TAKING SIDES

Clashing Views on Controversial

Issues in World Politics

TWELFTH EDITION

TAKING SIDES

Clashing Views on Controversial

Issues in World Politics

TWELFTH EDITION

Selected, Edited, and with Introductions by

John T. Rourke
University of Connecticut

McGraw-Hill/Dushkin
A Division of The McGraw-Hill Companies

For my son and friend—John Michael

Cover image: Photos.com

Cover Acknowledgment
Maggie Lytle

Manufactured in the United States of America

Twelfth Edition

123456789DOCDOC98765

Library of Congress Cataloging-in-Publication Data
Main entry under title:
Taking sides: clashing views on controversial issues in world politics/selected, edited, and with
introductions by John T. Rourke.—12th ed.
Includes bibliographical references and index.
1. World Politics—1989–. I. Rourke, John T., *comp.*
909.82

0-07-352716-5
ISSN: 1094-754X

Printed on Recycled Paper

Preface

In the first edition of *Taking Sides: Clashing Views on Controversial Issues in World Politics,* I wrote of my belief in informed argument: [A] book that debates vital issues is valuable and necessary. . . . [It is important] to recognize that world politics is usually not a subject of absolute rights and absolute wrongs and of easy policy choices. We all have a responsibility to study the issues thoughtfully, and we should be careful to understand all sides of the debates.

It is gratifying to discover, as indicated by the success of *Taking Sides* over 11 editions, that so many of my colleagues share this belief in the value of a debate-format text.

The format of this edition follows a formula that has proved successful in acquainting students with the global issues that we face and in generating discussion of those issues and the policy choices that address them. This book addresses 20 issues on a wide range of topics in international relations. Each issue has two readings, one pro and one con. Each is also accompanied by an issue *introduction*, which sets the stage for the debate, provides some background information on each author, and generally puts the issue into its political context. Each issue concludes with a *postscript* that summarizes the debate, gives the reader paths for further investigation, and suggests additional readings that might be helpful. I have also provided relevant Internet site addresses (URLs) in each postscript and on the *On the Internet* page that accompanies each part opener. At the back of the book is a listing of all the *contributors to this volume,* which will give you information on the political scientists and other commentators whose views are debated here.

I have continued to emphasize issues that are currently being debated in the policy sphere. The authors of the selections are a mix of practitioners, scholars, and noted political commentators.

Changes to this edition   The dynamic, constantly changing nature of the world political system and the many helpful comments from reviewers have brought about significant changes to this edition. Of the 40 readings in this edition, 26 are new, with 14 readings being carried over from the previous edition.

The kaleidoscopic dynamism of the international system is also evident in the high turnover in issues from one edition to the next of this reader. More than half of the issues have changed entirely or in part. Only 6 (30 percent) of the 20 issues and their readings are carried over directly from the previous edition. In contrast, 10 issues (50 percent) and their readings are completely new. They are: *Does Globalization Threaten Cultural Diversity?* (Issue 2); *Should the United States Decrease Its Global Presence?* (Issue 4); *Is Russian Foreign Policy Taking an Unsettling Turn?* (Issue 6); *Does a Strict "One China" Policy Still Make Sense?* (Issue 7); *Are Strict Sanctions on Cuba Warranted?* (Issue 11); *Is Capitalism the Best Model for the Global Economy?* (Issue 12); *Is Preemptive War an Unacceptable Doctrine?* (Issue 14); *Is the War on Terrorism Succeeding?* (Issue 15); *Is the United Nations*

Fundamentally Flawed? (Issue 17); and *Is the Convention on the Elimination of All Forms of Discrimination Against Women Worthy of Support?* (Issue 19). Another 4 issues (20 percent) that were included in the last edition have one or more new readings to reflect changes in their ongoing and dynamic nature. These debates are: *Is Economic Globalization a Positive Trend?* (Issue 1); *Should the United States Continue to Encourage a United Europe?* (Issue 5); *Should North Korea's Nuclear Arms Program Evoke a Hard-line Response?* (Issue 8), and *Was War with Iraq Justified?* (Issue 10).

It is important to note that the changes to this edition from the last should not disguise the fact that most of the issues address enduring human concerns, such as global political organization, arms and arms control, justice, development, and the environment. Also important is the fact that many of the issues have both a specific and a larger topic. For instance, Issue 17 is about the specific topic of reforming the United Nations, but it is also about more general topics. These include the proper role of international organizations in the global system and the degree to which countries should subordinate their sovereignty to them.

A word to the instructor An *Instructor's Manual With Test Questions* (multiple-choice and essay) is available through the publisher for instructors using *Taking Sides* in the classroom. A general guidebook, *Using Taking Sides in the Classroom,* which discusses methods and techniques for integrating the pro-con approach into any classroom setting, is also available. An online version of *Using Taking Sides in the Classroom* and a correspondence service for *Taking Sides* adopters can be found at http://www.dushkin.com/usingts/. *Taking Sides: Clashing Views on Controversial Issues in World Politics* is only one title in the Taking Sides series. If you are interested in seeing the table of contents for any of the other titles, please visit the Taking Sides Web site at http://www.dushkin.com/takingsides/.

A note especially for the student reader You will find that the debates in this book are not one-sided. Each author strongly believes in his or her position. And if you read the debates without prejudging them, you will see that each author makes cogent points. An author may not be "right," but the arguments made in an essay should not be dismissed out of hand, and you should work to remain tolerant of those who hold beliefs that are different from your own. There is an additional consideration to keep in mind as you pursue this debate approach to world politics. To consider divergent views objectively does not mean that you have to remain forever neutral. In fact, once you are informed, you ought to form convictions. More important, you should try to influence international policy to conform better with your beliefs. Write letters to policymakers; donate to causes you support; work for candidates who agree with your views; join an activist organization. *Do* something, whichever side of an issue you are on!

Acknowledgments I received many helpful comments and suggestions from colleagues and readers across the United States and Canada. Their suggestions have markedly enhanced the quality of this edition of *Taking Sides*. If as you read this book you are reminded of a selection or an issue that could be included in a future edition, please write to me in care of McGraw-Hill/Dushkin with your recommendations or e-mail them to me at john.rourke@uconn.edu.

My thanks go to those who responded with suggestions for the 12th edition. I would also like to thank Jill Peter, my editor for this volume, for her help in refining this edition.

John T. Rourke
University of Connecticut

Contents In Brief

Contents

Anne O. Krueger, first deputy managing director of the International Monetary Fund, asserts that the growth of economic globalization is the best approach to improving the economies of Africa and, by extension, other countries as well. José Bové, a French farmer and anti-globalization activist, contends that multinational corporations, government leaders, and others are engaged in a propaganda campaign to sell the world on the false promise of economic globalization.

Julia Galeota of McLean, Virginia, who was seventeen years old when she wrote her essay that won first place for her age category in the 2004 *Humanist* Essay Contest for Young Women and Men of North America, contends that many cultures around the world are gradually disappearing due to the overwhelming influence of corporate and cultural America. Philippe Legrain, chief economist of Britain in Europe, an organization supporting the adoption by Great Britain of the euro as its currency, counters that it is a myth that globalization involves the imposition of Americanized uniformity, rather than an explosion of cultural exchange.

Professor of international relations Stephen D. Krasner contends that the nation-state has a keen instinct for survival and will adapt to globalization and other challenges to sovereignty. Kimberly Weir, an assistant professor of political science, maintains that the tide of history is running against the sovereign state as a governing principle, which will soon go the way of earlier, now-discarded forms of governance, such as empire.

PART 2 REGIONAL AND COUNTRY ISSUES 51

YES: Louis Janowski, from "Neo-Imperialism and U.S. Foreign Policy," *Foreign Service Journal* (May 2004) 54

NO: Niall Ferguson, from "A World Without Power," *Foreign Policy* (July/August 2004) 62

Louis Janowski, a former U.S. diplomat with service in Vietnam, France, Ethiopia, Saudi Arabia, and Kenya, maintains that the view that the 9/11 attacks ushered in a new geo-strategic reality requiring new foreign policy approaches is based on a false and dangerous premise and is leading to an age of American neo-imperialism. Niall Ferguson, Herzog Professor of History at New York University's Stern School of Business and senior fellow at the Hoover Institution at Stanford University, contends that a U.S. retreat from global power would result in an anarchic nightmare of a new Dark Age.

YES: A. Elizabeth Jones, from Testimony Before the Subcommittee on Europe, Committee on International Relations, U.S. House of Representatives (March 13, 2002) 72

NO: John C. Hulsman, from "Laying Down Clear Markers: Protecting American Interests from a Confusing European Constitution," *The Heritage Foundation Backgrounder* (December 12, 2003) 78

A. Elizabeth Jones, assistant secretary of state for European and Eurasian affairs, maintains that the United States looks forward to working cooperatively with such exclusively or mostly European institutions as the European Union, the Organization for Cooperation and Security in Europe, and the North Atlantic Treaty Organization. John C. Hulsman, a research fellow for European affairs in the Kathryn and Shelby Cullom Davis Institute for International Studies at the Heritage Foundation, argues that the United States should support European countries on a selective basis but not be closely tied to Europe as a whole.

YES: Ariel Cohen and Yevgeny Volk, from "Recent Changes in Russia and Their Impact on U.S.-Russian Relations," *The Heritage Foundation Backgrounder* (March 9, 2004) 86

NO: Leon Aron, from Testimony During Hearings on "U.S.-Russia Relations in Putin's Second Term," Committee on International Relations, U.S. House of Representatives (March 18, 2004) 94

Ariel Cohen, research fellow in Russian and Eurasian studies in the Kathryn and Shelby Cullom Davis Institute for International Studies at The Heritage Foundation, and Yevgeny Volk, The Heritage Foundation's Moscow office director, write that the revival of statism and nationalism has seriously diminished Russia's chances of being regarded as a close and reliable partner that is clearly committed to democratic values. Leon Aron, director of Russian studies at the American Enterprise Institute, recognizes that there are pressures within Russia to try to take a more confrontational stance but believes that the forces for moderation are stronger.

Vice President Richard Cheney argues that Saddam Hussein's drive to acquire weapons of mass destruction, links with terrorists, and brutal dictatorship warranted U.S. action to topple his regime. West Virginia Senator Robert Byrd criticizes the decision to invade Iraq in the first place as ill-founded and further contends that the consequences have been too costly.

The Commission for Assistance to a Free Cuba, which President George W. Bush established on October 10, 2003, and charged with making recommendations about how to hasten a transition to democracy in Cuba, argues in its report to the president that the U.S. government should take stronger measures to undermine the Castro regime and to promote conditions that will help the Cuban people hasten the end of President Fidel Castro's dictatorial regime. William Ratliff, a research fellow at the Hoover Institution, argues that sanctions on Cuba only hurt the Cuban people because nothing the United States is doing today contributes significantly to the achievement of any change in the Castro regime.

PART 3   ECONOMIC ISSUES   171

Johan Norberg, a fellow at the Swedish think tank Timbro, portrays capitalism as the path to global economic prosperity and argues further that free markets and free trade mean free choices for individuals that transfer power to them at the expense of political institutions. Walden Bello, executive director of Focus on the Global South, the Bangkok, Thailand–based project of Chulalongkorn University's Social Research Institute, and professor of sociology and public administration at the University of the Philippines, contends that global capitalism is the source of societal and environmental destruction.

Romilly Greenhill, an economist with Jubilee Research at the New Economics Foundation, contends that if the world community is going to achieve its goal of eliminating world poverty by 2015, as stated in the UN's Millennium Declaration, then there is an urgent need to forgive the massive debt owed by the heavily indebted poor countries. William Easterly, a senior adviser in the Development Research Group at the World Bank, maintains that while debt relief is a popular cause and seems good at first glance, the reality is that debt relief is a bad deal for the world's poor.

The High Level Panel on Threats, Challenges, and Change, which was appointed by United Nations Secretary-General Kofi Annan in response to the global debate on the nature of threats and the use of force to counter them, concludes that in a world full of perceived potential threats, the risk to the global order posed by preemptive war is too great for its legality to be accepted. Colonel Steven L. Kenny argues in a research report he wrote at the U.S. Army War College, Carlisle Barracks, Pennsylvania, that substantial support from the acceptability of preemptive war results from such factors as the failure of the UN to enforce its charter, customary international law, and the growing threat of terrorists and weapons of mass destruction.

Douglas J. Feith, U.S. undersecretary of defense for policy, tells his audience that in the global war on terrorism, the United States is succeeding in defeating the terrorist threat to the American way of life and argues that the terrorists are on the run, that the world is safer and better for what has been accomplished, and that Americans have much of which to be proud. John Gershman, who is co-director of Foreign Policy in Focus for the Interhemispheric Resource Center and teaches at the Robert F. Wagner School for Public Service at New York University, contends that the "war on terrorism" being waged by the administration of President George W. Bush reflects a major failure of leadership and makes Americans more vulnerable rather than more secure.

Bruce Berkowitz, a research fellow at the Hoover Institution at Stanford University, argues that while government-directed political assassinations are hard to accomplish and are not a reliably effective political tool, there are instances where targeting and killing an individual is both prudent and legitimate. Margot Patterson, a senior writer for *National Catholic Reporter,* contends that assassinations are morally troubling, often counterproductive, and have a range of other drawbacks.

PART 5 INTERNATIONAL LAW AND ORGANIZATION ISSUES 265

Brett D. Schaefer, the Jay Kingham Fellow in International Regulatory Affairs in the Center for International Trade and Economics at The Heritage Foundation, contends that the UN is not doing as well as it should in championing the principles set forth in its charter and that, therefore, fundamental UN reform is required. Mary Robinson, the United Nations high commissioner for human rights and a former president of Ireland, argues that despite all the United Nations' shortcomings and criticism, the UN is as relevant now as it was when created.

The Lawyers Committee for Human Rights, in a statement submitted to the U.S. Congress, contends that the International Criminal Court (ICC) is an expression, in institutional form, of a global aspiration for justice. John R. Bolton, senior vice president of the American Enterprise Institute in Washington, D.C., contends that support for an international criminal court is based largely on naive emotion and that adhering to its provisions is not wise.

Harold Hongju Koh, the Gerard C. and Bernice Latrobe Smith Professor of International Law at Yale University and former assistant secretary of state for human rights and democracy, contends that the United States cannot be a global leader championing progress for women's human rights around the world unless it is also a party to the global women's treaty. Christina P. Hoff-Sommers, resident scholar at the American Enterprise Institute, Washington, D.C., tells Congress that the United States can and should help women everywhere to achieve the kind of equity American women have. She maintains, however, that ratifying the CEDAW is the wrong way to pursue that goal.

Professor of statistics Bjørn Lomborg argues that it is a myth that the world is in deep trouble on a range of environmental issues and that drastic action must be taken immediately to avoid an ecological catastrophe. Fred Krupp, president of Environmental Defense, asserts that although Lomborg's message is alluring because it says we can relax, the reality is that there are serious problems that, if not addressed, will have a deleterious effect on the global environment.

Introduction

World Politics and the Voice of Justice

John T. Rourke

Some years ago, the Rolling Stones recorded "Sympathy With the Devil." If you have never heard it, go find a copy. It is worth listening to. The theme of the song is echoed in a wonderful essay by Marshall Berman, "Have Sympathy for the Devil" (*New American Review,* 1973). The common theme of the Stones' and Berman's works is based on Johann Goethe's *Faust.* In that classic drama, the protagonist, Dr. Faust, trades his soul to gain great power. He attempts to do good, but in the end he commits evil by, in contemporary paraphrase, "doing the wrong things for the right reasons." Does that make Faust evil, the personification of the devil Mephistopheles among us? Or is the good doctor merely misguided in his effort to make the world better as he saw it and imagined it might be? The point that the Stones and Berman make is that it is important to avoid falling prey to the trap of many zealots who are so convinced of the truth of their own views that they feel righteously at liberty to condemn those who disagree with them as stupid or even diabolical.

It is to the principle of rational discourse, of tolerant debate, that this reader is dedicated. There are many issues in this volume that appropriately excite passion—for example, Issue 9 on whether or not Israel should agree to an independent Palestinian state, or Issue 16 which examines whether or not government-ordered assassinations are an acceptable practice. Few would deny, for example, that it is wrong to kill the leader of an enemy country during the normal course of military operations. But is it different to specifically target another country's leaders and conduct an overt or covert operation to kill them?

In other cases, the debates you will read diverge on goals. In Issue 5 Elizabeth Jones argues that the U.S. goal should be to encourage further European integration. John Hulsman disagrees. He worries that a united Europe will become a powerful rival of the United States rather than a powerful friend.

As you will see, each of the authors in all the debates strongly believes in his or her position. If you read these debates objectively, you will find that each side makes cogent points. They may or may not be right, but they should not be dismissed out of hand. It is important to repeat that the debate format does not imply that you should remain forever neutral. In fact, once you are informed, you *ought* to form convictions, and you should try to act on those convictions and try to influence international policy to conform better with your beliefs. Ponder the similarities in the views of two very different leaders, a very young president in a relatively young democracy and a very old emperor in a very old country: In 1963 President John F. Kennedy, in recalling the words of the author of the epic poem *The Divine Comedy* (1321), told a West German audience, "Dante once said that the hottest places in hell are reserved for those who in a period of moral crisis maintain their neutrality." That very same year, while speaking to the United Nations, Ethiopia's emperor Haile Selassie (1892–1975) said, "Throughout history it has been the inaction of those who could have acted, the indifference of those who should have known better, the silence of the voice of justice when it mattered most that made it possible for evil to triumph."

The point is: Become Informed. Then *do* something! Write letters to policy-makers, donate money to causes you support, work for candidates with whom you agree, join an activist organization, or any of the many other things that you can do to make a difference. What you do is less important than that you do it.

Approaches to Studying International Politics

As will become evident as you read this volume, there are many approaches to the study of international politics. Some political scientists and most practitioners specialize in *substantive topics,* and this reader is organized along topical lines. Part 1 (Issues 1 through 3) features debates on the evolution of the international system in the direction of greater globalization. In Issue 1, Anne Krueger and French dissident José Bové debate the general topic of economic globalization. Krueger is optimistic, maintaining that economic globalization is a positive trend; Bové is pessimistic. Issue 2 addresses the accompanying phenomenon of cultural global-ization. American student Julia Galeota fears that American culture is wiping away cultural diversity in the world, while British analyst Phillippe Legrain rejects these charges and points to the positive aspects of cultural. Then in Issue 3 Stephen D. Krasner and Kimberly Weir engage in a debate over whether countries will continue to maintain their sovereignty in the future or whether they will become at least partially subordinate to regional and global organizations.

Part 2 (Issues 4 through 11) focuses on regional and country-specific issues, including the role of the United States in the international system, the future of U.S.-European relations, trends in Russian foreign policy, what to do about the complex status of Taiwan, North Korea's nuclear weapons program, the possibility of a Palestinian state, the 2003 U.S.-led war against Iraq, and the ongoing U.S. sanctions against Cuba.

Part 3 (Issues 12 and 13) deals with specific concerns of the international economy, a topic introduced more generally in Issue 1. Issue 12 takes up the value of the capitalist approach to the global economy. Swedish analyst Johan Norberg argues that capitalism has been responsible for the prosperity of the economically developed countries and is the best way to bring prosperity to all countries. Filipino analyst Walden Bello condemns capitalism for enriching the few at the expense of the many and other deleterious impacts. In Issue 13 the debate turns to the huge financial debt owed by poor countries to international financial organizations such as the International Monetary Fund (IMF) and to the governments, banks, and other investors of the wealthiest countries. Romilly Greenhill argues that the debt should be forgiven; William Easterly disagrees.

Part 4 (Issues 14 through 16) examines violence in the international system. The past few decades have witnessed a rising concern about terrorism and the spread of nuclear, chemical, biological, and radiological weapons of mass destruction, not only among countries, but also to terrorist groups. Partly in response to these changes, President George W. Bush proclaimed that the United States had the right to take preemptive military action against threats before any direct or even immediately impending attack against Americans. The report of a committee appointed by the UN secretary general rejects preemptive action, while a report authored at the U.S. Army War College supports it. Issue 15 turns to what has become a high-profile threat: terrorism. Douglas Feith argues that the war on terrorism being led by the Bush administration in Washington is being won. John Gershman counters that the efforts to date reflect a failure of leadership. Issue 16 debates whether or not government-orchestrated assassinations are an effective and moral way of conducting foreign policy. Although it was generally shunned in the past, in the current era of increased terrorism, assassination has become more acceptable to some people.

Part 5 (Issues 17 through 19) addresses controversies related to international law and organizations. The ability of the United Nations to deploy effective peacekeeping forces is severely constrained by a number of factors. Similarly, the UN has not been able to resolve the world's many social and economic inequities. Moreover, the UN has suffered some mismanagment. Is the organization a failure in need of fundamental change, as Brett Schaefer argues in Issue 17? Mary Robinson, the former president of Ireland, does not think so. Issue 18 evaluates the wisdom of establishing a permanent international criminal court to punish those who violate the law of war. It is easy to advocate such a court as long as it is trying and sometimes punishing alleged war criminals from other countries. But one has to understand that one day a citizen of one's own country could be put on trial. The third debate in Part 5 takes up the Convention on the Elimination All Forms of Discrimination Against Women in Issue 19. The United States is one of the very few countries that has not agreed to the treaty. Harold Hongju Koh argues that is a mistake that should be corrected. Christina Hoff-Sommers contends that doing so would be an error that would bedevil the United States and make no contribution to global women's rights.

Part 6, which consists of Issue 20 addresses the environment issue. Over the past few decades there has been a growing concern that population growth; the discharge of liquid, gaseous, and solid waste into the air, water, and ground; the overconsumption of natural resources; and other human activities are severely, even irreparably, damaging the Earth's ecosphere. From this perspective, significant changes have to be made to avert further damage and an environmental disaster. Bjørn Lomborg of Denmark argues that the cries of alarm are vastly overstated and that whatever ills do exist can be addressed without radical change. Fred Krupp takes the concerned view, arguing that action needs to be taken to ensure a protected environment.

Political scientists also approach their subject from differing *methodological perspectives*. You will see, for example, that world politics can be studied from different *levels of analysis*. The question is, What is the basic source of the forces that shape the conduct of politics? Possible answers are world forces, the individual political processes of the specific countries, or the personal attributes of a country's leaders and decision makers. Various readings will illustrate all three levels.

Another way for students and practitioners of world politics to approach their subject is to focus on what is called the realist versus the idealist (or liberal) debate. Realists tend to assume that the world is permanently flawed and therefore advocate following policies in their country's narrow self-interests. Idealists take the approach that the world condition can be improved substantially by following policies that, at least in the short term, call for some risk or self-sacrifice. This divergence is an element of many of the debates in this book.

Dynamics of World Politics

The action on the global stage today is vastly different from what it was a few decades ago, or even a few years ago. *Technology* is one of the causes of this change. Technology has changed communications, manufacturing, health care, and many other aspects of the human condition. Technology has given humans the ability to create biological, chemical, and nuclear compounds and other material that in relatively small amounts have the ability to kill and injure huge numbers of people. Another negative byproduct of technology may be the vastly increased consumption of petroleum and other natural resources and the global environmental degradation that has been caused by discharges of waste products, deforestation, and a host of other technology-enhanced human activities.

Another dynamic aspect of world politics involves the *changing axes* of the world system. For about 40 years after World War II ended in 1945, a bipolar system existed, the primary axis of which was the *East-West* conflict, which pitted the United States and its allies against the Soviet Union and its allies. Now that the Cold War is over, one broad debate is over what role the United States should play. In Issue 4, former U.S. diplomat Louis Janowski argues that the United States should reduce its global presence; Niall Ferguson believes doing so will harm global stability. A related issue is whether or not there are potential enemies to the United States and its allies and, if so, who they are. As for potential rivals to U.S. hegemony, Issue 5 deals with Europe and its increasing uneasiness with U.S. power. Could a united Europe become a rival instead of an ally? More conventionally, Issue 7 deals with China, a Cold War antagonist of the United States. Some people believe that China is the next superpower and that it will pose a threat to U.S. security and interests. A crucial issue between Washington and Beijing is the status of Taiwan (Issue 7). The future of that island could bring great tension, even war, between China and the United States. Then there is the former chief antagonist of the United States during the Cold War. Issue 6 examines whether Russia, the successor state of the collapsed Soviet Union, is beginning to follow policies that could bring increasing confrontations in the future.

Technological changes and the shifting axes of international politics also highlight the *increased role of economics* in world politics. Economics have always played a role, but traditionally the main focus has been on strategic-political questions—especially military power. This concern still strongly exists, but now it shares the international spotlight with economic issues. One important change in recent decades has been the rapid growth of regional and global markets and the promotion of free trade and other forms of international economic interchange. As Issue 1 on economic interdependence indicates, many people support these efforts and see them as the wave of the future. But there are others who believe that free economic globalization and interdependence undermine sovereignty and the ability of governments to control their destinies. One related topic, which is taken up in Issue 12 is whether or not capitalism offers the best economic approach for ensuring continued prosperity among the developed countries and for improving the circumstances of the less economically developed countries.

Another change in the world system has to do with the main *international* actors. At one time states (countries) were practically the only international actors on the world stage. Now, and increasingly so, there are other actors. Some actors are regional. Others, such as the United Nations, are global actors. Issue 3 discusses the future of countries as principal and sovereign actors in the international system. Turning to the most notable international organization, Issue 17 examines the call for fundamentally reforming the UN and, by extension, the role of that world organization and the proper approach of member countries to it and to global cooperation. Issue 18 focuses on whether or not a supranational criminal court should be established to take over the prosecution and punishment of war criminals from the domestic courts and ad hoc tribunals that have sometimes dealt with such cases in the past.

Perceptions Versus Reality

In addition to addressing the general changes in the world system outlined above, the debates in this reader explore the controversies that exist over many of the fundamental issues that face the world.

One key to these debates is the differing *perceptions* that protagonists bring to them. There may be a reality in world politics, but very often that reality is obscured. Many observers, for example, are alarmed by the seeming rise in radical actions by Islamic fundamentalists. However, the image of Islamic radicalism is not a fact but a perception; perhaps correct, perhaps not. In cases such as this, though, it is often the perception, not the reality, that is more important because policy is formulated on what decision makers *think*, not necessarily on what *is*. Thus, perception becomes the operating guide, or *operational reality*, whether it is true or not. Perceptions result from many factors. One factor is the information that decision makers receive. For a variety of reasons, the facts and analyses that are given to leaders are often inaccurate or represent only part of the picture. The conflicting perceptions of Israelis and Palestinians, for example, make the achievement of peace in Israel very difficult. Many Israelis and Palestinians fervently believe that the conflict that has occurred in the region over the past 50 years is the responsibility of the other. Both sides also believe in the righteousness of their own policies. Even if both sides are well-meaning, the perceptions of hostility that each holds means that the operational reality often has to be violence. These differing perceptions are a key element in the debate in Issue 9.

A related aspect of perception is the tendency to see oneself differently than some others do. Specifically, the tendency is to see oneself as benevolent and to perceive rivals as sinister. This reverse image is partly at issue in the debate on preemptive war (Issue 14). Many Americans see themselves as threatened by terrorists and rogue nations and support preemptive war as a legitimate defensive strategy. Other countries worry that the Bush Doctrine is an excuse for the United States to try to dominate others and attack those who offer serious opposition to U.S. imperialism. For example, one aspect of Issue 8 on how to deal with North Korea's nuclear weapons program centers on whether the North Koreans are trying to increase their power for aggressive purposes or are trying to deter any chance of a U.S. attack on them. Perceptions, then, are crucial to understanding international politics. It is important to understand objective reality, but it is also necessary to comprehend subjective reality in order to be able to predict and analyze another country's actions.

Levels of Analysis

Political scientists approach the study of international politics from different levels of analysis. The most macroscopic view is *system-level analysis*. This is a top-down approach that maintains that world factors virtually compel countries to follow certain foreign policies. Governing factors include the number of powerful actors, geographic relationships, economic needs, and technology. System analysts hold that a country's internal political system and its leaders do not have a major impact on policy. As such, political scientists who work from this perspective are interested in exploring the governing factors, how they cause policy, and how and why systems change.

After the end of World War II, the world was structured as a *bipolar* system, dominated by the United States and the Soviet Union. Furthermore, each superpower was supported by a tightly organized and dependent group of allies. For a variety of reasons, including changing economics and the nuclear standoff, the bipolar system has faded. Some political scientists argue that the bipolar system is being replaced by a *multipolar* system. In such a configuration, those who favor *balance-of-power* politics maintain that it is unwise to ignore power considerations. Or it may be that something like a one-power (unipolar) system exists, with the United States as that power, as taken up in Issue 4.

State-level analysis is the middle and most common level of analysis. Social scientists who study world politics from this perspective focus on how countries, singly or comparatively, make foreign policy. In other words, this perspective is concerned with internal political dynamics, such as the roles of and interactions between the executive and legislative branches of government, the impact of bureaucracy, the role of interest groups, and the effect of public opinion. This level of analysis is very much in evidence in Issue 19 and the explanation of why the United States has not ratified the treaty enumerating women's rights. The dangers to the global environment, which are debated in Issue 20 extend beyond rarified scientific controversy to important issues of public policy. For example, should the United States and other industrialized countries adopt policies that are costly in terms of economics and lifestyle to significantly reduce the emission of carbon dioxide and other harmful gases? This debate pits interest groups against one another as they try to get the governments of their respective countries to support or reject the steps necessary to reduce the consumption of resources and the emission of waste products. To a large degree, it is the environmentalists versus the business groups.

A third level of analysis, which is the most microscopic, is *human-level analysis*. This approach focuses, in part, on the role of individual decision makers. This technique is applied under the assumption that individuals make decisions and that the nature of those decisions is determined by the decision makers' perceptions, predilections, and strengths and weaknesses. Part of Issue 11 on U.S. sanctions on Cuba is based on conjectures about whether Cuba will experience a basic change in its political system once long-time President Fidel Castro is no longer in power.

The Political and Ecological Future

Future *world alternatives* are discussed in many of the issues in this volume. Abraham Lincoln once said, "A house divided against itself cannot stand." One suspects that the 16th president might say something similar about the world today if he were with us. Issue 1, for example, debates whether growing globalization is a positive or negative trend. The world has responded to globalization by creating and strengthening the UN, the IMF, the World Bank, the World Trade Organization, and many other international organizations to try to regulate the increasing number of international interactions. There can be little doubt that the role of global governance is growing, and this reality is the spark behind the debate in Issue 3 over whether or not the traditional sovereignty of states will persist in a time of increasing globalization. More specific debates about the future are taken up in many of the selections that follow the three debates in Part 1. Far-reaching alternatives to a state-centric system based on sovereign countries include international organizations' (Issue 17) taking over some (or all) of the sovereign responsibilities of national governments, such as the prosecution of international war criminals (Issue 18). The global future also involves the ability of the world to prosper economically while not denuding itself of its natural resources or destroying the environment. This is the focus of Issue 20 on the environment.

The Axes of World Division

The world is politically dynamic, and the nature of the political system is undergoing profound change. As noted, the once-primary axis of world politics, the East-West confrontation, has broken down. Yet a few vestiges of the conflict on that axis remain. These can be seen in Issue 7 about future relations with still-communist and increasingly powerful China and Issue 6 about relations with the former heart of the so-called communist bloc, Russia/the Soviet Union. In what could be an ironic turn, it is even arguable that the growing tensions between the United States and Europe, as discussed in Issue 5, could lead to a newly configured East-West divide, with the Atlantic Ocean rather than the Iron Curtain separating the antagonists.

In contrast to the moribund East-West axis, the *North-South axis* has increased in importance and tension. The wealthy, industrialized countries (North) are at one end, and the poor, less developed countries (LDCs, South) are at the other extreme. Economic differences and disputes are the primary dimension of this axis, in contrast to the military nature of the East-West axis. Issue 13 explores these differences and debates one approach of the North giving economic aid to the South.

Something of military relations between the North and the South are present in the war with Iraq and the crisis with North Korea, which are taken up in Issues 10 and 8, respectively. There are some, especially in the South, who believe that the negative U.S. reactions to Iraqi and North Korean power, especially their possible possession of weapons of mass destruction, constitute evidence that the powerful countries are determined to maintain their power by keeping weak countries from obtaining the same types of weapons that the United States and other big powers have.

Increased Role of Economics

As the growing importance of the North-South axis indicates, economics are playing an increased role in world politics. The economic reasons behind the decline of the East-West axis is further evidence. Economics have always played a part in international relations, but the traditional focus has been on strategic-political affairs, especially questions of military power.

Political scientists, however, are increasingly focusing on the international political economy, or the economic dimensions of world politics. International trade, for instance, has increased dramatically, expanding from an annual world export total of $20 billion in 1933 to $7.9 trillion in 2002. The impact has been profound. The domestic economic health of most countries is heavily affected by trade and other aspects of international economics. Since World War II there has been an emphasis on expanding free trade by decreasing tariffs and other barriers to international commerce. In recent years, however, a downturn in the economies of many of the industrialized countries has increased calls for more protectionism. Yet restrictions on trade and other economic activity can also be used as diplomatic weapons. The intertwining of economies and the creation of organizations to regulate them, such as the World Trade Organization, is raising issues of sovereignty and other concerns. This is a central matter in the debate in Issue 1 over whether or not the trend toward global economic integration is desirable and in Issue 12 on the economic model the global economy should follow.

Conclusion

Having discussed many of the various dimensions and approaches to the study of world politics, it is incumbent on this editor to advise against your becoming too structured by them. Issues of focus and methodology are important both to studying international relations and to understanding how others are analyzing global conduct. However, they are also partially pedagogical. In the final analysis, world politics is a highly interrelated, perhaps seamless, subject. No one level of analysis, for instance, can fully explain the events on the world stage. Instead, using each of the levels to analyze events and trends will bring the greatest understanding.

Similarly, the realist-idealist division is less precise in practice than it may appear. As some of the debates indicate, each side often stresses its own standards of morality. Which is more moral: defeating a dictatorship or sparing the sword and saving lives that would almost inevitably be lost in the dictator's overthrow? Furthermore, realists usually do not reject moral considerations. Rather, they contend that morality is but one of the factors that a country's decision makers must consider. Realists are also apt to argue that standards of morality differ when dealing with a country as opposed to an individual. By the same token, most idealists do not completely ignore the often dangerous nature of the world. Nor do they argue that a country must totally sacrifice its short-term interests to promote the betterment of the current and future world. Thus, realism and idealism can be seen most accurately as the ends of a continuum—with most political scientists and practitioners falling somewhere between, rather than at, the extremes. The best advice, then, is this: think broadly about international politics. The subject is very complex, and the more creative and expansive you are in selecting your foci and methodologies, the more insight you will gain. To end where we began, with Dr. Faust, I offer his last words in Goethe's drama, "*Mehr licht*," . . . More light! That is the goal of this book.

On the Internet . . .

The Ultimate Political Science Links Page

Under the editorship of Professor P. S. Ruckman, Jr., at Rock Valley College in Rockford, Illinois, this site provides a gateway to the academic study of not just world politics but all of political science. It includes links to journals, news, publishers, and other relevant resources.

http://www.rvc.cc.il.us/faclink/
pruckman/PSLinks.htm

Poly-Cy: Internet Resources for Political Science

This is a worthwhile gateway to a broad range of political science resources, including some on international relations. It is maintained by Robert D. Duval, director of graduate studies at West Virginia University.

http://www.polsci.wvu.edu/polycy/

The WWW Virtual Library: International Affairs Resources

Maintained by Wayne A. Selcher, professor of international studies at Elizabethtown College in Elizabethtown, Pennsylvania, this site contains approximately 2,000 annotated links relating to a broad spectrum of international affairs. The sites listed are those that the Webmaster believes have long-term value and that are cost-free, and many have further links to help in extended research.

http://www.etown.edu/vl/

The Globalization Website

The goals of this site are to shed light on the process of globalization and contribute to discussions of its consequences, to clarify the meaning of globalization and the debates that surround it, and to serve as a guide to available sources on globalization.

http://www.emory.edu/SOC/globalization/

PART 1

Globalization

*T*he most significant change that the international system is experiencing is the trend toward globalization. Countries are becoming interdependent, the number of international organizations and their power are increasing, and global communications have become widespread and almost instantaneous. As reflected in the issues that make up this part, these changes and others have led to considerable debate about the value of globalization and what it will mean with regard to human governance.

- Is Economic Globalization a Postitive Trend?

- Does Globalization Threaten Cultural Diversity?

- Will State Sovereignty Survive Globalism?

ISSUE 1

Is Economic Globalization a Positive Trend?

YES: Anne O. Krueger, from "Expanding Trade and Unleashing Growth: The Prospects for Lasting Poverty Reduction," Remarks at the International Monetary Fund Seminar on Trade and Regional Integration, Dakar, Senegal (December 6, 2004)

NO: José Bové, from "Globalisations Misguided Assumptions," *OECD Observer* (September 2001)

ISSUE SUMMARY

YES: Anne O. Krueger, first deputy managing director of the International Monetary Fund, asserts that the growth of economic globalization is the best approach to improving the economies of Africa and, by extension, other countries as well.

NO: José Bové, a French farmer and anti-globalization activist, contends that multinational corporations, government leaders, and others are engaged in a propaganda campaign to sell the world on the false promise of economic globalization.

Globalization is a process that is diminishing many of the factors that divide the world. Advances in travel and communication have made geographical distances less important, people around the world increasingly resemble one another culturally, and the United Nations and other international organizations have increased the level of global governance. Another aspect, economic integration, is the most advanced of any of the strands of globalization. Tariffs and other barriers to trade have decreased significantly since the end of World War II. As a result, all aspects of international economic exchange have grown rapidly. For example, global trade, measured in the value of exported goods and services, grew over 1,200 percent during the last half of the twentieth century and now comes to more than $8 trillion annually. International investment in real estate and stocks and bonds in other countries, and in total, now exceeds $20 trillion. The flow of currencies is so massive that there is no accurate measure, but it certainly is more that $1.5 trillion a day.

In this liberalized atmosphere, huge multinational corporations (MNCs) have come to dominate global commerce. Just the top 100 MNCs have combined annual sales of over $4.4 trillion. The impact of all these changes is that the economic prosperity of almost all countries and the individuals within them is heavily dependent on what they import and export, the flow of investment in and out of each country, and the exchange rates of the currency of each country against the currencies of other countries.

The issue here is whether this economic globalization and integration is a positive or negative trend. For about 60 years, the United States has been at the center of the drive to open international commerce. The push to reduce trade barriers that occurred during and after World War II was designed to prevent a recurrence of the global economic collapse of the 1930s and the war of the 1940s. Policymakers believed that protectionism had caused the Great Depression, that the ensuing human desperation provided fertile ground for the rise of dictators who blamed scapegoats for what had occurred and who promised national salvation, and that these fascist dictators had set off World War II. In sum, policymakers thought that protectionism caused economic depression, which caused dictators, which caused war. They believed that free trade, by contrast, would promote prosperity, democracy, and peace.

Based on these political and economic theories, American policymakers took the lead in establishing a new international economic system, including helping to found such leading global economic organizations as the International Monetary Fund (IMF), the World Bank, and the World Trade Organization (WTO). During the entire latter half of the twentieth century, the movement toward economic globalization has been strong, and there have been few influential voices opposing it. Most national leaders, business leaders, and other elites continue to support economic interdependence. The people in various countries have largely followed the path set by their leaders.

In the following selection, Anne O. Krueger, a ranking official of the IMF, focuses on the benefits of economic globalization for Africa as part of her general view and that of the IMF that globalization not only benefits economically developed countries but is also a key to the economic development of poorer countries. Not everyone agrees, though, and in recent years the idea that globalization is necessarily beneficial has come under increasing scrutiny and has met increasing resistance. Within countries, globalization has benefited some, while others—usually those at the bottom of the economic ladder—have lost jobs to imports and suffered other negative consequences. Similarly, some countries, notably those in sub-Saharan Africa, have not prospered. Reflecting this uneven impact, one line of criticism of globalization comes from those who believe that the way global politics work is a function of how the world is organized economically. These critics contend that people within countries are divided into "haves" and "have-nots" and that the world is similarly divided into have and have-not countries. Moreover, these critics believe that, both domestically and internationally, the wealthy haves are using globalization to keep the have-nots weak and poor in order to exploit them. Representing this view, French political activist José Bové argues in the second selection that economic globalization benefits the few at the expense of the many.

3

Anne O. Krueger  **YES**

Expanding Trade and Unleashing Growth: The Prospects for Lasting Poverty Reduction

$\mathbf{A}$ll of us here are concerned to ensure that there is more rapid progress in reducing poverty and raising living standards in Africa. I have no doubt that it is perfectly possible for African countries to achieve the rapid rates of growth and of poverty reduction that countries elsewhere in the world have attained. The evidence is clear that economic agents in Africa respond in exactly the same way to incentives as do agents in other economies. As one African central bank governor put it recently, the supply curve does not slope downwards in Africa.

There is now almost universal agreement on what the principal economic objectives should be—and, I would argue, on how to achieve them. All developing economies need more rapid and sustained rates of growth that will in turn permit large scale and lasting poverty reduction and rising living standards for all.

Trade has a central role in helping achieve more rapid growth. So I want in my remarks this morning to say something about the link between trade, growth and poverty reduction; and to explain what I see as the contribution that regional trade integration can make.

Trade and Global Prosperity

Let me start by putting our discussions into a broad context. A healthy rapidly growing world economy is desirable for everyone—and it is vital if we are to see more rapid growth in Africa. It is vital, too, for African countries to be full participants in the global economy. No country has achieved the sustained rapid growth needed to reduce poverty without opening up its trade with the rest of the world. Of course, the more rapid growth of intra-African trade is, of itself, desirable. But it will never be a substitute for the growth of trade between Africa and all other countries.

Trade liberalization is closely linked with economic growth. We can go right back to the days of trade around the Mediterranean to see that the countries that traded most actively were also the most prosperous. The industrial revolution of the eighteenth and nineteenth centuries led to a surge in trade as manufacturers sought new export markets and scoured the world for inputs that enabled them fully to exploit new technological processes.

Yet the experience of the past sixty years is, even by historical standards, remarkable. The rapid growth of the world economy that followed the Second World War was unprecedented. And the surge in global growth led directly to improvements in material welfare and in the quality of life. Most people in most countries shared in the benefits of this. Some countries grew much more rapidly than others, of course. And the evidence shows that the best performers were those where policy reforms aimed at creating macroeconomic stability, liberalizing trade and encouraging enterprise.

Postwar global growth was accompanied, indeed driven by, an unprecedented expansion in world trade. In 1950, world merchandise exports amounted to about 8% of world GDP [gross domestic product, a measure of a country's economic productivity]. Today that figure is close to 26%. In the IMF's [International Monetary Fund] latest *World Economic Outlook*, world growth this year is expected to be 5%—an impressive number. But we estimate that world trade will have grown by 8.5% this year—an even more impressive number.

Trade is an engine of global growth because it brings competition. It enables producers to have access to inputs at the lowest possible prices. No modern industry can compete without access to the world market for inputs. Trade ensures that scarce resources are used in the most effective possible way.

One important feature of the structure of trade today is the extent to which the "value-added chain" has been chopped up as transport and other costs have fallen, enabling firms in different countries to specialize, importing parts and intermediate goods, and often exporting products that will be further processed or assembled in yet other countries. Trade also enables firms and individuals to move up the value-added chain as they accumulate experience.

The benefits from trade are large and diverse. The challenge is therefore to provide the framework in which trade can thrive.

The Role of Trade Liberalization

Trade liberalization is the key. The progressive dismantling of trade barriers has been at the heart of the multilateral economic framework established in the aftermath of the Second World War. It is the process of multilateral trade liberalization—by which I mean all countries acting together to lower their trade barriers against each other—that has created the environment in which trade has grown rapidly, and driven global growth.

And I am talking here about the reduction and removal of both tariff and non-tariff barriers. All barriers act as an impediment to trade—indeed, that is their primary purpose. All lead to inefficient allocation of resources, because they introduce distortions into price signals. Trade barriers raise costs, both to final consumers and to the users of intermediate inputs. They enable vested interests—the firms in protected industries, and the workers in those firms—to benefit at the expense of the majority.

Firms can make what sounds like a plausible case for protection—without some kind of barrier, either a higher tariff or a quantitative restriction, they can argue that they would lose business to a foreign competitor and have to lay off workers. But the evidence has repeatedly shown that such protection only delays necessary adjustments, because it enables firms to avoid the need to cut costs and become competitive. Eventually no amount of protection will keep an inefficient firm in business. And while some workers in protected firms may keep their jobs for longer than they might have otherwise, others will be deprived of jobs in new or growing firms that would be able to perform better in the absence of the distortions that protection brings. And, interestingly, many firms and industries seeking, but failing to achieve, protection often survive and even prosper despite their pessimistic view of their prospects.

Multilateral liberalization provides the greatest gains. The benefits from trade liberalization in one country are larger when others liberalize at the same time. But that is not an argument for waiting until others are ready. The benefits from trade liberalization are significant even when it is done unilaterally, on a most favored nation (MFN) basis [putting tariff rates on a country that are no higher than those on a country with the lowest rates]. The countries that have experienced phases of really rapid growth have all been ones where policymakers recognized the gains to be had from unilateral liberalization of their trade with all countries and have acted accordingly. Korea and Chile are both obvious examples of the rewards of unilateral trade liberalization. Both economies grew rapidly as their trade barriers were lowered.

In this context, let me say something about the trade barriers of the rich industrial countries. I know that policymakers in many developing countries argue that it is up to the world's more prosperous economies to take the lead in trade liberalization.

It is true that the trade restrictions of industrial countries, especially in agriculture where average tariffs remain high, impose costs on developing countries. It is clear that the industrial countries should lower tariffs and other trade barriers, and improve access to their markets. They themselves would benefit greatly by removing these barriers.

But action by the industrial countries would not, by itself, bring many of the gains that many in the developing world expect. Barriers imposed by developing countries on each other are far higher than those imposed on them by industrial countries. The gains from lowering developing country trade barriers are accordingly greater. And developing countries have much to gain from unilateral liberalization. Yes those benefits will be greater if there is subsequently further multilateral liberalization. But developing countries will not be in a position to gain the maximum possible benefits from multilateral liberalization if they do not liberalize themselves, and open their trade to all countries.

The evidence is incontrovertible. Multilateral trade liberalization is the most desirable policy option. That is why a successful Doha round [of general talks to lower global trade barriers] is so important. That is why the IMF has worked hard to support the WTO [World Trade Organization] during the Doha process.

But this is not a justification to wait. The evidence of the gains from unilateral liberalization is, as I said, unequivocal.

There will always be some who lose as a result of trade liberalization. The fact that not everyone will gain in the short-term is not an excuse for postponing liberalization, far from it.

We need to bear several important factors in mind when assessing the short-term impact of trade liberalization. First, of course, is that trade liberalization of itself results in more rapid economic growth. This brings new business opportunities for firms, and new employment opportunities for individuals. As I have already noted, economic growth is the principal route to lasting poverty reduction.

Second, those who are disadvantaged as a result of lowering trade barriers are far fewer than people anticipate *ex ante* [before an occurrence]. Those who think they are going to lose out often do not. Fear of change is the source of much opposition to trade liberalization.

And third, remember that those who may lose have a far better idea of who they are than those who are likely to gain. Those in heavily protected but highly inefficient industries or firms might well be right to fear competition, although more firms and industries find ways of adjusting than is generally realized. But it is almost impossible to predict where the new business and employment opportunities will come ahead of time, because we cannot know in advance which sectors and firms will best respond to the new opportunities that trade liberalization brings. What we can be certain of is that those new opportunities will materialize—and that somebody will seize on them.

Let me illustrate this with a true story. During the negotiations for the North American Free Trade Area [NAFTA], one of the principal opponents of Mexican participation was the main manufacturer of Mexican refrigerators. In those days, Mexican refrigerators were of such poor quality that they barely counted as durable goods, at least in the eyes of most Mexican consumers. Our manufacturer was convinced that his business would be wiped out when, under NAFTA, American manufacturers, offering a much better quality product, gained free access to his home market. He did not see how he would be able to compete.

He could hardly have been more wrong, and his misguided reasoning is a textbook illustration of the benefits that trade can bring. It turned out that the fatal flaw in Mexican refrigerators was the compressor. Mexican-made compressors were just not up to the job, with the result that the refrigerators had a short lifespan. Once NAFTA came into effect, our manufacturer was able to obtain much better-made American compressors at competitive prices.

The result was that far from losing business, he gained a great deal of new business, much of it in the form of exports. His company became the largest supplier of smaller, apartment-sized refrigerators to the US market.

So as trade barriers were lowered, Mexican consumers gained access to much higher quality refrigerators at a competitive price; the Mexican manufacturer was able greatly to improve his product by gaining access to competitively-priced and high-quality inputs; additional workers were employed; and he was as a result able to exploit a large new export market.

I tell this story not because it is unusual but because it is typical. The bene-fits of trade liberalization aren't confined to some theoretic textbook—they are there for all to see, and for firms and individuals to experience for themselves.

I am not suggesting that we should be unconcerned for the plight of those who lose their jobs in firms that go out of business as a result of competition and who for one reason or another are unable to find another job. But denying opportunities to a larger number in order to protect a small minority—often at considerable expense both to consumers and taxpayers—is not the way to respond to concern. Properly targeted and effective social safety nets are a better, fairer and far cheaper way to help those who lose out as economies adjust to the beneficial changes that trade liberalization brings.

Regional Integration

When we come to focus on issues concerning regional integration, then, we must not lose sight of the bigger picture. Economies need to open up to the rest of the world as fast as possible. Full participation in a healthy growing world economy is essential if countries are to experience the rapid growth needed to alleviate poverty on a significant scale. Multilateral trade liberalization—all countries reducing their trade barriers with each other—is undoubtedly the best way to achieve this opening, but unilateral liberalization still offers very considerable benefits—far more than waiting for others to lower their trade barriers.

We should not look on the debate as one between unilateral and multilat-eral. It is rather a question of the large benefits that unilateral opening, on an MFN basis, will bring, or the even larger ones that will flow from multilateral trade liberalization. Those countries that open unilaterally will still reap the benefits of multilateral liberalization when it happens.

So it is essential that regional integration does not act as a distraction from this wider picture. Regional trade agreements (RTAs) must not be used as a cover for protectionist behavior. It is not enough to lower tariff barriers among the members of an RTA. This will severely limit the benefits that trade liberalization can bring: these will pale in comparison to the gains to be had from full-scale trade liberalization.

The experience of what is now the European Union [EU] is important in this context. Recall the context in which the European Economic Community [EEC] was founded in the 1957. The original six members of the EEC as it then was experienced rapid growth over a long period and experienced a consider-able degree of economic integration. They did so in large part because the EEC was established at a time of rapid multilateral liberalization. Intra-European tar-iff barriers fell rapidly to zero from in excess of 40%; but the EEC's external tar-iff was also reduced from the same 40% plus level to a weighted average of just over 3% now (excluding estimated ad valorem equivalents [AVEs, of specific tariffs); 6.5% if AVEs are included. [An ad avalorem tax is a tax based on the assessed value of real estate or merchandise.] Quantitative restrictions were also eliminated. RTAs are most effective when they are a complement to, and not a substitute for, multilateral liberalization.

The economic success of the EU over a long period of time was the result *both* of the removal of internal trade barriers and the growth of intra-European trade; *and* of trade liberalization with the rest of the world. And the EU's trade framework, by locking in the principle of tariff reduction, acted as a disincentive to protectionist lobbying.

There is a great danger that external trade restrictions associated with regional trade agreements will harm the economic prospects of member countries. For efficiency reasons any would-be importer should have incentives to choose the lowest-cost source. Systems that ignore this fundamental precept simply store up trouble for the future—even if the arrangements are meant to be temporary and only a route to further liberalization. Regional trade agreements give higher cost producers an advantage that distorts the market.

And to the extent that RTAs are inward-looking, and give lower priority to trade and economic engagement with the rest of the world, global trade will be lower than it otherwise might be and the global economy will grow less rapidly than it otherwise could.

Regional trade agreements have long proliferated in Africa. Yet as both the IMF paper tabled for this seminar and a recent study by the World Bank conclude the economic benefits have been at best questionable. Intra-African trade actually fell in the 1970s and only recovered to its 1970 levels in the mid-1990s. It remains at around 10% of total African trade.

Tariff and non-tariff barriers to trade are a serious obstacle to trade within Africa and, more importantly, to trade between African countries and the rest of the world. There are also huge practical impediments to trade on the continent. These include high costs, partly a result of poor infrastructure. As the IMF paper notes, it costs around $5,000 to ship a car from Addis Ababa to Abidjan—more than three times the cost of shipping the same car from Japan.

But these practical obstacles also include inefficient customs procedures and other bureaucratic hurdles than often hamper business activity in general. These problems need to be tackled, whether policymakers choose to focus on trade opening at the regional or global level. Without action, such practical obstacles to the distribution and transfer of goods will continue to place limits on the benefits to be had from trade liberalization.

I know that one of the recurring concerns about lowering trade barriers in Africa is the potential revenue losses that governments face. We in the [IMF] understand those fears: we recognize the importance of infrastructure investment in the appropriate macroeconomic framework. But we also remain convinced that revenue concerns should not be an excuse for postponing trade liberalization. Removing quantitative restrictions, which arbitrarily distort market signals, and switching to tariffs will, in the short term boost customs revenues. It is also important to remember that as lower barriers stimulate trade, the revenues from lower tariffs may be greater than those from higher tariffs.

Most important, though, is the need to broaden the tax base and improve tax collection rates. This is the best way of ensuring that government revenues and macroeconomic stability are not undermined. The Fund is ready to provide technical assistance to help in this process.

Conclusion

Let me conclude. Africa has great economic potential. The fruits of macroeconomic reforms are already beginning to show in several countries, with growth picking up and inflation falling.

Trade liberalization has to be a central ingredient in any economic program if lasting high rates of growth are to be achieved. Without progress on lowering trade barriers, economic growth will, at best, be lower than it otherwise would.

It is all too easy to identify the problems associated with opening up an economy to trade and, as a consequence, all too easy to lose sight of the enormous gains that will flow from trade liberalization—not just at the regional level but, more importantly, with the rest of the world economy.

NO

José Bové

Globalisation's Misguided Assumptions

Humanity is grappling with a formidable creed, which, like so many others, is totalitarian and planetary in scope, namely free trade. The gurus and zealous servants of this doctrine ("responsible" people) are saying that the Market is the only god, and that those who want to combat it are heretics ("irresponsible" people). So we find ourselves faced with a modern-day obscurantism—a new opium on which the high priests and traffickers are sure they can make populations dependent. Recent articles in the international press supporting new trade rounds and the like are quite clear on the dogma that some people would like to impose on the men and women of this planet.

More and more people are coming out against the free market credo advocated by the WTO [World Trade Organization], the damage inflicted by it being so plain to see, and the falsehoods on which it is based so blatant.

The first falsehood is the market's self-regulating virtues, which form the basis of the dogma, but this ideological mystification is belied by the facts. In the field of agriculture, for example, since 1992 the major industrialised countries have embraced global markets with open arms—the United States enacted the FAIR (Federal Agriculture Improvement and Reform) Act, a policy instrument that did away with direct production subsidies, instead "decoupling" aid and allowing farmers to produce with no restrictions whatsoever—but this has done nothing to calm the wild swings in the markets.

It has, in fact, done quite the opposite, since markets have experienced unprecedented instability since the trade agreements [modifying and extending the General Agreement on Tariffs and Trade] signed in Marrakech in 1995. The most spectacular effect of this American "decoupling" has been the explosion of emergency direct subsidies to offset declining prices. These subsidies reached a record high of more than $23 billion in 2000 (four times more than the amount budgeted in the 1996 Farm Act).

So, contrary to free-marketers' assertions, markets are inherently unstable and chaotic. Government intervention is needed to regulate markets and adjust price trends, to guarantee producers' incomes and thus ensure that farming activity is sustained.

The second blatant untruth is that competition generates wealth for everyone. Competition is meaningful only if competitors are able to survive. This is especially true for agriculture, where labour productivity varies by a factor of a thousand to one between a grain farmer on the plains of the Middle West and a spade-wielding peasant in the heart of the Sahel.

To claim that the terms of competition will be healthy and fair, and thus tend towards equilibrium if farm policy does not interfere with the workings of a free market, is hypocritical. How can there be a level playing field in the same market between a majority of 1.3 billion farm workers who harvest the land with their hands or with harnessed animals, and a tiny minority of 28 million mechanised farmers formidably equipped for export? How can there be "fair" competition when the most productive farmers of rich countries receive emergency subsidies and multiple guarantees against falling prices on top of their direct and indirect export bonuses?

The third falsehood is that world market prices are a relevant criterion for guiding output. But these prices apply to only a very small fraction of global production and consumption. The world wheat market accounts for only 12% of overall output and international trade takes place at prices that are determined not by aggregate trade, but by the prices of the most competitive exporting country.

The world price of milk and dairy products is determined by production costs in New Zealand, while New Zealand's share of global milk production averaged only 1.63% between 1985 and 1998. The world price of wheat itself is pegged to the price in the United States, which accounted for only 5.84% of aggregate world output from 1985 to 1998.

What is more, these prices are nearly always tantamount to dumping (i.e., to selling below production costs in the producing and importing countries) and are only economically viable for the exporters thanks to the substantial aid they receive in return.

The fourth falsehood is that free trade is the engine of economic development. For free marketers, customs protection schemes are the root of all evil: they claim that such systems stifle trade and economic prosperity, and even hinder cultural exchanges and vital dialogue between peoples. Yet who would dare to claim that decades of massive northbound coffee, cocoa, rice and banana exports have enriched or improved the living standards of farmers in the south? Who would dare to make such a claim, looking these poverty-stricken farmers straight in the eyes? And who would dare to tell African breeders, bankrupted by competition from subsidised meat from Europe, that it is for their own good that customs barriers are falling?

To achieve their ends, the proponents of free trade exploit science in the name of so-called "modernism," asserting that the development of any scientific discovery constitutes progress—as long as it is economically profitable. They cannot bear the idea that life can reproduce on its own, free of charge, whence the race for patents, licences, profits and forcible expropriation.

Obviously, when talking about agriculture, it is impossible not to evoke the farce of GMOs [genetically modified organisms]. Nobody is asking for them, yet they must be the answer to everyone's dreams! There is pressure on us to concede that genetically modified rice (cynically dubbed "golden rice") is going to nourish people who are dying of hunger and protect them from all sorts of diseases, thanks to its new Vitamin A-enriched formula. But this will not solve the problems of vitamin deficiencies, because a person would have to eat three kilograms of dry rice every day, whereas the normal ration is no more than 100 grams.

The way to fight malnutrition, which affects nearly a third of humanity, is to diversify people's diets. This entails rethinking the appalling state of society, underpinned by free market economics which strives to keep wages in southern countries as low as possible in order to maximise profits. It is therefore a good idea to throw some vitamins into the rice that is sold to poor people, so that they don't die too quickly and can continue working for low wages, rather than helping them build a freer and fairer society. Jacques Diouf, Director-General of FAO [Food and Agriculture Organization of the United Nations], recently pointed out that "to feed the 8 003 million people who are hungry, there is no need for GMOs" (*Le Monde*, 10 May). No wonder the Indian farmers of Via Campesina, an international small farmers' movement, destroy fields of genetically modified rice.

The FAO is not the only international institution to question some of the certainties and radical WTO positions regarding the benefits of free markets. The highly free-market OECD [Organization for Economic Co-operation and Development] acknowledges in a recent report entitled *The Well-being of Nations* that the preservation and improvement of government services (healthcare, education) are a key factor underlying the economic success of nations.

We therefore have every reason to oppose the dangerous myth of free trade. Judging by the substantial social and environmental damage free trade has inflicted, before anything else, it is necessary for all of us—farmers and nonfarmers alike—to make it subject to three fundamental principles: food sovereignty—the right of peoples and of countries to produce their food freely, and to protect their agriculture from the ravages of global "competition"; food safety—the right to protect oneself from any threat to one's health; and the preservation of bio-diversity.

Along with adherence to these principles must come a goal of solidarity-based development, via the institution of economic partnership areas among neighbouring countries, including import protection for such groups of countries having uniform structures and levels of development.

The WTO wants to take its free-market logic even further. Next November [2002], in the seclusion of a monarchy that outlaws political parties and demonstrations—Qatar—it will attempt to attain its goals. But if major international institutions are becoming increasingly critical and are casting doubt on these certainties, then mobilised citizens can bring their own laws to bear on trade.

Between the absolute sovereign attitude of nationalists and the proponents of free trade, there are other roads. To echo the theme of the World Social Forum that took place in Porto Alegre last January [2001], "another world is possible!"—a world that respects different cultures and the particularities of each, in a spirit of openness and understanding. We are happy and proud to be part of its emergence.

POSTSCRIPT

Is Economic Globalization a Positive Trend?

Globalization is both old and new. It is old in that the efforts of humans to overcome distance and other barriers to increased interchange have long existed. The first canoes and signal fires are part of the history of globalization. The first event in true globalization occurred between 1519 and 1522, when Ferdinand Magellan circumnavigated the globe. MNCs have existed at least since 1600 when a group of British merchants formed the East India Company. In the main, though, globalization is mostly a modern phenomenon. The progress of globalization until the latter half of the 1800s might be termed "creeping globalization." There were changes, but they occurred very slowly. Since then, the pace of globalization has increased exponentially. A brief introduction to globalization is available in Jurgen Osterhammel, Niels P. Petersson, and Dona Geyer, *Globalization: A Short History* (Princeton University Press, 2005).

The recent rapid pace of globalization has sparked an increasing chorus of criticism against many of its aspects including economic interdependence. Now it is not uncommon for massive protests to occur when the leaders of the world countries meet to discuss global or regional economics or when the WTO, IMF, or World Bank hold important conferences.

One of the oddities about globalization, economic or otherwise, is that it often creates a common cause between those of marked conservative and marked liberal views. More than anything, conservatives worry that their respective countries are losing control of their economies and, thus, a degree of their independence. Echoing this view, archconservative political commentator Patrick Buchanan has warned that unchecked globalization threatens to turn the United States into a "North American province of what some call The New World Order."

Some liberals share the conservatives' negative views of globalization but for different reasons. This perspective is less concerned with sovereignty and security; it is more concerned with workers and countries being exploited and the environment being damaged by MNCs that shift their operations to other countries to find cheap labor and to escape environmental regulations. Referring to violent anti-WTO protests that had recently occurred, U.S. labor leader John J. Sweeney told reporters that they were "just the beginning" and that "If globalization brings more inequality, then it will generate a violent reaction that [these protests] will look tame." One widely read critique of economic globalization is Joseph E. Stiglitz, *Globalization and Its Discontents* (W. W. Norton 2003).

Despite the upsurge of criticism, globalization continues to have many supporters. Most national leaders, especially among the industrialized countries, continue to support free economic interchange. In the United States, both President Bill Clinton and George W. Bush took that stand, although they sometimes also applied protectionist practices to shield U.S. business and workers. The support for globalization is also strong among economists, including Jagdish Bhagwati, *In Defense of Globalization* (Oxford University Press, 2004).

What many analysts argue is that globalization is not good or bad, as such. Rather, it is how it is applied that makes the difference. Two books that look toward reform and discuss new ways of thinking are Manfred B. Steger, ed., *Rethinking Globalism* (Roman & Littlefield, 2003) and Joseph S. Tulching and Gary Bland, *Getting Globalization Right* (Lynne Reiner, 2005). The Internet provides further resources that examine globalization. A good site is The Globalization Web site, http://www.emory.edu/SOC/globalization/, hosted by Emory University. It is even-handed and is particularly meant to support undergraduates.

ISSUE 2

Does Globalization Threaten Cultural Diversity?

YES: Julia Galeota, from "Cultural Imperialism: An American Tradition," *The Humanist* (May/June 2004)

NO: Philippe Legrain, from "In Defense of Globalization," *The International Economy* (Summer 2003)

ISSUE SUMMARY

YES: Julia Galeota of McLean, Virginia, who was seventeen years old when she wrote her essay that won first place for her age category in the 2004 *Humanist* Essay Contest for Young Women and Men of North America, contends that many cultures around the world are gradually disappearing due to the overwhelming influence of corporate and cultural America.

NO: Philippe Legrain, chief economist of Britain in Europe, an organization supporting the adoption by Great Britain of the euro as its currency, counters that it is a myth that globalization involves the imposition of Americanized uniformity, rather than an explosion of cultural exchange.

Globalization is often thought of in terms of economic integration, but it is a much broader phenomenon. Another important aspect of globalization is the spread of national cultures to other countries, regions, and, indeed, the world. One impetus for cultural globalization is economic globalization, as products spread around the world and as huge multinational corporations establish global operations. Additionally, cultural globalization is a product of advances in transportation that allow an increasing number of people to travel to other countries and of radio, television, the Internet, and other advances in communications that permit people to interact passively or actively with others around the world.

To a degree, the culture of many nations is spreading, with Japanese shushi bars now a common site in the United States, Europe, and elsewhere. Cultural globalization also involves a certain amount of cultural amalgamation, with influences merging to create new cultural realities. A third possibility, and

16

the one that is at the heart of this debate, is when the spread of one culture is far greater than the spread of others. Arguably, that is what is currently occurring, with American "cultural exports" much greater than those of any other country.

There is significant evidence of the spread of Western, particularly American, cultural. Casual dress around the world is more apt to include jeans, T-shirts, and sneakers than traditional dress. Young people everywhere listen music by Beyonce, Jay-Z, and others artists, and fast-food hamburgers, fries, and milk shakes are consumed around the world. Adding to the spread of American culture, U.S. movies are everywhere, earning the majority of all film revenues in Japan, Europe, and Latin America, and U.S. television programming is increasingly omnipresent, with, for instance, about two-thirds of the market in Latin America.

Another indication of the spread of American culture is that English is increasingly the common language of business, diplomacy, communications, and even culture. Among Europeans, for instance, nearly all school children receive English instruction, and two-thirds of younger Europeans speak at least some English compared to less than 20 percent of retirement-age Europeans.

It is important to not trivialize cultural globalization even though it involves, in part, fast food, sneakers, rock music, and other elements of pop culture. Some scholars argue the elimination of culture differences will help reduce conflicts as people become more familiar with one another and, indeed, more similar to each other.

Others, however, believe the cultural globalization has negative aspects. One argument is that it is causing a backlash as people face the loss of their own cultures. Some analysts contend that the growth of religious fundamentalism and even terrorism is a reaction to cultural threats. Other analysts believe that defense of cultural traditionalism could even lead to culture wars in the future, with the world dividing itself into antagonist cultural groups. A third worry is represented by Julia Galeota in the first reading, who worries that the spread of American culture amounts for cultural imperialism that is destroying the rich cultural variety that has heretofore marked human society. Galeota suggests that the spread of American culture is a product of American economic and other forms of power, rather than the result of the superiority of American culture and its attractiveness to others. As such, she depicts the Americanization of global culture as cultural imperialism. Philippe Legrain is much more at ease with cultural globalization. He contends that it reflects new realities and is making many contributions, such as giving people the freedom to adopt whatever language, style of dress, or other cultural aspect that they find most compatible with their tastes and needs.

Julia Galeota

 **YES**

Cultural Imperialism: An American Tradition

T ravel almost anywhere in the world today and, whether you suffer from habitual Big Mac cravings or cringe at the thought of missing the newest episode of MTV's *The Real World*, your American tastes can be satisfied practically everywhere. This proliferation of American products across the globe is more than mere accident. As a byproduct of globalization, it is part of a larger trend in the conscious dissemination of American attitudes and values that is often referred to as *cultural imperialism*. In his 1976 work *Communication and Cultural Domination*, Herbert Schiller defines cultural imperialism as:

> The sum of the processes by which a society is brought into the modern world system, and how its dominating stratum is attracted, pressured, forced, and sometimes bribed into shaping social institutions to correspond to, or even to promote, the values and structures of the dominant center of the system.

Thus, cultural imperialism involves much more than simple consumer goods; it involves the dissemination of ostensibly American principles, such as freedom and democracy. Though this process might sound appealing on the surface, it masks a frightening truth: many cultures around the world are gradually disappearing due to the overwhelming influence of corporate and cultural America.

The motivations behind American cultural imperialism parallel the justifications for U.S. imperialism throughout history: the desire for access to foreign markets and the belief in the superiority of American culture. Though the United States does boast the world's largest, most powerful economy, no business is completely satisfied with controlling only the American market; American corporations want to control the other 95 percent of the world's consumers as well. Many industries are incredibly successful in that venture. According to the *Guardian*, American films accounted for approximately 80 percent of global box office revenue in January 2003. And who can forget good old Micky D's? With over 30,000 restaurants in over one hundred countries, the ubiquitous golden arches of McDonald's are now, according to Eric Schlosser's *Fast Food*

From *The Humanist* by Julia Galeota, vol. 64, no. 3, May/June 2004, pp. 22–24, 46. Copyright © 2004 by American Humanist Association. Reprinted by permission.

Nation, "more widely recognized than the Christian cross." Such American domination inevitably hurts local markets, as the majority of foreign industries are unable to compete with the economic strength of U.S. industry. Because it serves American economic interests, corporations conveniently ignore the detrimental impact of American control of foreign markets.

Corporations don't harbor qualms about the detrimental effects of "Americanization" of foreign cultures, as most corporations have ostensibly convinced themselves that American culture is superior and therefore its influence is beneficial to other, "lesser" cultures. Unfortunately, this American belief in the superiority of U.S. culture is anything but new; it is as old as the culture itself. This attitude was manifest in the actions of settlers when they first arrived on this continent and massacred or assimilated essentially the entire "savage" Native American population. This attitude also reflects that of the late nineteenth-century age of imperialism, during which the jingoists attempted to fulfill what they believed to be the divinely ordained "manifest destiny" of American expansion. Jingoists strongly believe in the concept of social Darwinism: the stronger, "superior" cultures will overtake the weaker, "inferior" cultures in a "survival of the fittest." It is this arrogant belief in the incomparability of American culture that characterizes many of our economic and political strategies today.

It is easy enough to convince Americans of the superiority of their culture, but how does one convince the rest of the world of the superiority of American culture? The answer is simple: marketing. Whether attempting to sell an item, a brand, or an entire culture, marketers have always been able to successfully associate American products with modernity in the minds of consumers worldwide. While corporations seem to simply sell Nike shoes or Gap jeans (both, ironically, manufactured *outside* of the United States), they are also selling the image of America as the land of "cool." This indissoluble association causes consumers all over the globe to clamor ceaselessly for the same American products.

Twenty years ago, in his essay "The Globalization of Markets," Harvard business professor Theodore Levitt declared, "The world's needs and desires have been irrevocably homogenized." Levitt held that corporations that were willing to bend to local tastes and habits were inevitably doomed to failure. He drew a distinction between weak multinational corporations that operate differently in each country and strong global corporations that handle an entire world of business with the same agenda.

In recent years, American corporations have developed an even more successful global strategy: instead of advertising American conformity with blonde-haired, blue-eyed, stereotypical Americans, they pitch diversity. These campaigns—such as McDonald's new international "I'm lovin' it" campaign—work by drawing on the United State's history as an ethnically integrated nation composed of essentially every culture in the world. An early example of this global marketing tactic was found in a Coca Cola commercial from 1971 featuring children from many different countries innocently singing, "I'd like to teach the world to sing in perfect harmony/I'd like to buy the world a Coke to keep it company." This commercial illustrates an attempt to portray a U.S. goods as a product capable of transcending political, ethnic, religious, social, and economic differences to unite the world (according to the Coca-Cola Company, we can achieve world peace through consumerism).

More recently, Viacon's MTV has successfully adapted this strategy by integrating many different Americanized cultures into one unbelievably influential American network (with over 280 million subscribers worldwide). According to a 1996 "New World Teen Study" conducted by DMB&B's BrainWaves division, of the 26,700 middle-class teens in forty-five countries surveyed, 85 percent watch MTV every day. These teens absorb what MTV intends to show as a diverse mix of cultural influences but is really nothing more than manufactured stars singing in English to appeal to American popular taste.

If the strength of these diverse "American" images is not powerful enough to move products, American corporations also appropriate local cultures into their advertising abroad. Unlike Levitt's weak multinationals, these corporations don't bend to local tastes; they merely insert indigenous celebrities or trends to present the facade of a customized advertisement. MTV has spawned over twenty networks specific to certain geographical areas such as Brazil and Japan. These specialized networks further spread the association between American and modernity under the pretense of catering to local taste. Similarly, commercials in India in 2000 featured Bollywood stars Hrithik Roshan promoting Coke and Shahrukh Khan promoting Pepsi (Sanjeev Srivastava, "Cola Row in India." BBC News Online). By using popular local icons in their advertisements, U.S. corporations successfully associate what is fashionable in local cultures with what is fashionable in America. America essentially samples the world's cultures, repackages them with the American trademark of materialism, and resells them to the world.

Critics of the theory of American cultural imperialism argue that foreign consumers don't passively absorb the images America bombards upon them. In fact, foreign consumers do play an active role in the reciprocal relationship between buyer and seller. For example, according to Naomi Klein's *No Logo,* American cultural imperialism has inspired a "slow food movement" in Italy and a demonstration involving the burning of chickens outside of the first Kentucky Fried Chicken outlet in India. Though there have been countless other conspicuous and inconspicuous acts of resistance, the intense, unrelenting barrage of American cultural influence continues ceaselessly.

Compounding the influence of commercial images are the media and information industries, which present both explicit and implicit messages about the very real military and economic hegemony of the United States. Ironically, the industry that claims to be the source for "fair and balanced" information plays a large role in the propagation of American influence around the world. The concentration of media ownership during the 1990s enabled both American and British media organizations to gain control of the majority of the world's news services. Satellites allow over 150 million households in approximately 212 countries and territories worldwide to subscribe to CNN, a member of Time Warner, the world's largest media conglomerate. In the words of British sociologist Jeremy Tunstall, "When a government allows news importation, it is in effect importing a piece of another country's politics—which is true of no other import." In addition to politics and commercials, networks like CNN also present foreign countries with unabashed accounts of the military and economic superiority of the United States.

The Internet acts as another vehicle for the worldwide propagation of American influence. Interestingly, some commentators cite the new "information economy" as proof that American cultural imperialism is in decline. They argue that the global accessibility of this decentralized medium has decreased the relevance of the "core and periphery" theory of global influence. This theory describes an inherent imbalance in the primarily outward flow of information and influence from the stronger, more powerful "core" nations such as the United States. Additionally, such critics argue, unlike consumers of other types of media, Internet users must actively seek out information; users can consciously choose to avoid all messages of American culture. While these arguments are valid, they ignore their converse: if one so desires, anyone can access a wealth of information about American culture possibly unavailable through previous channels. Thus, the Internet can dramatically increase exposure to American culture for those who desire it.

Fear of the cultural upheaval that could result from this exposure to new information has driven governments in communist China and Cuba to strictly monitor and regulate their citizens' access to websites (these protectionist policies aren't totally effective, however, because they are difficult to implement and maintain). Paradoxically, limiting access to the Internet nearly ensures that countries will remain largely the recipients, rather than the contributors, of information on the Internet.

Not all social critics see the Americanization of the world as a negative phenomenon. Proponents of cultural imperialism, such as David Rothkopf, a former senior official in Clinton's Department of Commerce, argue that American cultural imperialism is in the interest not only of the United States but also of the world at large. Rothkopf cites Samuel Huntington's theory from *The Clash* of Civilizations and the Beginning of the World Order that, the greater the cultural disparities in the world, the more likely it is that conflict will occur. Rothkopf argues that the removal of cultural barriers through U.S. cultural imperialism will promote a more stable world, one in which American culture reigns supreme as "the most just, the most tolerant, the most willing to constantly reassess and improve itself, and the best model for the future." Rothkopf is correct in one sense: Americans are on the way to establishing a global society with minimal cultural barriers. However, one must question whether this projected society is truly beneficial for all involved. Is it worth sacrificing countless indigenous cultures for the unlikely promise of a world without conflict?

Around the world, the answer is an overwhelming "No!" Disregarding the fact that a world of homogenized culture would not necessarily guarantee a world without conflict, the complex fabric of diverse cultures around the world is a fundamental and indispensable basis of humanity. Throughout the course of human existence, millions have died to preserve their indigenous culture. It is a fundamental right of humanity to be allowed to preserve the mental, physical, intellectual, and creative aspects of one's society. A single "global culture" would be nothing more than a shallow, artificial "culture" of materialism reliant on technology. Thankfully, it would be nearly impossible to create one bland culture in a world of over six billion people. And nor should we want to. Contrary to Rothkopf's (and George W. Bush's) belief that, "Good and evil, better and worse coexist in this world," there are no such absolutes in this world. The United States should not be able to relentlessly force other nations to accept its definition of what is "good" and "just" or even "modern."

Fortunately, many victims of American cultural imperialism aren't blind to the subversion of their cultures. Unfortunately, these nations are often too weak to fight the strength of the United States and subsequently to preserve their native cultures. Some countries—such as France, China, Cuba, Canada, and Iran—have attempted to quell America's cultural influence by limiting or prohibiting access to American cultural programming through satellites and the Internet. However, according to the UN Universal Declaration of Human Rights, it is a basic right of all people to "seek, receive, and impart information and ideas through any media and regardless of frontiers," Governments shouldn't have to restrict their citizens' access to information in order to preserve their native cultures. We as a world must find ways to defend local cultures in a manner that does not compromise the rights of indigenous people.

The prevalent proposed solutions to the problem of American cultural imperialism are a mix of defense and compromise measures on behalf of the endangered cultures. In *The Lexus and the Olive Tree,* Thomas Friedman advocates the use of protective legislation such as zoning laws and protected area laws, as well as the appointment of politicians with cultural integrity, such as those in agricultural, culturally pure Southern France. However, many other nations have no voice in the nomination of their leadership, so those countries need a middle-class and elite committed to social activism. If it is utterly impossible to maintain the cultural purity of a country through legislation, Friedman suggests the country attempt to "glocalize," that is:

> To absorb influences that naturally fit into and can enrich [a] culture, to resist those things that are truly alien and to compartmentalize those things that, while different, can nevertheless be enjoyed and celebrated as different.

These types of protective filters should help to maintain the integrity of a culture in the face of cultural imperialism. In *Jihad vs. McWorld,* Benjamin Barber calls for the resuscitation of nongovernmental, noncapitalist spaces—to the "civic spaces"—such as village greens, places of religious worship, or community schools. It is also equally important to focus on the education of youth in their native values and traditions. Teens especially need a counterbalance images of American consumerism they absorb from the media. Even if individuals or countries consciously choose to become "Americanized" or "modernized," their choice should be made freely and independently of the coercion and influence of American cultural imperialism.

The responsibility for preserving cultures shouldn't fall entirely on those at risk. The United States must also recognize that what is good for its economy isn't necessarily good for the world at large. We must learn to put people before profits. The corporate and political leaders of the United States would be well advised to heed these words of Gandhi:

> I do not want my house to be walled in on all sides and my windows to be stuffed. I want the culture of all lands to be blown about my house as freely as possible. But I refuse to be blown off my feet by any.

The United States must acknowledge that no one culture can or should reign supreme, for the death of diverse cultures can only further harm future generations.

In Defense of Globalization

Fears that globalization is imposing a deadening cultural uniformity are as ubiquitous as Coca-Cola, McDonald's, and Mickey Mouse. Many people dread that local cultures and national identifies are dissolving into a crass all-American consumerism. That cultural imperialism is said to impose American values as well as products, promote the commercial at the expense of the authentic, and substitute shallow gratification for deeper satisfaction.

Thomas Friedman, columnist for the *New York Times* and author of *The Lexus and the Olive Tree*, believes that globalization is "globalizing American culture and American cultural icons." Naomi Klein, a Canadian journalist and author of *No Logo*, argues that "Despite the embrace of polyethnic imagery, market-driven globalization doesn't want diversity; quite the opposite. Its enemies are national habits, local brands, and distinctive regional tastes."

But it is a myth that globalization involves the imposition of Americanized uniformity, rather than an explosion of cultural exchange. And although—as with any change—it can have downsides, this cross-fertilization is overwhelmingly a force for good.

The beauty of globalization is that it can free people from the tyranny of geography. Just because someone was born in France does not mean they can only aspire to speak French, eat French food, read French books, and so on. That we are increasingly free to choose our cultural experiences enriches our lives immeasurably. We could not always enjoy the best the world has to offer.

Globalization not only increases individual freedom, but also revitalizes cultures and cultural artifacts through foreign influences, technologies, and markets. Many of the best things come from cultures mixing: Paul Gauguin painting in Polynesia, the African rhythms in rock 'n' roll, the great British curry. Admire the many-colored faces of France's World Cup-winning soccer team, the ferment of ideas that came from Eastern Europe's Jewish diaspora, and the cosmopolitan cities of London and New York.

Fears about an Americanized uniformity are overblown. For a start, many "American" products are not as all-American as they seem; MTV in Asia promotes Thai pop stars and plays rock music sung in Mandarin. Nor are American products all-conquering. Coke accounts for less than two of the 64 fluid ounces that the typical person drinks a day. France imported a mere $620 million in

From *The International Economy* by Philippe Legrain, vol. 17, no. 3, Summer 2003, pp. 62–65.

food from the United States in 2000, while exporting to America three times that. Worldwide, pizzas are more popular than burgers and Chinese restaurants sprout up everywhere.

In fashion, the ne plus ultra is Italian or French. Nike shoes are given a run for their money by Germany's Adidas, Britain's Reebok, and Italy's Fila. American pop stars do not have the stage to themselves. According to the IFPI, the record-industry bible, local acts accounted for 68 percent of music sales in 2000, up from 58 percent in 1991. And although nearly three-quarters of television drama exported worldwide comes from the United States, most countries' favorite shows are homegrown.

Nor are Americans the only players in the global media industry. Of the seven market leaders, one is German, one French, and one Japanese. What they distribute comes from all quarters: Germany's Bertelsmann publishes books by American writers; America's News Corporation broadcasts Asian news; Japan's Sony sells Brazilian music.

In some ways, America is an outlier, not a global leader. Baseball and American football have not traveled well; most prefer soccer. Most of the world has adopted the (French) metric system; America persists with antiquated British Imperial measurements. Most developed countries have become intensely secular, but many Americans burn with fundamentalist fervor—like Muslims in the Middle East.

Admittedly, Hollywood dominates the global movie market and swamps local products in most countries. American fare accounts for more than half the market in Japan and nearly two-thirds in Europe. Yet Hollywood is less American than it seems. Top actors and directors are often from outside America. Some studios are foreign-owned. To some extent, Hollywood is a global industry that just happens to be in America. Rather than exporting Americana, it serves up pap to appeal to a global audience.

Hollywood's dominance is in part due to economics: Movies cost a lot to make and so need a big audience to be profitable; Hollywood has used America's huge and relatively uniform domestic market as a platform to expand overseas. So there could be a case for stuffing subsidies into a rival European film industry, just as Airbus was created to challenge Boeing's near-monopoly. But France's subsidies have created a vicious circle whereby European film producers fail in global markets because they serve domestic demand and the wishes of politicians and cinematic bureaucrats.

Another American export is also conquering the globe: English. By 2050, it is reckoned, half the world will be more or less proficient in it. A common global language would certainly be a big plus—for businessmen, scientists, and tourists—but a single one seems far less desirable. Language is often at the heart of national culture, yet English may usurp other languages not because it is what people prefer to speak, but because, like Microsoft software, there are compelling advantages to using it if everyone else does.

But although many languages are becoming extinct, English is rarely to blame. People are learning English as well as—not instead of—their native tongue, and often many more languages besides. Where local languages are dying, it is typically national rivals that are stamping them out. So although, within the United States, English is displacing American Indian tongues, it is not doing away with Swahili or Norwegian.

Even though American consumer culture is widespread, its significance is often exaggerated. You can choose to drink Coke and eat at McDonald's without becoming American in any meaningful sense. One newspaper photo of Taliban fighters in Afghanistan showed them toting Kalashnikovs—as well as a sports bag with Nike's trademark swoosh. People's culture—in the sense of their shared ideas, beliefs, knowledge, inherited traditions, and art—may scarcely be eroded by mere commercial artifacts that, despite all the furious branding, embody at best flimsy values.

The really profound cultural changes have little to do with Coca-Cola. Western ideas about liberalism and science are taking root almost everywhere, while Europe and North America are becoming multicultural societies through immigration, mainly from developing countries. Technology is reshaping culture: Just think of the Internet. Individual choice is fragmenting the imposed uniformity of national cultures. New hybrid cultures are emerging, and regional ones re-emerging. National identity is not disappearing, but the bonds of nationality are loosening.

Cross-border cultural exchange increases diversity within societies—but at the expense of making them more alike. People everywhere have more choice, but they often choose similar things. That worries cultural pessimists, even though the right to choose to be the same is an essential part of freedom.

Cross-cultural exchange can spread greater diversity as well as greater similarity: more gourmet restaurants as well as more McDonald's outlets. And just as a big city can support a wider spread of restaurants than a small town, so a global market for cultural products allows a wider range of artists to thrive. If all the new customers are ignorant, a wider market may drive down the quality of cultural products: Think of tourist souvenirs. But as long as some customers are well informed (or have "good taste"), a general "dumbing down" is unlikely. Hobbyists, fans, artistic pride, and professional critics also help maintain (and raise) standards.

A bigger worry is that greater individual freedom may undermine national identity. The French fret that by individually choosing to watch Hollywood films they might unwittingly lose their collective Frenchness. Yet such fears are overdone. Natural cultures are much stronger than people seem to think. They can embrace some foreign influences and resist others. Foreign influences can rapidly become domesticated, changing national culture, but not destroying it. Clearly, though, there is a limit to how many foreign influences a culture can absorb before being swamped. Traditional cultures in the developing world that have until now evolved (or failed to evolve) in isolation may be particularly vulnerable.

In *The Silent Takeover*, Noreena Hertz describes the supposed spiritual Eden that was the isolated kingdom of Bhutan in the Himalayas as being defiled by such awful imports as basketball and Spice Girls T-shirts. But is that such a bad thing? It is odd, to put it mildly, that many on the left support multiculturalism in the West but advocate cultural purity in the developing world—an attitude they would tar as fascist if proposed for the United States. Hertz appears to want people outside the industrialized West preserved in unchanging but supposedly pure poverty. Yet the Westerners who want this supposed paradise preserved in aspic rarely feel like settling there. Nor do most people in developing countries want to lead an "authentic" unspoiled life of isolated poverty.

In truth, cultural pessimists are typically not attached to diversity per se but to designated manifestations of diversity, determined by their preferences. Cultural pessimists want to freeze things as they were. But if diversity at any point in time is desirable, why isn't diversity across time? Certainly, it is often a shame if ancient cultural traditions are lost. We should do our best to preserve them and keep them alive where possible. Foreigners can often help, by providing the new customers and technologies that have enabled reggae music, Haitian art, and Persian carpet making, for instance, to thrive and reach new markets. But people cannot be made to live in a museum. We in the West are forever casting off old customs when we feel they are no longer relevant. Nobody argues that Americans should ban nightclubs to force people back to line dancing. People in poor countries have a right to change, too.

Moreover, some losses of diversity are a good thing. Who laments that the world is now almost universally rid of slavery? More generally, Western ideas are reshaping the way people everywhere view themselves and the world. Like nationalism and socialism before it, liberalism is a European philosophy that has swept the world. Even people who resist liberal ideas, in the name of religion (Islamic and Christian fundamentalists), group identity (communitarians), authoritarianism (advocates of "Asian values") or tradition (cultural conservatives), now define themselves partly by their opposition to them.

Faith in science and technology is even more widespread. Even those who hate the West make use of its technologies. Osama bin Laden plots terrorism on a cellphone and crashes planes into skyscrapers. Antiglobalization protesters organize by e-mail and over the Internet. China no longer turns its nose up at Western technology: It tries to beat the West at its own game.

Yet globalization is not a one-way street. Although Europe's former colonial powers have left their stamp on much of the world, the recent flow of migration has been in the opposite direction. There are Algerian suburbs in Paris, but not French ones in Algiers. Whereas Muslims are a growing minority in Europe, Christians are a disappearing one in the Middle East.

Foreigners are changing America even as they adopt its ways. A million or so immigrants arrive each year, most of them Latino or Asian. Since 1990, the number of foreign-born American residents has risen by 6 million to just over 25 million, the biggest immigration wave since the turn of the 20th century. English may be all-conquering outside America, but in some parts of the United States, it is now second to Spanish.

The upshot is that national cultures are fragmenting into a kaleidoscope of different ones. New hybrid cultures are emerging. In "Amexica" people speak Spanglish. Regional cultures are reviving. The Scots and Welsh break with British monoculture. Estonia is reborn from the Soviet Union. Voices that were silent dare to speak again.

Individuals are forming new communities, linked by shared interests and passions, that cut across national borders. Friendships with foreigners met on holiday. Scientists sharing ideas over the Internet. Environmentalists campaigning together using e-mail. Greater individualism does not spell the end of community. The new communities are simply chosen rather than coerced, unlike the older ones that communitarians hark back to.

So is national identity dead? Hardly. People who speak the same language, were born and live near each other, face similar problems, have a common experience, and vote in the same elections still have plenty in common. For all our awareness of the world as a single place, we are not citizens of the world but citizens of a state. But if people now wear the bonds of nationality more loosely, is that such a bad thing? People may lament the passing of old ways. Indeed, many of the worries about globalization echo age-old fears about decline, a lost golden age, and so on. But by and large, people choose the new ways because they are more relevant to their current needs and offer new opportunities.

The truth is that we increasingly define ourselves rather than let others define us. Being British or American does not define who you are: It is part of who you are. You can like foreign things and still have strong bonds to your fellow citizens. As Mario Vargas Llosa, the Peruvian author, has written: "Seeking to impose a cultural identity on a people is equivalent to locking them in a prison and denying them the most precious of liberties—that of choosing what, how, and who they want to be."

POSTSCRIPT

Does Globalization Threaten Cultural Diversity?

Cultural globalization, dominated by the spread of Western, primarily American, culture is likely to continue into the foreseeable future. For example, English may not be a common global language, but the possibility of that occurring is given some credence by a survey of people in 42 countries that recorded the vast majority in every region agreed with the statement, "Children need to learn English to succeed in the world today."

Attitudes toward cultural globalization are less clear-cut and are even contradictory. A global survey found that, on average, three-quarters of all people thought culture imports were good. Regionally, that favorable response ranged from 61 percent in the Middle East to 86 percent in western Europe. At the same time, though, an approximately equal percentage of people thought cultural imports were eroding their traditional way of life, with Africans, at 86 percent, the most likely to think so. Not surprisingly, this sense of cultural threat also leads to a desire to protect traditional cultures. The survey found that about 70 percent of its respondents felt that their way of life needed protection from foreign influence. At 79 percent each, people in Africa and the Middle East were most likely to feel their traditional cultures needed protection; western Europeans (56 percent) were the least insecure. Whether this sense of cultural loss is all due to globalization is unclear. It may well be that the changes that are unsettling most people worldwide are also part of the even broader phenomenon of rapid technological modernization that is spurring globalization.

An oddity of the cultural globalization phenomenon is that American attitudes are not much different from those of other people, despite the worry that American culture is becoming dominant. The poll showed that the overwhelming majority of Americans were favorable to the increased availability of goods, music, films, and other cultural imports. Yet two-thrids of all Americans replied to the survey that their traditional way of life was being lost, and a similar percentage responded that they believed that their way of life needed to be protected against foreign influences.

For more on the progress of globalization, visit the British Broadcasting Corporation's Web site at http://www.bbc.co.uk/worldservice/programmes/globalisation/ and Randolph Kluver and Wayne Fu, "The Cultural Globalization Index," posted in February 2004 to the Web site of *Foreign Policy* at http://www.foreignpolicy. For a mostly negative view of cultural globalization, see John Tomlinson, *Globalization and Culture* (University of Chicago Press, 1999). F. Jan Nederveen Pieterse, *Globalization and Culture* (Rowman & Littlefield, 2003) views the process as a cultural hybridization rather than Americanization. Of all countries, the less-developed ones are the most strongly impacted by cultural globalization, as discussed by Jeff Haynes in *Religion, Globalization, and Political Culture in the Third World* (St. Martin's Press, 1999). A discussion of maintaining cultural distinctiveness can be found in Harry Redner, *Conserving Cultures: Technology, Globalization, and the Future of Local Cultures* (Rowman & Littlefield, 2003).

ISSUE 3

Will State Sovereignty Survive Globalism?

YES: Stephen D. Krasner, from "Sovereignty," *Foreign Policy* (January/February 2001)

NO: Kimberly Weir, from "The Waning State of Sovereignty," An Original Essay Written for This Volume (2002)

ISSUE SUMMARY

YES: Professor of international relations Stephen D. Krasner contends that the nation-state has a keen instinct for survival and will adapt to globalization and other challenges to sovereignty.

NO: Kimberly Weir, an assistant professor of political science, maintains that the tide of history is running against the sovereign state as a governing principle, which will soon go the way of earlier, now-discarded forms of governance, such as empire.

There are political, economic, and social forces that are working to break down the importance and authority of states and that are creating pressures to move the world toward a much higher degree of political, economic, and social integration. Whatever one may think of international organizations and their roles, the increasing number and importance of them provide evidence of the trend toward globalization.

Many have questioned whether or not the growth of international law and norms, international organizations, and other transnational phenomena are lessening the sovereignty of states. The debate concerning this issue takes the discussion over the international political system yet another step by asking what the future holds for states. Will the state persist as the principal actor in the international system or be eclipsed?

To grasp this debate it is crucial to understand that countries (states) are not a natural political order. Instead, there are two things to bear in mind. One is that states as we know them have not always existed as a form of political organization. Logically, this means that states need not always exist. Second, states, or any form of governance, are best regarded as tools. They are vehicles to serve the interests of their citizens and arguably only deserve loyalty as long as they provide benefits.

The modern state is largely a Western creation. For almost a millennium after the universalistic Roman Empire fell in 476, political power rested at two levels of authority—one universal, the other local. On the universal level authority existed in the form of the Roman Catholic Church in an era when kings and other secular leaders were subordinate in theory, and often in practice, to popes. Later, the broad authority of the Catholic Church was supplemented and, in some cases, supplanted by the Holy Roman Empire and other multieth-nic empires that exercised control over many different peoples. Most of the people within these empires had little if any sense of loyalty to the empire.

The second level was local feudal authority in principalities, dukedoms, baronies, and other such fiefdoms in political units that were smaller than most modern states. Here, too, the common people were not citizens, as we know the concept today. Instead, the minor royalty came close to owning the peasantry.

For a variety of economic, military, and other reasons, this old system failed, and a new system based on territorially defined sovereign states slowly evolved.

States came to dominate the international system not because of any ideological reason but simply because they worked better than the models of governance that had failed or any of the other models (such as a rebirth of city-states like Venice) that were tried. The Treaty of Westphalia (1648), which divided Europe into Catholic and Protestant states and marked the end of the pan-European dominance of the Holy Roman Empire, is often used to symbolize the establishment of the state. The treaty was certainly important, but, in fact, that establishment of the state and the growth of the concept of citizenry and patriotism evolved over centuries.

Once this occurred, all of the basic parameters for the modern state that exists today were in place. Yet in the ceaseless ebb and flow of world forces, pressures were beginning to build that would work to undermine the modern states. Those economic, military, and other forces have, in the view of some, built up rapidly since the mid-twentieth century and cast doubt on the future role, even existence, of states as sovereign entities at the center of the international system.

There are some observers who contend that globalization of trade, communications, and other processes do not and should not threaten the existence of the state as a fundamental political unit. Analysts of this persuasion believe that countries can, will, and should continue to exist as the sovereign entities they are today. Stephen D. Krasner takes this point of view in the following selection.

Other analysts believe that states have run their course as a model for governance and are becoming outmoded in a world that is increasingly inter-dependent. They note that the world has changed greatly since the rise of the state to political dominance centuries ago. The analysts ask whether or not it is reasonable to assume that a model of governance that worked hundreds of years ago is the best model for the future. Their answer is no. Kimberly Weir represents this view in the second of the following selections.

YES

Stephen D. Krasner

Sovereignty

The idea of states as autonomous, independent entities is collapsing under the combined onslaught of monetary unions, CNN, the Internet, and non-governmental organizations [NGOs]. But those who proclaim the death of sovereignty misread history. The nation-state has a keen instinct for survival and has so far adapted to new challenges—even the challenge of globalization.

The Sovereign State Is Just About Dead

Very wrong Sovereignty was never quite as vibrant as many contemporary observers suggest. The conventional norms of sovereignty have always been challenged. A few states, most notably the United States, have had autonomy, control, and recognition for most of their existence, but most others have not. The polities of many weaker states have been persistently penetrated, and stronger nations have not been immune to external influence. China was occupied. The constitutional arrangements of Japan and Germany were directed by the United States after World War II. The United Kingdom, despite its rejection of the euro, is part of the European Union [EU].

Even for weaker states—whose domestic structures have been influenced by outside actors, and whose leaders have very little control over transborder movements or even activities within their own country—sovereignty remains attractive. Although sovereignty might provide little more than international recognition, that recognition guarantees access to international organizations and sometimes to international finance. It offers status to individual leaders. While the great powers of Europe have eschewed many elements of sovereignty, the United States, China, and Japan have neither the interest nor the inclination to abandon their usually effective claims to domestic autonomy.

In various parts of the world, national borders still represent the fault lines of conflict, whether it is Israelis and Palestinians fighting over the status of Jerusalem, Indians and Pakistanis threatening to go nuclear over Kashmir, or Ethiopia and Eritrea clashing over disputed territories. Yet commentators nowadays are mostly concerned about the erosion of national borders as a consequence of globalization. Governments and activists alike complain that multilateral institutions such as the United Nations, the World Trade Organization, and the International Monetary Fund overstep their authority by promoting universal standards for everything from human rights and the environment to monetary policy and immigration. However, the most important impact of economic globalization and transnational norms will be to alter the scope of state authority rather than to generate some fundamentally new way to organize political life.

Sovereignty Means Final Authority

Not anymore, if ever   When philosophers Jean Bodin and Thomas Hobbes first elaborated the notion of sovereignty in the 16th and 17th centuries, they were concerned with establishing the legitimacy of a single hierarchy of domestic authority. Although Bodin and Hobbes accepted the existence of divine and natural law, they both (especially Hobbes) believed the word of the sovereign was law. Subjects had no right to revolt. Bodin and Hobbes realized that imbuing the sovereign with such overweening power invited tyranny, but they were predominately concerned with maintaining domestic order, without which they believed there could be no justice. Both were writing in a world riven by sectarian strife. Bodin was almost killed in religious riots in France in 1572. Hobbes published his seminal work, *Leviathan,* only a few years after parliament (composed of Britain's emerging wealthy middle class) had executed Charles I in a civil war that had sought to wrest state control from the monarchy.

This idea of supreme power was compelling, but irrelevant in practice. By the end of the 17th century, political authority in Britain was divided between king and parliament. In the United States, the Founding Fathers established a constitutional structure of checks and balances and multiple sovereignties distributed among local and national interests that were inconsistent with hierarchy and supremacy. The principles of justice, and especially order, so valued by Bodin and Hobbes, have best been provided by modern democratic states whose organizing principles are antithetical to the idea that sovereignty means uncontrolled domestic power.

If sovereignty does not mean a domestic order with a single hierarchy of authority, what does it mean? In the contemporary world, sovereignty primarily has been linked with the idea that states are autonomous and independent from each other. Within their own boundaries, the members of a polity are free to choose their own form of government. A necessary corollary of this claim is the principle of nonintervention: One state does not have a right to intervene in the internal affairs of another.

More recently, sovereignty has come to be associated with the idea of control over transborder movements. When contemporary observers assert that the sovereign state is just about dead, they do not mean that constitutional structures are about to disappear. Instead, they mean that technological change has made it very difficult, or perhaps impossible, for states to control movements across their borders of all kinds of material things (from coffee to cocaine) and not-so-material things (from Hollywood movies to capital flows). Finally, sovereignty has meant that political authorities can enter into international agreements. They are free to endorse any contract they find attractive. Any treaty among states is legitimate provided that it has not been coerced. . . .

Universal Human Rights Are an Unprecedented Challenge to Sovereignty

Wrong   The struggle to establish international rules that compel leaders to treat their subjects in a certain way has been going on for a long time. Over the centuries the emphasis has shifted from religious toleration, to minority rights (often focusing on specific ethnic groups in specific countries), to human rights (emphasizing rights enjoyed by all or broad classes of individuals). In a few instances states have voluntarily embraced international supervision, but generally the weak have acceded to the preferences of the strong: The Vienna settlement following the Napoleonic wars guaranteed religious toleration for Catholics in the Netherlands. All of the successor states of the Ottoman Empire, beginning with Greece in 1832 and ending with Albania in 1913, had to accept provisions for civic and political equality for religious minorities as a condition for international recognition. The peace settlements following World War I included extensive provisions for the protection of minorities. Poland, for instance, agreed to refrain from holding elections on Saturday because such balloting would have violated the Jewish Sabbath. Individuals could bring complaints against governments through a minority rights bureau established within the League of Nations.

But as the Holocaust tragically demonstrated, interwar efforts at international constraints on domestic practices failed dismally. After World War II, human, rather than minority, rights became the focus of attention. The United Nations Charter endorsed both human rights and the classic sovereignty principle of nonintervention. The 20-plus human rights accords that have been signed during the last half century cover a wide range of issues including genocide, torture, slavery, refugees, stateless persons, women's rights, racial discrimination, children's rights, and forced labor. These U.N. agreements, however, have few enforcement mechanisms, and even their provisions for reporting violations are often ineffective.

The tragic and bloody disintegration of Yugoslavia in the 1990s revived earlier concerns with ethnic rights. International recognition of the Yugoslav successor states was conditional upon their acceptance of constitutional provisions guaranteeing minority rights. The Dayton accords established externally controlled authority structures in Bosnia, including a Human Rights Commission (a majority of whose members were appointed by the Western European states). NATO [North Atlantic Treaty Organization] created a de facto protectorate in Kosovo.

The motivations for such interventions—humanitarianism and security—have hardly changed. Indeed, the considerations that brought the great powers into the Balkans following the wars of the 1870s were hardly different from those that engaged NATO and Russia in the 1990s.

Globalization Undermines State Control

No   State control could never be taken for granted. Technological changes over the last 200 years have increased the flow of people, goods, capital, and ideas—but the problems posed by such movements are not new. In many ways, states are better able to respond now than they were in the past.

The impact of the global media on political authority (the so-called CNN effect) pales in comparison to the havoc that followed the invention of the printing press. Within a decade after Martin Luther purportedly nailed his 95 theses to the Wittenberg church door, his ideas had circulated throughout Europe. Some political leaders seized upon the principles of the Protestant Reformation as a way to legitimize secular political authority. No sovereign monarch could contain the spread of these concepts, and some lost not only their lands but also their heads. The sectarian controversies of the 16th and 17th centuries were perhaps more politically consequential than any subsequent transnational flow of ideas.

In some ways, international capital movements were more significant in earlier periods than they are now. During the 19th century, Latin American states (and to a lesser extent Canada, the United States, and Europe) were beset by boom-and-bust cycles associated with global financial crises. The Great Depression, which had a powerful effect on the domestic politics of all major states, was precipitated by an international collapse of credit. The Asian financial crisis of the late 1990s was not nearly as devastating. Indeed, the speed with which countries recovered from the Asian flu reflects how a better working knowledge of economic theories and more effective central banks have made it easier for states to secure the advantages (while at the same time minimizing the risks) of being enmeshed in global financial markets.

In addition to attempting to control the flows of capital and ideas, states have long struggled to manage the impact of international trade. The opening of long-distance trade for bulk commodities in the 19th century created fundamental cleavages in all of the major states. Depression and plummeting grain prices made it possible for German Chancellor Otto von Bismarck to prod the landholding aristocracy into a protectionist alliance with urban heavy industry (this coalition of "iron and rye" dominated German politics for decades). The tariff question was a basic divide in U.S. politics for much of the last half of the19th and first half of the 20th centuries. But, despite growing levels of imports and exports since 1950, the political salience of trade has receded because national governments have developed social welfare strategies that cushion the impact of international competition, and workers with higher skill levels are better able to adjust to changing international conditions. It has become easier, not harder, for states to manage the flow of goods and services.

Globalization Is Changing the Scope of State Control

Yes   The reach of the state has increased in some areas but contracted in others. Rulers have recognized that their effective control can be enhanced by walking away from issues they cannot resolve. For instance, beginning with the Peace of Westphalia [1648 treaty often cited as the political big bang that created the modern system of autonomous states], leaders chose to surrender their control over religion because it proved too volatile. Keeping religion within the scope of state authority undermined, rather than strengthened, political stability. Monetary policy is an area where state control expanded and then ultimately contracted. Before the 20th century, states had neither the administrative competence nor the inclination to conduct independent monetary policies. The mid-20th-century effort to control monetary affairs, which was associated with Keynesian economics, has now been reversed due to the magnitude of short-term capital flows and the inability of some states to control inflation. With the exception of Great Britain, the major European states have established a single monetary authority. Confronting recurrent hyperinflation, Ecuador adopted the U.S. dollar as its currency in 2000.

Along with the erosion of national currencies, we now see the erosion of national citizenship—the notion that an individual should be a citizen of one and only one country, and that the state has exclusive claims to that person's loyalty. For many states, there is no longer a sharp distinction between citizens and noncitizens. Permanent residents, guest workers, refugees, and undocumented immigrants are entitled to some bundle of rights even if they cannot vote. The ease of travel and the desire of many countries to attract either capital or skilled workers have increased incentives to make citizenship a more flexible category.

Although government involvement in religion, monetary affairs, and claims to loyalty has declined, overall government activity, as reflected in taxation and government expenditures, has increased as a percentage of national income since the 1950s among the most economically advanced states. The extent of a country's social welfare programs tends to go hand in hand with its level of integration within the global economy. Crises of authority and control have been most pronounced in the states that have been the most isolated, with sub-Saharan Africa offering the largest number of unhappy examples.

NGOs Are Nibbling at National Sovereignty

To some extent Transnational nongovernmental organizations (NGOs) have been around for quite awhile, especially if you include corporations. In the 18th century, the East India Company possessed political power (and even an expeditionary military force) that rivaled many national governments. Throughout the 19th century, there were transnational movements to abolish slavery, promote the rights of women, and improve conditions for workers. The number of transnational NGOs, however, has grown tremendously, from around 200 in 1909 to over 17,000 today. The availability of inexpensive and very fast communications technology has made it easier for such groups to organize and make an impact on public policy and international law—the international agreement banning land mines being a recent case in point. Such groups prompt questions about sovereignty because they appear to threaten the integrity of domestic decision making. Activists who lose on their home territory can pressure foreign governments, which may in turn influence decision makers in the activists' own nation.

But for all of the talk of growing NGO influence, their power to affect a country's domestic affairs has been limited when compared to governments, international organizations, and multinational corporations. The United Fruit Company had more influence in Central America in the early part of the 20th century than any NGO could hope to have anywhere in the contemporary

world. The International Monetary Fund and other multilateral financial institutions now routinely negotiate conditionality agreements that involve not
only specific economic targets but also domestic institutional changes, such as
pledges to crack down on corruption and break up cartels.

Smaller, weaker states are the most frequent targets of external efforts to
alter domestic institutions, but more powerful states are not immune. The
openness of the U.S. political system means that not only NGOs, but also foreign governments, can play some role in political decisions. (The Mexican government, for instance, lobbied heavily for the passage of the North American
Free Trade Agreement [NAFTA].) In fact, the permeability of the American polity
makes the United States a less threatening partner; nations are more willing to
sign on to U.S.-sponsored international arrangements because they have some
confidence that they can play a role in U.S. decision making.

Sovereignty Blocks Conflict Resolution

Yes, sometimes   Rulers as well as their constituents have some reasonably clear
notion of what sovereignty means—exclusive control within a given territory—
even if this norm has been challenged frequently by inconsistent principles
(such as universal human rights) and violated in practice (the U.S.- and British-
enforced no-fly zones over Iraq). In fact, the political importance of conventional sovereignty rules has made it harder to solve some problems. There is, for
instance, no conventional sovereignty solution for Jerusalem, but it doesn't
require much imagination to think of alternatives: Divide the city into small
pieces; divide the Temple Mount vertically with the Palestinians controlling the
top and the Israelis the bottom; establish some kind of international authority;
divide control over different issues (religious practices versus taxation, for
instance) among different authorities. Any one of these solutions would be better for most Israelis and Palestinians than an ongoing stalemate, but political
leaders on both sides have had trouble delivering a settlement because they are
subject to attacks by counterelites who can wave the sovereignty flag.

Conventional rules have also been problematic for Tibet. Both the Chinese
and the Tibetans might be better off if Tibet could regain some of the autonomy
it had as a tributary state within the traditional Chinese empire. Tibet had extensive local control, but symbolically (and sometimes through tribute payments)
recognized the supremacy of the emperor. Today, few on either side would even
know what a tributary state is, and even if the leaders of Tibet worked out some
kind of settlement that would give their country more self-government, there
would be no guarantee that they could gain the support of their own constituents.

If, however, leaders can reach mutual agreements, bring along their constituents, or are willing to use coercion, sovereignty rules can be violated in inven-
tive ways. The Chinese, for instance, made Hong Kong a special administrative
region after the transfer from British rule, allowed a foreign judge to sit on the
Court of Final Appeal, and secured acceptance by other states not only for Hong
Kong's participation in a number of international organizations but also for separate visa agreements and recognition of a distinct Hong Kong passport. All of
these measures violate conventional sovereignty rules since Hong Kong does not

have juridical independence. Only by inventing a unique status for Hong Kong, which involved the acquiescence of other states, could China claim sovereignty while simultaneously preserving the confidence of the business community.

The European Union Is a New Model for Supranational Governance

Yes, but only for the Europeans   The European Union (EU) really is a new thing, far more interesting in terms of sovereignty than Hong Kong. It is not a conventional international organization because its member states are now so intimately linked with one another that withdrawal is not a viable option. It is not likely to become a "United States of Europe"—a large federal state that might look something like the United States of America—because the interests, cultures, economies, and domestic institutional arrangements of its members are too diverse. Widening the EU to include the former communist states of Central Europe would further complicate any efforts to move toward a political organization that looks like a conventional sovereign state.

The EU is inconsistent with conventional sovereignty rules. Its member states have created supranational institutions (the European Court of Justice, the European Commission, and the Council of Ministers) that can make decisions opposed by some member states. The rulings of the court have direct effect and supremacy within national judicial systems, even though these doctrines were never explicitly endorsed in any treaty. The European Monetary Union created a central bank that now controls monetary affairs for three of the union's four largest states. The Single European Act and the Maastricht Treaty provide for majority or qualified majority, but not unanimous, voting in some issue areas. In one sense, the European Union is a product of state sovereignty because it has been created through voluntary agreements among its member states. But, in another sense, it fundamentally contradicts conventional understandings of sovereignty because these same agreements have undermined the juridical autonomy of its individual members.

The European Union, however, is not a model that other parts of the world can imitate. The initial moves toward integration could not have taken place without the political and economic support of the United States, which was, in the early years of the Cold War, much more interested in creating a strong alliance that could effectively oppose the Soviet Union than it was in any potential European challenge to U.S. leadership. Germany, one of the largest states in the European Union, has been the most consistent supporter of an institutional structure that would limit Berlin's own freedom of action, a reflection of the lessons of two devastating wars and the attractiveness of a European identity for a country still grappling with the sins of the Nazi era. It is hard to imagine that other regional powers such as China, Japan, or Brazil, much less the United States, would have any interest in tying their own hands in similar ways. (Regional trading agreements such as Mercosur and NAFTA have very limited supranational provisions and show few signs of evolving into broader monetary or political unions.) The EU is a new and unique institutional structure, but it will coexist with, not displace, the sovereign-state model.

Kimberly Weir  **NO**

The Waning State of Sovereignty

T hose who think that the sovereign state will reign supreme forever in the international system have been fooled. Like a torrential downpour, a multitude of things—ranging from the industrial revolution to the inception of the Internet, from the end of colonialism to the formation of the European Union [EU]—are at work eroding state sovereignty. Although they will try to weather the storm, states will eventually go the way of the Holy Roman Empire.

The Sovereign State Is Just About Dead

No, but they are dying. The state as a political unit is not on its immediate deathbed, but states are outdated institutions that have difficulty meeting the needs of their citizens. One argument is that states are too big to do the small things. The size of states and the scope of their governments leave many people complaining about "big government," where the multiple layers of bureaucracy complicate accomplishing any task. Securing a passport can take months, and implementing a new social welfare program can take years to get through the system.

It can also be said that states are too small to do the big things. We live in an era of global problems. Just a few of the concerns that affect the Earth and which are difficult or impossible for any state to address alone include: global warming; the depletion of the ozone layer; the increase of nuclear, biological, and chemical weapons of mass destruction which can be delivered intercontinentally by missiles or terrorists; the movement of trade, investment capital, and money across borders; transnational communications and travel; and the global spread of AIDS and other diseases. Can states any longer truly protect the health, wealth, and very lives of their citizens? The answer in many cases is, "not very well."

Then there is the problem of small states. Microstates are on the increase as more and more people seek national self-determination. Too often what that means, however, is the creation of sovereign entities that more resemble a small city than a country. Nauru is a good example. With a territory that is but eight square miles, a population of barely 10,000, and an economy that depends on the export of guano (seabird droppings) for fertlizer, Nauru is a sham state. Yes, it may take only a day to obtain a Nauruan passport, but no, Nauru cannot provide a secure future for its population.

In 1648, the Treaty of Westphalia marked the birth of the modern state. The states that formed and survived did so because at that time they were the most available means to provide people with security and resources. A competition ensued between the states that formed in Europe. They sought to secure trade relations and to stake claims on territories throughout the world to create empires. This process continued until World War II. The atrocities brought about by the fascist ideology driving the conflict pushed the victors of the war towards working to preserve human dignity and identity. To facilitate this goal of self-determinism, the colonial empires were broken up, and the former colonies were granted their independence and recognized as states. As a result, the number of states increased exponentially during the mid-twentieth century. It was expected that through a process of decolonization, the former empires would endow the colonies with the institutions they needed to become successful and independent states. Instead, a division has emerged between the wealthy North (most of the former imperial countries) and poor South (most of the former colonies) that structures the international system. The geographical description of these countries does not fully illustrate the economic, political, social, and ecological disparities that separate these states.

Indeed many factors have contributed to the developmental problems encountered by the former colonies. Not the least of these is rather abstractly-drawn political borders that failed to take into consideration the indigenous settlements established long before colonization. The intrastate (internal) ethnic conflict that now plagues the majority of all developing states—particularly in Africa, the region most randomly dissected—constitutes the majority of all conflict in the post–cold war era.

One of the main functions of the state is that it is supposed to provide security for its citizens. Yet the millions of people who have been killed as a result of conflict since the formation of states raises questions as to the ability of states to provide security. Considering that weapons are increasingly powerful, and that biological, chemical, and weapons of mass destruction know no political boundaries, the state does not appear to be a very effective means of protection. In most cases, it is the state that incites war and violence. Furthermore, there are countless examples where states even do this against their own citizens who [are] the very ones the state is supposed to protect.

Those who argue that states are still strong sovereign units are missing the reality of state existence. In theory, all states are equal. Any state recognized as sovereign by the international community earns the same status as other states. This is true regardless of the state's resources, its level of development, or its ability to sustain itself and its population. International recognition also almost automatically earns the state a seat in the United Nations (UN) General Assembly, gives the state the opportunity to make treaties with other states, and otherwise gives the state legal standing as one among equals with other states. This theory is largely fiction, however. According all states the same status masks the reality of the situation. It appears that states possess all of the qualities necessary to act as sovereign units, when, indeed, many states fail to provide economic security for their own citizens. Many people in such states would suffer without international assistance offered by nongovernmental organizations (NGOs) and intergovernmental organizations (IGOs).

So, those who argue that state sovereignty is not in decline ignore the fact that a majority of the states have very limited sovereignty to begin with. Recognizing the inconsistencies between the entities collectively called states prompted scholar Robert Jackson to coin the term 'quasi-state' to refer to states that are recognized as sovereign, yet do not meet the criteria for actual statehood. Even more discouraging is the fact that Jackson was also moved to coin the term 'failed-state' to describe states like Somalia, Chad, Liberia, and Afghanistan, whose infrastructures have disintegrated almost beyond repair. Despite their inability to function, however, these states continue to be recognized as legitimate actors because the international system does not permit states to "quit" or decide they are no longer going to be states.

Sovereignty Means Final Authority

Not anymore, if ever.   Many, even most, states have never been free to [do] whatever they want, and some small states with powerful neighbors have been on very short leashes. That continues to occur. Furthermore, most developing states still struggle to lay the foundation necessary to hold free and fair elections, keep government corruption in check, and maintain a state of order sufficient enough to withstand meddling by other states and the international community.

Though having sovereign recognition gives states the ability to enter into agreements and treaties, enables them to become members of IGOs, makes them eligible for international aid, and is supposed to protect states from unwanted foreign intervention, these "benefits" also compromise state authority. A closer look at sovereignty reveals that it is not necessarily what it appears to be.

While no government wants to relinquish any of its sovereignty and most claim that no other authority supersedes its own, the international system is pushing states more and more to abide by decisions made by IGOs. Membership in an international organization necessarily requires a state to relinquish at least some of its authority for the organization to function effectively. Changes in the international system (the rising importance, number, and influence of NGOs, emergent transnational issues, and the increasing number of overall actors) have required states and other actors to cooperate through international organizations. The primary incentive for states to cooperate is to minimize the effects of the changes and the challenges to state sovereignty. China took considerable heat during talks over its application for admission to the World Trade Organization (WTO) from the United States and the European Union (EU) because of its human rights record. And the WTO members, including both the powerful and less powerful members alike, have been subject to rulings by the WTO that require these states to change their domestic policies to be consistent with those at the international level.

Most states attempting to meet the basic needs of their citizens require international assistance. However, that aid usually has stipulations attached, ranging from revamping state budgets to downsizing bureaucracy size to requiring that states adopt a democratic form of government. States are then evaluated on a regular basis and may have their aid withdrawn if they are noncompliant.

While the notion of nonintervention is considered sacred in the international community, the reality of the situation is that intervention, both overt and covert, happens regularly. Instances abound, from the justifiable drive to end apartheid in South Africa to the equally understandable wish to end ethnic cleansing in Kosovo, where the international community, as well as individual states, have interfered in one or another state's domestic affairs.

Universal Human Rights Are an Unprecedented Challenge to Sovereignty

Indeed they are. The Universal Declaration of Human Rights, as negotiated by the UN member-states in 1948 following the atrocities of World War II, invites interference in states' domestic affairs in the name of humanitarian intervention. In the post–cold war era, the number of conflicts throughout the world has exploded, and the Somalias, Bosnias, Rwandas, Yugoslavias, and Haitis of the world have prompted the international community to intervene for humanitarian reasons. However, the sovereignty of these states is depleted as foreigners intervene in the internal affairs of these countries. This is not to say that the massacre and violence taking place is justified, but rather that events are indicators of how states have become obsolete.

Not only is the sovereignty of domestic states diminished by the Universal Declaration of Human Rights, but the authority of Eastern states is undermined by the Western-oriented values that are the foundation of this so-called universal declaration. The difference is that the Western idea of rights is individually-based while the Eastern concept of rights is communally-based. Many Eastern cultures feel that Western values predominated in defining what a "universal" set of human rights should encompass. Thus, states like China are constantly criticized for violating the Universal Declaration of Human Rights, even though their practices coincide with their own values of the good of the community taking precedence over that of the individual.

Although some argue that UN agreements regarding the various human rights issues are often ineffective, more people are now conscious of these issues. Oppression, hate, slavery, abuse, and violence have not been wiped out by these agreements, but international agreements have spread awareness about these issues. Furthermore, the establishment of international war crime tribunals now provides a way to evaluate and determine whether or not actions warrant punishment. For those found guilty of unnecessarily inhumane practices during a time of war, the international system now has a means for seeking retribution. NGOs such as Amnesty International and WorldWatch monitor state action as well as lobby, protest, and boycott states that violate human rights. At least, their efforts have put states' human rights violations in the spotlight. Their perseverance has helped to push countries to amend their policies, with South Africa as the most stark example of a state's authority being undermined because of its human rights violations.

Globalization Undermines State Control

Yes.   While states undeniably remain significant actors in the international system, the effects of globalization are steadily eroding away at state authority. Trade, communications, technology, and travel all serve to undermine state control.

The increase in international trade has had devastating effects throughout various sectors of economically advanced economies. One does not have to see [the film] *The Full Monty* to be reminded of the power that MNCs [multinational corporations] wield over states. A visit to Bethlehem, Pennsylvania, or Birmingham, England, finds huge factories abandoned, houses vacant due to bank foreclosures, and unemployment rates through the sky. Whole cities have lost their main source of jobs when MNCs decided to utilize cheap foreign labor and have moved abroad, leaving both blue-collar workers and middle-managers jobless. Consider that U.S. labor union members—fighting to keep jobs at home—constituted the largest number of protestors at the Seattle Millennium Round of the WTO talks in 1999.

States have been forced into devising social welfare strategies in attempts to salvage their economies from the effects of globalization. Indeed, it would appear that developing countries benefit from the industry losses of the industrially advanced countries. Instead, however, studies indicate that in most cases the high trade-offs, such as land, tax incentives, and capital, for developing countries offer only a few small returns. In the end, it is the MNCs that both contribute to and benefit from waning state sovereignty.

Communications have been chipping away at the state since the printing press was invented. Since that time, countless communications innovations, from Radio Free Europe to televised satellite broadcasts, have whittled away at government control. Undoubtedly the most recent of these is the Internet. Despite the many opportunities and convenience it brings, surfing the web diminishes state control. Citizens are moving from interacting with one another in civil society to virtual reality. The more time citizens spend alone at home in front of their computers, the less likely they are to take part in community or civic activities, play on a softball team, or sing in the church choir. Furthermore, the Internet provides the freedom to move beyond political boundaries. People have endless opportunities to voice their points of view or network globally to challenge states' authority on issues like environmental practices or human rights records. States, including China and Singapore, have attempted to regulate Internet access in order to preserve their authority. But, it has been an uphill battle as more people gain access to information outside of what their governments report.

Just as traditional forms of terrorism challenge a state's authority over its territory and/or nationals, cyberterrorism also undermines state control. The number of cyberterrorist attacks increases daily, with infiltrators planting viruses that destroy entire databanks, as has happened in the United States, or posting slanderous information, as occurred in Japan. As governments increasingly rely on the Internet to disseminate information and to process passports, driver's licenses, and welfare benefits, states open themselves to cyberterrorist attacks.

There is yet another way that technology presents problems for governments, especially those that do not have it. The increasing technology gap between the North and South only serves to undermine the authority of developing states as they struggle to advance. Falling farther behind the cutting edge of technology decreases job opportunities, chances for development, and hopes for improving living standards. As people become frustrated with the lack of services provided by the state, the possibility for social unrest increases tremendously.

The technological advances of the agricultural industry also affect state sovereignty. Though genetically modified organisms (GMOs) are touted for the fortified grains and vegetables that can be produced and the plants genetically encoded with built-in pesticides, there is a downside to altering genes. Biotech companies can also produce seeds that do not reproduce, forcing poor countries to buy new seeds every season, rather than storing seeds from the last crop for the next year. GMOs undermine state development programs by keeping poor countries dependent upon international aid to feed their people.

More efficient travel presents a plethora of problems for the state. To begin, travel facilitates terrorist activities. It provides terrorists with an easy way to enter a foreign country or to take hostage a country's nationals who are traveling abroad. Transportation also moves migrants. Consider the effect of the break-up of Yugoslavia and the consequent conflict. Tens of thousands of people abandoned their homes, seeking refuge. They spread across Europe, only to be faced with neofascist movements protesting their arrival. More affordable travel also undermines state sovereignty because it carries people abroad where, for any number of reasons, they choose to remain illegally. In many cases, these people rely on the social welfare of their host country to help to meet their needs, thus depleting the state's resources. Finally, more efficient travel means more efficient spread of disease. Thanks to mobility, AIDS has spread like wildfire across Africa, now threatening into Asia. The situation has become so horrific that many states now consider AIDS to be a threat to their national security because states cannot guarantee their citizens' protection from this threat.

Globalization Is Changing the Scope of State Control

Yes.   Globalization is forcing states to alter their authority. States have not voluntarily changed the amount or level of control they hold. Rather, the nature of the Global Village requires states to change if they want to remain players in the international system. This argument, however, diminishes the fact that the world is not a static thing and that, as Darwin concluded, all things must evolve if they are to survive. It is not enough to argue that state sovereignty is not waning but just trying to survive by transforming. Instead, what needs to be considered is that the significance of the state to its people is in decline, regardless of the changes that happen to a state as an evolving entity.

States are being challenged by international organizations. These include both the intergovernmental bodies that the states created and are members [of] and the myriad private membership nongovernmental organizations (NGOs) that have sprung up. If states did not concede at least some of their authority to the international organizations they formed, it would defeat the purpose of creating them, i.e., to facilitate relations between them.

NGOs Are Nibbling at National Sovereignty

To a large extent. This is especially if, as is appropriate, one considers corporations to be NGOs.

MNCs have power, mobility, and resources. They are very mobile; if a state is not willing to grant incentives to MNCs to set up or to stay, corporations will move. In a global business environment where states compete for industries that provide revenue and jobs, MNCs have the upper hand over many small developing economies. MNCs seek out places like the *maquiladoras* (factories near the U.S. border) in Mexico that draws in businesses because of the low cost of labor, or havens in the Caribbean that attract banks with lax financial regulations. And it is not just developing states that are willing to bargain away their control over tax and environmental regulations to entice businesses to build new factories in their backyards; indeed developed states are just as tenacious in offering enticements and incentives with hopes of boosting their economies.

Big businesses are not the only NGOs that nibble on states. The recent growth spurt of private citizen organizations can be attributed to a phenomenon called "post-materialism" by Professor Ronald Inglehart. Enough citizens of the North have achieved sufficient economic security that they are willing to spend time and money on substantive issues like the environment and human rights. These post-materialist trends have brought people with similar concerns together to create, join, and support NGOs. The result is people uniting across political boundaries to form transnational networks that challenge state authority. Many of the groups that protested the Seattle Millennium Round of the WTO or the United Nations Council on Trade and Development (UNCTAD) in Bangkok are NGOs that demonstrated to press governments to make responsible decisions regarding enforcement of stricter environmental codes or labor regulations. Their efforts at the WTO talks have recently been rewarded, as the major IGO that determines and governs trade policy between states is now trying to work with NGOs rather than exclude them, as had previously been the case.

As the overall number of NGOs increases, so does the number of NGOs moving into developing countries in need of assistance. In many instances, these groups provide essential social welfare services that struggling economies and fractured governments cannot supply. Increasingly, industrially advanced states and IGOs are using NGOs to distribute aid and evaluate progress because they have a local knowledge of the situation. Though countless people benefit from these services, the consequence is that the legitimacy of these developing states is even further undermined because people come to depend on the NGOs, rather than the states. The net effect is to put the NGOs in a position of power over those governments.

Sovereignty Blocks Conflict Resolution

Increasingly less. It appears that conflict resolution is based more on issues of self-identity and expression than on preserving state sovereignty. Before Lebanon crumbled to pieces, the Jews, Christians, and Arabs lived peaceably together under a well-functioning government. But Lebanon was eventually affected by neighboring religious conflict. As long as each of these ethnically diverse groups was recognized, the state prospered. When groups attempted to oppress one another, conflict broke out.

A majority of conflict occurring in the post–cold war era is a result of ethnic intrastate clashes in developing countries. Oppressing minorities without fair representation chips away at state sovereignty. Rwanda's Tutsi population was massacred as a result of ethnic hatred. Somalia is still divided by clan rivalries. Kurds throughout Turkey, Iran, and Iraq are persecuted daily. Though they express their desires in very different ways, the Zapatista rebels in the southern Mexican state of Chiapas and many of the French-heritage Quebecois in Canada continue to fight for separation seeking representation.

When the international community does intervene in these conflicts, most efforts are channeled through regional organizations and IGOs. The North Atlantic Treaty Organization (NATO) member-states as well as the UN forces intervened in the Balkans in attempts to save Yugoslavia from itself because of ethnic intrastate differences.

The European Union Is a New Model for Supranational Governance

Yes.   It provides a model of what is likely to come.

Given the success of the EU, particularly in the global market, it is not only desirable, but also necessary for other countries to emulate its success. Though the laws of capitalism are based on competition, cooperation has proven to be successful for the member-countries of the EU. The strengthening of the EU has even prompted other industrially advanced countries like the United States to join trade groups like the North American Free Trade Agreement (NAFTA) and seek even larger multilateral trade organizations, such as the Free Trade Agreements of the Americas that, if it comes into being as planned, will include virtually all the countries in the Western Hemisphere.

Globalization is pressuring individual states into regional blocs in order to compete with other regionally based economic organizations that have formed. Across the globe, few states do not belong to at least one regional economic organization. Despite resistance to external political and economic forces, even previously closed countries such as Burma/Myanmar have realized the benefits of regional cooperation.

Stephen Krasner argues that expanding the EU to include more Central and Eastern European countries would make it difficult to "move toward a political organization that looks like a conventional sovereign state." But, that's the point—the conventional states are becoming obsolete as regional and international organizations supplant them by providing more of what citizens need. The formation and relative success of the EU provided the inspiration for other regional economic organizations, such as MERCOSUR (the Southern Cone Common Market in South America), the Association of Southeast Asian Nations (ASEAN), and the Southern Africa Development Community (SADC), that hope to emulate the EU's successes. By joining together within regions, memberstates have already increased their potential in terms of a larger market for their products and possibilities for joint ventures, such as the hydroelectric plants being built in the Southern Cone region. Rather than competing against international powerhouses in a highly heterogeneous international market, Latin American, Asian, and African states open their markets regionally, thus exponentially increasing the size of their markets for their domestic businesses by targeting similarly situated consumers and economies. Just as the EU started out with joint economic ventures, these regional economic organizations have begun to lay the same foundations that have proven successful for the advancement of the EU.

What Can We Conclude About the Future of the Sovereign State?

Its prospects are poor.   Stephen Krasner is whistling past the proverbial graveyard of states. The growth of separatist movements and the general rising tide of frustration that people feel about the national governments are evidence that states have gotten too big to tend efficiently to the needs of their people. States also cannot hope to deal with global pollution, the spread of weapons of mass destruction, the rocketing level of economic interdependence, and a host of other problems that ignore national borders.

Those who cannot see the end of the state coming also seem oblivious to the impact that rapid international travel, almost instantaneous international communications, the Internet, the increasing homogenization of culture, and a host of other transnational trends are having on obliterating national distinctions. "What is now must always be," may seem comforting, but it is self-delusion.

POSTSCRIPT

Will State Sovereignty Survive Globalism?

Many people find it difficult to debate the issue of the survival of the sovereign state. For a general view of the role of states in the contemporary international system, see Walter C. Opello, Jr., and Stephen Rosow, *The Nation-State and Global Order: A Historical Introduction to Contemporary Politics* (Lynne Rienner, 2003). Nationalism, the identity link between people and their country, is such a powerful force that some find it almost treasonous to suggest that their country—whether the United States, Canada, Zimbabwe, or any other country—is destined to be eclipsed and perhaps to cease to exist as the dominant unit of the international political system.

Yet it could happen. Since states as we know them did not always exist as a political unit, there is little reason to believe that they will necessarily persist. Instead, as economic, military, and other factors change, it is reasonable to suspect that the form of governance best suited to address these new realities will also change. It may be that just as feudal units once proved to be economically nonviable, states may also fail to meet the tests of a global economy. So, too, just as small units could not provide adequate security amid new weapons and tactics, critics say that states provide little protection from weapons of mass destruction (WMDs) and terrorism. Maryann Cusimano Love, ed., *Beyond Sovereignty: Issues for a Global Agenda* (Wadsworth, 2002) discusses the problems that are beyond the ability of any single state to solve.

Krasner and Weir both make thoughtful arguments, but it is unclear which scholar the verdict of history will ultimately uphold on the question of whether or not states should continue to dominate the political system and to be the principal focus of political identity. Clearly, states continue to exercise great political strength and remain most people's main focus of political identification. But it is also the case that states exist in a rapidly changing political environment and that, ultimately, they need to adapt and serve the needs of their people. A good Web site to begin further exploration on the future of sovereignty is that of the Global Policy Forum at `http://www.globalpolicy.org`. Click the box that says "Nations & States." Taking a middle ground in the debate and providing a good perspective is Georg Sørenson, *The Transformation of the State: Beyond the Myth of Retreat* (Palgrave Macmillan, 2004).

More reading from Krasner's perspective is available in his edited volume *Problematic Sovereignty* (Columbia University Press, 2001). For a different perspective, see Gerard Kreijen, ed., *State, Sovereignty, and International Governance* (Oxford University Press, 2002), which investigates the response of the global community to states that collapse and that fail to observe the basic principles of international law.

Country Indicators for Foreign Policy (CIFP)

Hosted by Carlton University in Canada, the Country Indicators for Foreign Policy project represents an ongoing effort to identify and assemble statistical information conveying the key features of the economic, political, social, and cultural environments of countries around the world.

http://www.carleton.ca/cifp/

U.S. Department of State

The information on this site is organized into categories based on countries, topics, and other criteria. "Background Notes," which provide information on regions and specific countries, can be accessed through this site.

http://www.state.gov/index.cfm

http://www.state.gov/countries/

PART 2

Regional and Country Issues

*T*he issues in this section deal with countries that are major regional powers. In this era of interdependence among nations, it is important to understand the concerns that these issues address and the actors involved because they will shape the world and will affect the lives of all people.

- Should the United States Decrease Its Global Presence?

- Should the United States Continue to Encourage a United Europe?

- Is Russian Foreign Policy Taking an Unsettling Turn?

- Does a Strict "One China" Policy Still Make Sense?

- Should North Korea's Arms Program Evoke a Hard-Line Response?

- Would It Be an Error to Establish a Palestinian State?

- Was War With Iraq Justified?

- Are Strict Sanctions on Cuba Warranted?

ISSUE 4

Should the United States Decrease Its Global Presence?

YES: Louis Janowski, from "Neo-Imperialism and U.S. Foreign Policy," *Foreign Service Journal* (May 2004)

NO: Niall Ferguson, from "A World Without Power," *Foreign Policy* (July/August 2004)

ISSUE SUMMARY

YES: Louis Janowski, a former U.S. diplomat with service in Vietnam, France, Ethiopia, Saudi Arabia, and Kenya, maintains that the view that the 9/11 attacks ushered in a new geo-strategic reality requiring new foreign policy approaches is based on a false and dangerous premise and is leading to an age of American neo-imperialism.

NO: Niall Ferguson, Herzog Professor of History at New York University's Stern School of Business and senior fellow at the Hoover Institution at Stanford University, contends that a U.S. retreat from global power would result in an anarchic nightmare of a new Dark Age.

T here can be little doubt that the United States is the most powerful country in the world—no other country can rival the U.S. military or launch a successful conventional attack on the United States. Few could withstand a U.S. conventional attack. It is true that Americans are subject to terrorist attack, and the possibility of nuclear attack remains. Still, the U.S. nuclear arsenal overshadows that of any other country, with the aging Russian arsenal the closest rival, and terrorists find themselves under constant U.S. pressure. The United States is also in an unparalleled economic position, accounting for over 25 percent of the world's measured economic production. Scholars debate whether the world is structured as a unipolar system with just one dominant power (the United States in this case), whether there is a limited unpolar system, or whether a multipolar system with numerous power centers is emerging. Whatever the precise answer may be, there can be little doubt that the United States is currently the world dominant power, even if it has not achieved complete hegemony, or power dominance.

The issue is what the United States should do with its immense power. What role should the United States play in the world? The answers, at least in the view of the current U.S. administration, were given in 2002 when President George W. Bush issued a document entitled "The National Security Strategy of the United States of America." The report characterized the United States as possessing "unprecedented—and unequaled—strength and influence in the world" and went on to argue that "this position comes with unparalleled responsibilities, obligations, and opportunity. The great strength of this nation must be used to promote a balance of power that favors freedom." A number of the report's sections pledged that the United States would use its strength globally to promote human dignity, democracy, and free economic interchange, and the document also declared that the United States intends to develop cooperation with "other main centers of global power." These sections drew relatively little notice, although some analysts objected that, if taken literally, the pledges would lead the country into an unending series of draining efforts around the world. Much more controversial were the sections on threats to the United States and its allies, especially by terrorism and weapons of mass destruction (biological, chemical, nuclear, and radiological weapons). President Bush not only promised that attacks on the United States and its allies would bring devastating consequences to perpetrators, he also indicated that the United States would not wait to be attacked but might act preemptively to destroy perceived enemies and their capabilities before an attack could occur. The president's far-reaching position reflected the strength of neoconservative (neocon) views in his administration. This perspective argues that the United States has the right to act aggressively and unilaterally if necessary to protect its interests. Moreover, neocons believe that democracy, free enterprise (capitalism), and other aspects of the "American way" are superior to other approaches to governance and that the United States has the right and responsibility to promote and defend these approaches worldwide.

The U.S. invasion of Iraq in 2003 without UN support and in opposition to the views of most U.S. allies and other countries was the clearest expression of the Bush doctrine. That effort was not only designed to end what the neocons saw as the threat from Iraq, it also importantly included the determination to bring democracy to Iraq and, by extension, the rest of the Middle East.

The result of the Bush doctrine and its application, according to Louis Janowski in the first reading, has been a disastrous foreign policy that amounts to quasi-imperialism and that unduly taxes U.S. resources. Janowski concludes that the United States should greatly reduce its worldwide commitments. In the second reading, Niall Ferguson does not defend the Bush doctrine or the president's foreign policy as such. Ferguson argues that a U.S. retreat from the central role that it is playing in the world would create a power vacuum that would have dire consequences.

Louis Janowski

 **YES**

Neo-Imperialism and U.S. Foreign Policy

American foreign policy at its best combines a clear understanding of our national interests, the limits of our power, and the real and psychological needs of the American people. Effective foreign policy in our democracy has always been a combination of realpolitik and moral idealism. Pearl Harbor remains the classic example: a Japanese attack created the catalyst that allowed President Franklin Roosevelt to unite the American people behind moral and idealistic policies which successfully structured U.S. policies and advanced U.S. interests for the remainder of the 20th century.

Yet the U.S. foreign policy record over the past half-century has been mixed. All too often, our political leadership appears to suffer from attention deficit disorder and the dangerous, self-destructive behaviors that too often accompany ADD.

The Vietnam War failed the test of meeting a clearly defined and limited national interest. In addition, the realities of conducting guerrilla warfare meant that the average American perceived a nightmare rather than an idealistic and moral crusade for a better world. Both the Korean and Persian Gulf Wars had clear causes, limited objectives (recall President Harry Truman's dismissal of Gen. Douglas MacArthur over widening the scope of the Korean War) and wide global support. The Persian Gulf War was a good example of clear causes, limited objectives, morality, and broad international support. By contrast, Somalia was an example of unrealistic moral idealism combined with a lack of concrete national interest.

The 2003 invasion of Iraq failed to meet these criteria. It lacked virtually every element of this formula for success: a clearly defined *casus belli* [cause of war], an overriding national interest, limited goals, and international legitimacy. Indeed, the Bush administration was able to win popular support for the war only by pandering to the worst fears of the American public, conjuring up a link of terror between the secular nationalist Baathist rulers of Iraq and the diametrically opposed pan-Islamic religious fundamentalists of al-Qaida. The two represent essentially opposing ends of the political spectrum in the Middle East with little in common other than shared anti-Americanism.

Equally unbelievable was the portrait of an "axis of evil" linking Iran and Iraq (and North Korea). Saddam Hussein's invasion of Iran and the ensuing 1980–88 Iran-Iraq War render such a linkage a grotesque distortion of historical reality, as does the participation of several senior officials from the current Bush administration in the Reagan administration's efforts to cultivate Saddam during that period.

More generally, the Bush administration has attempted to argue that the terrorist attacks on the World Trade Center ushered in a new geo-strategic reality requiring new domestic and foreign policy approaches. This is a false premise. All that changed with 9/11 was a naive assumption that somehow the U.S.—unlike any other nation—could involve itself in ever-expanding external acts without potential negative or retaliatory responses on its territory.

In this regard, it is useful to recall that terrorism is specifically designed to cause overreaction. Perhaps terrorism's greatest success in the past century was Austria-Hungary's overreaction to the assassination of Archduke Ferdinand by a Serbian Pan-Slav "terrorist" which, in turn, led to World War I and the destruction of the Austro-Hungarian Empire.

Empire Building

The 9/11 attacks have been used to redefine U.S. foreign policy along neoconservative lines. The new policies emphasize unilateralism, unlimited objectives, and the use of military force as a primary adjunct to policy. This set of characteristics has little in common with historic U.S. policy, which until the 1940s emphasized isolationism, limited foreign policy objectives and an aversion to the use of military force outside the Western Hemisphere.

In one respect, the neoconservatives do harken back to the past in their approach to foreign policy. Unfortunately, they do so by invoking the now-obsolete political-military premises of the Cold War, such as a perceived need for overwhelming military superiority. The administrations proposed military budget of $401 billion for FY 2004–2005 is as great as those of the next six powers combined. Where is the threat to justify this expenditure? Ongoing efforts to expand the forward deployment of U.S. forces to areas such as Central Europe and South Asia can hardly he justified on the basis of a military threat to the territorial integrity or national existence of the United States or of our principal allies. There was a sound rationale for a forward projection of U.S. forces during the Cold War. But there is no basis for transforming forward defense into a strategy of unilateral global political-military imperialism, as we are in the process of doing.

President Dwight Eisenhower's farewell address, in which he warned of the dangers posed by the "military-industrial complex," was perhaps the last example of a leadership vision coupling an emphasis on adequate power with an understanding of the dangers that excessive power creates. Since Eisenhower, American political leadership has actively sought an ever-expanding role on the world stage and an expansion of military presence into far-flung regions of the world where U.S. interests are marginal at best.

For the sake of argument, however, let us assume that the only way for the United States to remain secure in the post-9/11 environment is to forge an empire. The basic ingredients for success at such an enterprise are: skillful diplomacy to forge strong alliances; the ability to formulate and implement rational decisions based on realistic threat assessments; sound decisions about when to use military force; and the wherewithal to support the demands of running and defending a global presence (e.g., a sound economic base, military hardware and human resources).

Keeping Bad Company

The long-term viability of any American empire will be based on the ability to make alliances with nations and leaders who support the long-term goals and values of American democracy while, to the extent possible, avoiding alliances of convenience with known bad actors. Yet in the case of Iraq, we reversed that formula.

Nearly all our major allies were strongly opposed to the war, and the few who stood with us did so despite strong domestic opposition. Thus, major by-products of the war have been a fundamental weakening of the NATO [North Atlantic Treaty Organization] alliance, rifts in the longstanding unity of the West, and the under-mining of pro-American governments in the Arab and Muslim worlds.

The war on terror demonstrates a similar inconsistency on the other side of the equation. In our zeal to acquire new allies against al-Qaida, the Bush administration seems willing to over-look the very same human rights violations and brutal suppression of democracy that the State Department details in its latest set of worldwide country reports. Countries like Uzbekistan, Turkmenistan and Pakistan were quick to learn that lesson, and others seem poised to follow in their footsteps.

This phenomenon is nothing new, regrettably. In Afghanistan, U.S. covert operations in support of Islamic fundamentalists fighting the Soviets two decades ago paved the way for the Taliban to fill the vacuum created when Moscow withdrew. When we eventually turned to tribal surrogates to help us oust the Taliban, we conveniently overlooked the fact that some of them were major players in the international drug trade. The result? Afghanistan today is the world's largest source of opium and heroin prices have fallen around the globe. It is, therefore, hard to make a convincing case that Afghanistan is any less a global danger to U.S. interests now.

Getting the Threat Right

Whether or not one believes that the Bush administration politicized the findings of the intelligence community concerning Saddam Hussein's alleged weapons of mass destruction programs, that debate underscores the need for sound analysis of often-ambiguous and incomplete indications regarding what our foes are doing and planning.

Understandably, the various components of the U.S. intelligence community frequently disagree among themselves when it comes to assessing the data, but in general, the most unrealistic threat assessments tend to come from the Defense Intelligence Agency and the other military intelligence services. This is so for several reasons. First, military commanders understandably want to ensure that they do not inadvertently endanger their troops by underestimating the forces they face. Second, DOD [Department of Defense] budgets are directly related to threat projections, while State [DDepartment], CIA [Central Intelligence Agency] and NSA [National Security Agency] budgets lack this seminal link. Third, DIA [Deense Intelligence Agency] assessments frequently ignore political, economic and cultural factors, and therefore misread both enemy intentions and capabilities. For example, the military threat the Soviet Union posed during the Cold War was never as serious as estimated. And in the post-Cold War period, our experiences in the Balkans, Iraq and elsewhere have clearly demonstrated the gaps between the military threat projected by DOD and actual conditions on the ground.

In the case of Iraq, Gen. Eric Shinseki [Armu chief of staff] and other combat-seasoned military officers were fully aware that our lack of cultural and language capability would seriously limit the utility of modern arms, particularly in non-traditional warfare. They therefore requested force levels higher than they otherwise would have. Yet the White House rejected the requests, citing the ability of our troops to destroy all conventional military resistance in Iraq. But the administration neglected to take into account the importance of destroying or forcing the surrender and disbanding of Iraqi units in place and securing weapons and ammunition dumps to ensure that most Iraqis perceived the likelihood of successful unconventional warfare as poor. It also ignored the reality that terrorism and unconventional warfare are the logical by-products of overwhelming military inferiority.

The other side of the threat assessment coin is formulating an appropriate response. Just as even the best analysts sometimes either overestimate or underestimate potential threats, policy-makers tend to favor the use of force to keep other countries from assessing U.S. decision-makers as weak or uncertain.

American military dominance has resulted in both the overuse of military force and errors in how we have applied it. Overuse is a natural result of being able to use military force in almost any scenario; as the saying goes, when you have a hammer, every problem looks like a nail. But it also reflects the desire for quick solutions to complex problems, and the political reality that the use of military force builds short- to midterm political support at the polls.

The potential for error exists in large part because there are major disconnects in our system between global political, economic, social and political-military knowledge and national decision-making power. America's foreign and strategic policy decision-making structures are so complex and multi-layered, and actual "on-the ground" knowledge is so far removed from those with decision-making authority, that serious mistakes are inevitable. The collapse of the Soviet Union and its empire from within is an excellent example of this type of structural problem. The Soviet centralized economic planning system worked reasonably well when the system it ran was a relatively simple one. But as the Soviet Union became economically mature and far more complex, centralized planning became incapable of meeting the varied tasks it faced. A similar reality is faced by American foreign policy today, with a potentially parallel outcome.

Then there is the problem of developing the human resources necessary for maintaining a global empire. After all, "smart" weapons systems are only as "smart" as those who operate them. It is exceptionally difficult to identify, track and destroy irregular forces and terrorists when you don't speak the local language or understand the local norms and mores—much less the broader culture and its complicated subcultures. What was true in Vietnam 35 years ago is just as true today in Iraq and Afghanistan. Yet no administration has been willing to commit the funds to ensure that U.S. diplomats, intelligence operatives and military forces have adequate linguistic, cultural and area-specific skills.

Despite our relative under-investment in these areas, our military, intelligence and diplomatic services have amassed an immense amount of knowledge (especially compared to what the political leadership of the day possesses). But power rivalries at both the political and bureaucratic levels and complex hierarchical structures work to keep knowledge and power apart. The longtime rivalry between the FBI and the CIA was one of the main factors that prevented solid intelligence about terrorist training in U.S. flight schools from cutting through multiple levels of bureaucracy and preventing the 9/11 terrorist attacks.

Following 9/11, top Defense Department officials chose to confine decision-making and intelligence assessment with respect to Iraq to a small group of like-thinking individuals (the "Office of Special Plans"). The result was that policy was made in secret by individuals with only a limited knowledge of the region, who never allowed their recommendations to face the open and ongoing scrutiny of the entire intelligence community (much less the political system). The outcome demonstrated manifold errors. There were no weapons of mass destruction. The assumption that Iraqi Arabs would warmly welcome the U.S., particularly given our longstanding support for Israel, failed to stand up in the light of day. Exiled Iraqis were not warmly welcomed upon their return. The assumption that Iraq's clan structure—where nepotism is a virtue, not a vice—is amenable to democracy appears to be either a misguided assumption or a cynical ploy.

Finally, the administration's insistence on requesting military force levels well below what the Joint Chiefs wanted, and its refusal to draw up clear-cut plans for occupation and exit, clearly created conditions more favorable to insurgency, costing hundreds of American lives.

Again, the lesson is that decision-makers need to have access to, and be willing to consult, those diplomats, analysts, troops and agents with first-hand knowledge of conditions on the ground. For this to occur, of course, a certain amount of humility is required by the political leadership as well as a fair amount of structural change to reduce the rigidity of our foreign affairs bureaucracies.

When he first arrived at State, Secretary [Colin] Powell stunned the bureaucracy by occasionally leaving the 7th floor and personally going to desk officers to seek out knowledge. Coming from a military background and drawing on his experiences in Vietnam, Powell was undoubtedly aware that had news is repeatedly filtered by multiple levels of bureaucracy before it reaches senior decision-makers. And, as any problem works its way through the system, more and more filtering is done by officials who, no matter how competent or capable, are less likely to have adequate knowledge of the realities on the ground. Furthermore, they have bureaucratic reasons not to disturb the status quo and existing chains of power and control. Few are prepared to appear disloyal by failing to cheerlead administration priorities and policies of the moment. Bureaucratic advancement is as much the result of the absence of perceived error as it is of actual accomplishment.

Paying the Tab

Finally, in an era of half-trillion dollar budget deficits, can we afford an empire?

One of the major consequences of seeking a global political-military—empire is the relegation of critical domestic and global economic, financial and other policy questions to secondary status rather than addressing them as key issues.

Except for the United States, almost every developed nation (and many aspiring to that status) has placed economic and financial policies, not political-military objectives, at the top of their respective agendas since the end of the Cold War. And, in large part, they did so precisely because they knew U.S. leadership—for its own reasons—was prepared to carry the burden for them. This was a rational decision both because of the lack of a pressing threat and in view of the lesson learned by other developed states from 1939 to 1989: namely, empires are expensive. The average citizen of a former colonial power may today regret the loss of his or her perceived superiority by association with an empire, but he or she certainly does not regret no longer having to pay the extravagant costs of maintaining an imperial system.

United States policy-makers desperately need to pay more attention to America's pressing financial and economic needs. Cordell Hull is almost completely forgotten today, but he deserves to be remembered not only for being the longest-serving Secretary of State in history (1933–1944) but for being the last one to concentrate on promoting U.S. economic interests. Admittedly, the collapse of the global economic and financial order left him and FOB with no alternative, but it is still disheartening to see how far we have gone in the opposite direction.

Today, U.S. policies seem to be assisting a collapse of the very international financial order we created in the aftermath of World War II. The disconnect between the United States and Western Europe on trade matters is growing. American restrictions on steel imports have caused greater harm to U.S. steel fabricators and consumers than the benefits they provided to domestic steel producers, and provoked threats of retaliatory measures from Europe, China and Japan. Coincidentally or not, Washington sharply increased subsidies to American agribusiness on the eve of global discussions on finding a way to increase and rationalize global agricultural trade, with predictable results.

If the United States were a minor player on the world stage instead of the major funding source for the International Monetary Fund [IMF] and the World Bank, those organizations would be demanding that the U.S. administration make major fiscal and economic policy changes. The United States' staggering trade deficit (5 percent of GDP) would have to be addressed, as would its addiction to foreign investment capital to finance that trade deficit. The IMF and the World Bank would also demand progress toward a balanced

U.S. budget. The rapid fall of the U.S. dollar relative to other major currencies over the past two years is a clear indication that investors worldwide are today far less comfortable with investing in the United States. Whatever else one may think of the policies of the Clinton administration, its fixation on a strong dollar and balanced budgets was, in part, based on a clear understanding of the need to assure foreign investors of the long-term strength of the U.S. economy.

A Return to Core Competencies

The impact of American unilateralism and its concentration on political-military matters is obvious. There is a vacuum of leadership in other international policy areas, be they political, economic, legal, or environmental. The vacuum exists by definition. If you have a unilateral policy, you can't lead, because you have been unwilling to make the compromises necessary for others to follow.

As with domestic issues, success in foreign policy means meeting the often-conflicting needs of concerned parties. For example, if we had shown some regard for the views of the United Nations and our traditional allies as we prepared for war (or even afterward), we might not be bearing the costs almost entirely alone—and the situation in Iraq, not to mention its prospects, would likely be considerably brighter. Compare the current mess with the handling of the Persian Gulf War. There, patient diplomacy ensured that the financial, political and human costs to the U.S. were minimal—as opposed to the open-ended costs of the present "Coalition of the Willing" in Iraq.

The unilateralist neoconservative policies of today are a badly mutated descendant of our isolationist heritage. Isolationism at least had the clear advantage of limited objectives, keeping the United States from entering two world wars until a national consensus existed for intervention. Our late entry into both conflicts spared the United States from most of the human, social and financial consequences of those two great conflicts.

Of course, isolationism is dead, buried by technology that makes it outmoded except in backwaters such as North Korea and Burma. But the concept of limiting commitments on the basis of national interest and real needs makes as much sense today as it always has. American foreign policy today needs to reexamine its commitments worldwide and redefine them. Our 60-year relationship with Europe is crumbling, and better solutions exist than moving U.S. military bases from Western to Central Europe.

U.S. foreign policy toward the Middle East has been a disaster since 1967. An even-handed policy with respect to the Israeli-Palestinian conflict would do more to reduce the threat of anti-American terrorism than any other step we could take. We also need to question why, in view of the end of the Cold War and a vastly changed energy situation world-wide over the past three decades, we need a military presence in the Persian Gulf.

In sum, the primary need of the United States today is to greatly reduce U.S. commitments world-wide. The existing U.S. decision-making and intelligence structures are no more capable of running a global empire (at least one in accord with the moral and democratic views of the American public) than the centralized

Soviet system was of controlling a far less complicated global equation. In the language of the business community, the United States needs to get back to its core competencies. In the 21st century, there is no reason for Americans to play the "Great Game" in the mode of 19th century European elites—particularly when no vital U.S. interests are at stake. To follow such a course is, in the words of Talleyrand, "Worse than wrong, monsieur. It is stupid."

Niall Ferguson

 **NO**

A World Without Power

. . . We tend to assume that power, like nature, abhors a vacuum. In the history of world politics, it seems, someone is always the hegemon [dominant power], or bidding to become it. Today, it is the United States; a century ago, it was the United Kingdom. Before that, it was France, Spain, and so on. The famed 19th-century German historian Leopold von Ranke, doyen of the study of statecraft, portrayed modern European history as an incessant struggle for mastery, in which a balance of power was possible only through recurrent conflict.

The influence of economics on the study of diplomacy only seems to confirm the notion that history is a competition between rival powers. In his best-selling 1987 work, *The Rise and Fall of the Great Powers: Economic Change and Military Conflict from 1500 to 2000*, Yale University historian Paul Kennedy concluded that, like all past empires, the U.S. and Russian superpowers would inevitably succumb to overstretch. But their place would soon be usurped, Kennedy argued, by the rising powers of China and Japan, both still unencumbered by the dead weight of imperial military commitments.

In his 2001 book, *The Tragedy of Great Power Politics*, University of Chicago political scientist John J. Mearsheimer updates Kennedy's account. Having failed to succumb to overstretch, and after surviving the German and Japanese challenges, he argues, the United States must now brace for the ascent of new rivals. "[A] rising China is the most dangerous potential threat to the United States in the early twenty-first century," contends Mearsheimer. "[T]he United States has a profound interest in seeing Chinese economic growth slow considerably in the years ahead." China is not the only threat Mearsheimer foresees. The European Union (EU) too has the potential to become "a formidable rival."

Power, in other words, is not a natural monopoly; the struggle for mastery is both perennial and universal. The "unipolarity" identified by some commentators following the Soviet collapse cannot last much longer, for the simple reason that history hates a hyperpower. Sooner or later, challengers will emerge, and back we must go to a multipolar, multipower world.

But what if these esteemed theorists are all wrong? What if the world is actually heading for a period when there is no hegemon? What if, instead of a balance of power, there is an absence of power?

Such a situation is not unknown in history. Although the chroniclers of the past have long been preoccupied with the achievements of great powers—whether civilizations, empires, or nation-states—they have not wholly overlooked eras when power receded.

Unfortunately, the world's experience with power vacuums (eras of "apolarity," if you will) is hardly encouraging. Anyone who dislikes U.S. hegemony should bear in mind that, rather than a multipolar world of competing great powers, a world with no hegemon at all may be the real alternative to U.S. primacy. Apolarity could turn out to mean an anarchic new Dark Age: an era of waning empires and religious fanaticism; of endemic plunder and pillage in the world's forgotten regions; of economic stagnation and civilization's retreat into a few fortified enclaves.

Pretenders to the Throne

Why might a power vacuum arise early in the 21st century? The reasons are not especially hard to imagine.

The clay feet of the U.S. colossus   Powerful though it may seem—in terms of economic output, military might, and "soft" cultural power—the United States suffers from at least three structural deficits that will limit the effectiveness and duration of its quasi-imperial role in the world. The first factor is the nation's growing dependence on foreign capital to finance excessive private and public consumption. It is difficult to recall any past empire that long endured after becoming so dependent on lending from abroad. The second deficit relates to troop levels: The United States is a net importer of people and cannot, therefore, underpin its hegemonic aspirations with true colonization. At the same time, its relatively small volunteer army is already spread very thin as a result of major and ongoing military interventions in Afghanistan and Iraq. Finally, and most critically, the United States suffers from what is best called an attention deficit. Its republican institutions and political traditions make it difficult to establish a consensus for long-term nation-building projects. With a few exceptions, most U.S. interventions in the past century have been relatively short lived. U.S. troops have stayed in West Germany, Japan, and South Korea for more than 50 years; they did not linger so long in the Philippines, the Dominican Republic, Haiti, or Vietnam, to say nothing of Lebanon and Somalia. Recent trends in public opinion suggest that the U.S. electorate is even less ready to sacrifice blood and treasure in foreign fields than it was during the Vietnam War.

"Old Europe" grows older.   Those who dream the EU might become a counterweight to the U.S. hyperpower should continue slumbering. Impressive though the EU's enlargement this year has been—not to mention the achievement of 12-country monetary union [using the euro as its common currency]—the reality is that demography likely condemns the EU to decline in international influence and importance. With fertility rates dropping and life expectancies rising, West European societies may, within fewer than 50 years, display median ages in the upper 40s. Europe's "dependency ratio" (the number of non-working-age

citizens for every working-age citizen) is set to become cripplingly high. Indeed, Old Europe will soon be truly old. By 2050, one in every three Italians, Spaniards, and Greeks is expected to be 65 or older, even allowing for ongoing immigration. Europeans therefore face an agonizing choice between Americanizing their economies, i.e., opening their borders to much more immigration, with the cultural changes that would entail, or transforming their union into a fortified retirement community. Meanwhile, the EU's stalled institutional reforms mean that individual European nation-states will continue exercising considerable autonomy outside the economic sphere, particularly in foreign and security policy.

China's coming economic crisis.   Optimistic observers of China insist the economic miracle of the past decade will endure, with growth continuing at such a sizzling pace that within 30 or 40 years China's gross domestic product will surpass that of the United States. Yet it is far from clear that the normal rules for emerging markets are suspended for Beijing's benefit. First, a fundamental incompatibility exists between the free-market economy, based inevitably on private property and the rule of law, and the Communist monopoly on power, which breeds corruption and impedes the creation of transparent fiscal, monetary, and regulatory institutions. As is common in "Asian tiger" economies, production is running far ahead of domestic consumption—thus making the economy heavily dependent on exports—and far ahead of domestic financial development. Indeed, no one knows the full extent of the problems in the Chinese domestic banking sector. Those Western banks that are buying up bad debts to establish themselves in China must remember that this strategy was tried once before: a century ago, in the era of the Open Door policy, when U.S. and European firms rushed into China only to see their investments vanish amid the turmoil of war and revolution.

Then, as now, hopes for China's development ran euphorically high, especially in the United States. But those hopes were dashed, and could be disappointed again. A Chinese currency or banking crisis could have earth-shaking ramifications, especially when foreign investors realize the difficulty of repatriating assets held in China. Remember, when foreigners invest directly in factories rather than through intermediaries such as bond markets, there is no need for domestic capital controls. After all, how does one repatriate a steel mill?

The fragmentation of Islamic civilization   With birthrates in Muslim societies more than double the European average, the Islamic countries of Northern Africa and the Middle East are bound to put pressure on Europe and the United States in the years ahead. If, for example, the population of Yemen will exceed that of Russia by 2050 (as the United Nations forecasts, assuming constant fertility), there must either be dramatic improvements in the Middle East's economic performance or substantial emigration from the Arab world to aging Europe. Yet the subtle Muslim colonization of Europe's cities—most striking in places like Marseille, France, where North Africans populate whole suburbs—may not necessarily portend the advent of a new and menacing "Eurabia." In fact, the Muslim world is as divided as ever, and not merely along the traditional fissure between Sunnis and Shiites. It is also split between those Muslims seeking a peaceful modus vivendi [way of living] with the West (an impulse embodied in the Turkish government's desire to join the EU) and those drawn to the revolutionary Islamic Bolshevism of renegades like al Qaeda leader Osama bin Laden. Opinion polls from Morocco to Pakistan suggest high levels of anti-American sentiment, but not unanimity. In Europe, only a minority expresses overt sympathy for terrorist organizations; most young Muslims in England clearly prefer assimilation to jihad. We are a long way from a bipolar clash of civilizations, much less the rise of a new caliphate that might pose a geopolitical threat to the United States and its allies.

In short, each of the potential hegemons of the 21st century—the United States, Europe, and China—seems to contain within it the seeds of decline; and Islam remains a diffuse force in world politics, lacking the resources of a superpower.

Dark and Disconnected

Suppose, in a worst-case scenario, that U.S. neoconservative hubris is humbled in Iraq and that the Bush administration's project to democratize the Middle East at gunpoint ends in ignominious withdrawal, going from empire to decolonization in less than two years. Suppose also that no aspiring rival power shows interest in filling the resulting vacuums—not only in coping with Iraq but conceivably also Afghanistan, the Balkans, and Haiti. What would an apolar future look like?

The answer is not easy, as there have been very few periods in world history with no contenders for the role of global, or at least regional, hegemon. The nearest approximation in modern times could be the 1920s, when the United States walked away from President Woodrow Wilson's project of global democracy and collective security centered on the League of Nations. There was certainly a power vacuum in Central and Eastern Europe after the collapse of the Romanov [Russia], Habsburg [Austria-Hungary], Hohenzollern [Germany], and Ottoman [Turkey] empires, but it did not last long. The old West European empires were quick to snap up the choice leftovers of Ottoman rule in the Middle East. The Bolsheviks had reassembled the czarist empire by 1922. And by 1936, German revanche [retaking old territory and power] was already far advanced.

One must go back much further in history to find a period of true and enduring apolarity; as far back, in fact, as the ninth and 10th centuries.

In this era, the remains of the Roman Empire—Rome and Byzantium—receded from the height of their power. The leadership of the West was divided between the pope, who led Christendom, and the heirs of Charlemagne, who divided up his short-lived empire under the Treaty of Verdun in 843. No credible claimant to the title of emperor emerged until Otto was crowned in 962, and even he was merely a German prince with pretensions (never realized) to rule Italy. Byzantium, meanwhile, was dealing with the Bulgar rebellion to the north.

By 900, the Abbasid caliphate initially established by Abu al-Abbas in 750 [in the Arabian peninsula] had passed its peak; it was in steep decline by the middle of the 10th century. In China, too, imperial power was in a dip between the T'ang and Sung dynasties. Both these empires had splendid capitals—Baghdad and Ch'ang-an—but neither had serious aspirations of territorial expansion.

The weakness of the old empires allowed new and smaller entities to flourish. When the Khazar tribe converted to Judaism in 740, their khanate occupied a Eurasian power vacuum between the Black Sea and the Caspian Sea. In Kiev, far from the reach of Byzantium, the regent Olga laid the foundation for the future Russian Empire in 957 when she converted to the Orthodox Church. The Seljuks—forebears of the Ottoman Turks—carved the Sultanate of Rum as the Abbasid caliphate lost its grip over Asia Minor. Africa had its mini-empire in Ghana; Central America had its Mayan civilization. Connections between these entities were minimal or nonexistent. This condition was the antithesis of globalization. It was a world broken up into disconnected, introverted civilizations.

One feature of the age was that, in the absence of strong secular polities, religious questions often produced serious convulsions. Indeed, religious institutions often set the political agenda. In the eighth and ninth centuries, Byzantium was racked by controversy over the proper role of icons in worship. By the 11th century, the pope felt confident enough to humble Holy Roman Emperor Henry IV during the battle over which of them should have the right to appoint bishops. The new monastic orders amassed considerable power in Christendom, particularly the Cluniacs, the first order to centralize monastic authority. In the Muslim world, it was the ulema (clerics) who truly ruled. This atmosphere helps explain why the period ended with the extraordinary holy wars known as the Crusades, the first of which was launched by European Christians in 1095.

Yet, this apparent clash of civilizations was in many ways just another example of the apolar world's susceptibility to long-distance military raids directed at urban centers by more backward peoples. The Vikings repeatedly attacked West European towns in the ninth century—Nantes in 842, Seville in 844, to name just two. One Frankish chronicler lamented "the endless flood of Vikings" sweeping southward. Byzantium, too, was sacked in 860 by raiders from Rus, the kernel of the future Russia. This "fierce and savage tribe" showed "no mercy," lamented the Byzantine patriarch. It was like "the roaring sea . . . destroying everything, sparing nothing." Such were the conditions of an anarchic age.

Small wonder that the future seemed to lie in creating small, defensible, political units: the Venetian republic—the quintessential city-state, which was conducting its own foreign policy by 840—or Alfred the Great's England, arguably the first thing resembling a nation-state in European history, created in 886.

Superpower Failure

Could an apolar world today produce an era reminiscent of the age of Alfred? It could, though with some important and troubling differences.

Certainly, one can imagine the world's established powers—the United States, Europe, and China—retreating into their own regional spheres of influence. But what of the growing pretensions to autonomy of the supranational bodies created under U.S. leadership after the Second World War? The United Nations, the International Monetary Fund, the World Bank, and the World Trade Organization (formerly the General Agreement on Tariffs and Trade) each considers itself in some way representative of the "international community." Surely their aspirations to global governance are fundamentally different from the spirit of the Dark Ages?

Yet universal claims were also an integral part of the rhetoric of that era. All the empires claimed to rule the world; some, unaware of the existence of other civilizations, maybe even believed that they did. The reality, however, was not a global Christendom, nor an all-embracing Empire of Heaven. The reality was political fragmentation. And that is also true today. The defining characteristic of our age is not a shift of power upward to supranational institutions, but downward. With the end of states' monopoly on the means of violence and the collapse of their control over channels of communication, humanity has entered an era characterized as much by disintegration as integration.

If free flows of information and of means of production empower multinational corporations and nongovernmental organizations (as well as evangelistic religious cults of all denominations), the free flow of destructive technology empowers both criminal organizations and terrorist cells. These groups can operate, it seems, wherever they choose, from Hamburg to Gaza. By contrast, the writ of the international community is not global at all. It is, in fact, increasingly confined to a few strategic cities such as Kabul and Pristina [capital of Kosovo]. In short, it is the nonstate actors who truly wield global power—including both the monks and the Vikings of our time.

So what is left? Waning empires. Religious revivals. Incipient anarchy. A coming retreat into fortified cities. These are the Dark Age experiences that a world without a hyperpower might quickly find itself reliving. The trouble is, of course, that this Dark Age would be an altogether more dangerous one than the Dark Age of the ninth century. For the world is much more populous—roughly 20 times more—so friction between the world's disparate "tribes" is bound to be more frequent. Technology has transformed production; now human societies depend not merely on freshwater and the harvest but also on supplies of fossil fuels that are known to be finite. Technology has upgraded destruction, too, so it is now possible not just to sack a city but to obliterate it.

For more than two decades, globalization—the integration of world markets for commodities, labor, and capital—has raised living standards throughout the world, except where countries have shut themselves off from the process through tyranny or civil war. The reversal of globalization—which a new Dark Age would produce—would certainly lead to economic stagnation and even depression. As the United States sought to protect itself after a second September 11 devastates, say, Houston or Chicago, it would inevitably become a less open society, less hospitable for foreigners seeking to work, visit, or do business. Meanwhile, as Europe's Muslim enclaves grew, Islamist extremists' infiltration of the EU would become irreversible, increasing trans-Atlantic tensions over the Middle East to the breaking point. An economic meltdown in China would plunge the Communist system into crisis, unleashing the centrifugal forces that undermined previous Chinese empires. Western investors would lose out and conclude that lower returns at home are preferable to the risks of default abroad.

The worst effects of the new Dark Age would be felt on the edges of the waning great powers. The wealthiest ports of the global economy—from New York to Rotterdam to Shanghai—would become the targets of plunderers and pirates.

With ease, terrorists could disrupt the freedom of the seas, targeting oil tankers, aircraft carriers, and cruise liners, while Western nations frantically concentrated on making their airports secure. Meanwhile, limited nuclear wars could devastate numerous regions, beginning in the Korean peninsula and Kashmir, perhaps ending catastrophically in the Middle East. In Latin America, wretchedly poor citizens would seek solace in Evangelical Christianity imported by U.S. religious orders. In Africa, the great plagues of aids and malaria would continue their deadly work. The few remaining solvent airlines would simply suspend services to many cities in these continents; who would wish to leave their privately guarded safe havens to go there?

For all these reasons, the prospect of an apolar world should frighten us today a great deal more than it frightened the heirs of Charlemagne. If the United States retreats from global hegemony—its fragile self-image dented by minor setbacks on the imperial frontier—its critics at home and abroad must not pretend that they are ushering in a new era of multipolar harmony, or even a return to the good old balance of power.

Be careful what you wish for. The alternative to unipolarity would not be multipolarity at all. It would be apolarity—a global vacuum of power. And far more dangerous forces than rival great powers would benefit from such a not-so-new world disorder.

POSTSCRIPT

Should the United States Decrease Its Global Presence?

Supporters of President George Bush greeted his reelection in November 2004 with chants of "four more years, four more years!" Just hours after Senator John Kerry had conceded the election, Bush told listeners "America has spoken," that the American people had given him a mandate to continue his domestic and foreign policies. "I earned capital in the campaign—political capital—and now I intend to spend it," he assured his audience.

In the aftermath of Bush's return to office, analysts debated how, if at all, his foreign policy would change. Some argued that the cost in American lives and treasure of trying to pacify defeated Iraq and install a self-reliant democracy there would serve to moderate the president's unilateralist and neoconserative (neocon) inclinations. Others contended that the departure of moderate Secretary of State Colin Powell, the continuance of neocon Secretary of Defense Donald Rumsfeld, as well as the reelection of neocon Vice President Dick Cheney, meant that, if anything, the president's application of the Bush doctrine would intensify, not abate. One such discussion is James Mann, "Four More Years," *Foreign Policy* (November 2004). For varying views of neocon beliefs, see Elizabeth Drew, "The Neocons in Power," *New York Review of Books* (June 12, 2003) and Zachary Seldon, "Neoconservatives and the American Mainstream," *Policy Review* (April 2004). For divergent views with a particular focus on foreign policy, see Lawrence F. Kaplan, "Springtime for Realism," *New Republic* (June 1, 2004) and Todd Gitlin, "America's Age of Empire," *Mother Jones* (January/February 2003). An expression of neocon views on foreign policy can also be found on the Web site of the Project for the New American Century at http://www.newamericancentury.org/.

Regardless of what degree neocon continues to shape the foreign policy during President Bush's second term, the United States will still be a dominant global power and will need to decide how to use its strength and influence. One study that specifically addresses the intersection of the U.S. superpower status and its role in international organizations is Rosemary Foot, S. Neil Macfarlane, and Michael Mastanduno, eds., *U.S. Hegemony in an Organized World: The United States and Multilateral Institutions* (Oxford University Press, 2002). Arguing for an assertive U.S. policy is Charles Krauthammer, "In Defense of Democratic Realism," *The National Interest* (Fall 2004). Championing a more restrained approach is Steven E. Lobell, "Historical Lessons to Extend America's Great Power Tenure," *World Affairs* (Spring 2004).

ISSUE 5

Should the United States Continue to Encourage a United Europe?

YES: A. Elizabeth Jones, from Testimony Before the Subcommittee on Europe, Committee on International Relations, U.S. House of Representatives (March 13, 2002)

NO: John C. Hulsman, from "Laying Down Clear Markers: Protecting American Interests from a Confusing European Constitution," *The Heritage Foundation Backgrounder* (December 12, 2003)

ISSUE SUMMARY

YES: A. Elizabeth Jones, assistant secretary of state for European and Eurasian affairs, maintains that the United States looks forward to working cooperatively with such exclusively or mostly European institutions as the European Union, the Organization for Cooperation and Security in Europe, and the North Atlantic Treaty Organization.

NO: John C. Hulsman, a research fellow for European affairs in the Kathryn and Shelby Cullom Davis Institute for International Studies at the Heritage Foundation, argues that the United States should support European countries on a selective basis but not be closely tied to Europe as a whole.

At the end of World War II, Europe was physically and economically devastated. World War II resulted in the deaths of tens of millions of Europeans and the destruction of a great deal of Europe's industry and infrastructure (transportation and communication networks, buildings, etc.). Soon after, the cold war between the United States and the Soviet Union added to Europe's problems by dividing the continent.

The United States took steps to strengthen Europe economically and politically. One path was by extending aid through the multibillion-dollar European Recovery Plan (also called the Marshall Plan) beginning in 1947. The United States aided European recovery in part because American economic health required Europe to once again become a strong trading partner and partly because Washington worried that left-wing movements in France, Italy, and elsewhere might result in communist influence in those countries' governments.

One of the strings that the United States tied to its aid was the demand that Europeans plan together how to best utilize it. This forced economic cooperation and the views of such Europeans as French statesman Jean Monnet (1888–1979) combined to promote the establishment that began in the 1950s and that evolved in the decades that followed into the European Union (EU). Today the EU has 15 member states, and it will soon be joined by an additional 13 eastern and southern European countries. In terms of the size of its population and gross domestic product, the EU rivals the United States.

On the security front, the United States encouraged the formation of the North Atlantic Treaty Organization (NATO) in 1949. The alliance consists of the United States, Canada, 13 Western European countries, and Turkey. The original purpose of NATO was to counter the threat that many in the West thought the Soviet Union posed. Other unspoken factors behind the founding of NATO were the desire to keep the United States engaged in Europe and to avoid any possibility that Germany might reemerge as a powerful, hostile country.

The cooperation in Europe, marked by NATO and what would become the EU, led to other joint organizations, such as the Organization for Security and Cooperation in Europe (OSCE). The OSCE includes 55 countries that are primarily European, but also Canada, the United States, and some states in Central Asia. The organization works through diplomats, observers, and others to prevent conflicts, manage crises, and facilitate post-conflict rehabilitation.

During the decades that followed the end of World War II, the United States served as the leader of the transatlantic bloc, and there was a great deal of unity, although some would argue that it was European subservience to Washington's direction.

However, the last decade or so has seen market changes in U.S.-European relations. One factor is the end of the cold war. The collapse of the Soviet Union and Soviet bloc (with Poland and some other former communist countries now belonging to NATO) erased any immediate military threat to Europe and, therefore, lessened the need for strategic cooperation. With no common enemy to bind Europe and the United States together, old disputes that had once been suppressed in the name of allied unity came to the fore and new disagreements became public and more acrimonious.

A second factor is the full recovery of Europe's economy and confidence. In particular, the expanded political and economic integration of the European Union have made Europeans increasingly willing to challenge U.S. hegemony on a range of political, economic, and social issues. Some in Europe even believe that its countries should work together to restrain U.S. power.

There are many people on both sides of the Atlantic Ocean who continue to believe that a strong, increasingly united Europe and strong transatlantic cooperation are in the mutual interests of the United States and Europe. That is the official U.S. position, which is reflected in the following selection by A. Elizabeth Jones. In the second selection, John C. Hulsman argues that such statements are papering over the distinct and growing division between the United States and its formal allies on the other side of the Atlantic. He contends that the United States should adopt a more realistic policy.

A. Elizabeth Jones  **YES**

U.S. and Europe: The Bush Administration and Transatlantic Relations

President [George W.] Bush said last August [2001] in Warsaw that the Administration seeks a Europe "whole, free, and at peace." This is even more vital to America's national security in the aftermath of September 11th. The imperative for closer coordination has opened up new opportunities to achieve our goals in Europe and Eurasia. We are cooperating more broadly to combat terrorism. We are pursuing a deeper relationship with Russia. We are advancing throughout the region respect for democracy, the rule of law, human rights, and free market economies.

We know who our friends are when the chips are down and we need help. By this measure, we have friends in Europe. Following September 11th our European partners offered critical assistance in military deployments to Afghanistan. They cracked down on terrorist activities in their territory. European and U.S. soldiers are working side-by-side in Afghanistan. . . . German and Danish troops suffered fatalities while trying to disarm abandoned ordnance in Kabul [Afghanistan]. Europe and the U.S. are partners in every sense.

Recently, a few European leaders have expressed concerns about U.S. "unilateralism." Some wonder about our long-term goals in the War on Terrorism and our intentions regarding pariah states such as Iraq. We take these concerns seriously. But we must put them in perspective. Europeans speak as our coalition partners. They are vulnerable to the same dangers that we are. As one European explained it: "September 11th was an attack on all of us. We want to be involved in the solution." As Secretary [of State Colin] Powell says constantly, the U.S. will continue to engage vigorously with our European partners. Our policies have not changed. We will remain in close touch. U.S.-European relations remain steadfast.

We are reinvigorating our partnership with the European Union [EU]. Counter-terrorism is front and center. In December Secretary Powell signed an agreement with EUROPOL [European Police Office]. We are aiming next for an agreement on judicial cooperation. There is potential for progress on nonproliferation, intelligence sharing, asset freezes, and uprooting terrorist networks. We are taking joint action against terrorist organizations.

From U.S. House of Representatives. Committee on International Relations. Subcommittee on Europe. *U.S. and Europe: The Bush Administration and Transatlantic Relations*. Hearing, March 13, 2002. Washington, D.C.: U.S. Government Printing Office, 2002.

The U.S. and EU economies are increasingly integrated. Trade and reciprocal foreign investment rise each year, doubling since 1990. The U.S. supports a fair, open international trading system. We worked with the EU on a successful launch of the new WTO [World Trade Organization] Round at Doha [Qatar]. We pursue vigorously the resolution of U.S.-EU trade disputes. We will continue to promote U.S. business and economic interests in resolving outstanding disagreements, not just on steel, but on Foreign Sales Corporation tax, biotechnology and beef hormones. Europeans have reacted strongly to the President's decision to impose temporary safeguards on steel. We will work with our European friends and other steel producing countries to address the heart of this problem: excess global capacity in steel production. Our goal is that transatlantic trade solidify all aspects of our relationship, including security.

Our European friends and allies share our concern about the need to accord recognition to surviving Holocaust victims within their lifetimes. In the past eight months, the German foundation "Remembrance, Responsibility and the Future" distributed more than $1.1 billion to 600,000 former slave and forced laborers as provided under the July 17, 2000 agreements. The payment of insurance claims is a difficult issue. We will continue to work with the International Commission on Holocaust Era Insurance Claims and other involved parties to resolve outstanding procedural problems. We are engaged on property restitution. In this regard, the International Task Force on Holocaust Education, Remembrance and Research is an important focus. The foundation's board of trustees is working on criteria for projects of the Future Fund. The interest on the endowment will be used to combat racism and hatred.

NATO [North Atlantic Treaty Organization] remains the cornerstone of transatlantic security. In the aftermath of September 11th, Allies invoked NATO's Article 5 collective defense commitment for the first time in history. Our Allies have provided invaluable support to the anti-terrorist effort. This includes force deployments, intelligence sharing, and extensive law enforcement assistance. Allies recognize that we must intensify this cooperation to address the threats of terrorism and Weapons of Mass Destruction. That is among our goals for the Prague Summit next November.

The September 11th attacks and continued terrorist threats have underscored the need for NATO to improve its ability to meet new challenges to our common security. Allies recognized this threat in the 1991 Strategic Concept. They reinforced it at the Washington Summit in 1999. When President Bush meets with Allied Leaders in Prague, NATO is expected to approve a program of action to enhance its ability to deal with these threats. It is vital that our European Allies, who have not followed through on all the commitments made in NATO's Defense Capabilities Initiative, refocus and reprioritize their efforts to address the growing capabilities gap within NATO. Thus, the development of new capabilities is one of our priorities for the Prague Summit next November.

A second key goal for Prague is the addition of new members to the Alliance. Continued NATO enlargement will reinforce the strength and cohesion of states committed to our values. It will bolster our own defense. We are looking closely at values issues among aspirant countries. We will evaluate candidates on their ability to further NATO's principles and contribute to the security of the North Atlantic area. An inter-agency team recently visited each of the nine countries participating in the Membership Action Plan for frank discussions of their progress toward these goals. As we approach these historic decisions, we look forward to a close dialogue with the Congress. Our goal is to forge a united U.S. approach to enlargement and a solid consensus within the Alliance.

We also hope to advance new relationships at the Prague Summit. Foremost among these is a constructive NATO-Russia relationship, which I will address later. NATO's continued outreach to Partnership for Peace [PfP] member states has overcome entrenched hostility and historical divisions. Through its unique Partnerships, NATO remains the only institution that can unite the continent in security cooperation. NATO remains the indispensable nexus for broadening and deepening Euro-Atlantic security, democracy, free markets, and the rule of law. At Prague, we intend to continue building closer links with Russia, Ukraine, and all of NATO's Partners.

As NATO further evolves, we will work to strengthen Alliance links between those Partners who are not yet ready or do not seek NATO membership. Many of our Partners, such as the Nordic countries and Ireland, have contributed significantly to NATO's efforts in the Balkans. They have reached out to the states of the former Soviet Union. We will continue to work closely with these Partners to improve interoperability and capabilities of all NATO's Partners.

Most recently, our Central Asian and Caucasus Partners have stepped forward to play critical roles in the anti-terrorist effort. We intend to energize all elements of the Partnership for Peace at NATO to engage Central Asian and Caucasus Partners. Working with our Allies and more advanced Partners, we hope to increase, coordinate and target assistance to the Central Asian and Caucasus states. We believe PfP programs should address issues that have the greatest appeal to these countries. These include terrorism, border security, and civil emergency planning. We will continue to support the development of democracy and market economic institutions to help ensure the viability of our security partnerships with these countries. We look to the OSCE [Organization for Security and Cooperation in Europe] to play an increasing role in this regard.

We continue to support a European Security and Defense Policy that strengthens NATO while increasing the EU's ability to act where NATO as a whole is not engaged. At the same time, the broader value of close NATO-EU cooperation is nowhere more evident than in Southeast Europe, where NATO and the EU have worked closely to prevent instability, overcome violence and begin to build a lasting peace. The Macedonia peace settlement is a model of our collective ability to draw on the unique strengths of these organizations in a common effort.

Key to a Europe "whole, free and at peace" is a more stable, democratic and prosperous Southeast Europe. Despite the region's great strides since the Dayton Peace Accords, governments still have much to do. Working in partnership with the U.S. and the Europeans, these nations must complete reform efforts and establish an environment conducive to prosperity. Corruption, insufficient border controls and weak export control regimes contribute to trafficking throughout the region—in arms, drugs and people. Work in these areas also contributes to our global counterterrorism efforts.

NATO and its partners in SFOR [Stabilization Force in Bosnia] and KFOR [Kosovo Security Force] still have a role to play, as does the German-led NATO "Task Force Fox" in Macedonia. Our vision is that the U.S. and the international community deal with this region "normally"—without troops on the ground and through trade and investment rather than aid. We are mindful that we came into this region together with our Allies and we should go out together.

Our engagement with Southeast Europe is changing. We continue to support economic reform and regional trade development, supported by a Southeast Europe Trade Preferences Act (SETPA). We are encouraging further integration of the region with Europe. We promote rule of law, cooperation with the International Criminal Tribunal, and ethnic tolerance. With success, our European partners and we have been able to reduce force levels in Bosnia. We anticipate that NATO Military Authorities will recommend further reductions in Bosnia and Kosovo. The EU will take over the UN's police mission in Bosnia at the end of [2002]. The international community recently agreed to a blueprint for streamlining and downsizing its presence in Bosnia. The creation of a government in Kosovo will allow the transfer of many responsibilities from the international community to local democratically elected authorities. In Macedonia, the close and continuing cooperation between the EU, NATO and the OSCE is a model for transatlantic cooperation in crisis management. Task Force Fox is small. It is of limited duration and made up almost entirely of Europeans.

A critical element of achieving the President's vision of a Europe "whole, free and at peace" is the resolution of regional and ethnic conflicts in Europe and neighboring Eurasia. We are pleased by progress in the Cyprus talks. We will encourage the leaders on the island to achieve a final settlement in the coming months. The Good Friday Accord is being implemented in Northern Ireland. We will work to solidify the role of the police force there. Cooperation among all factions is crucial. In Northern Europe, we will continue to work with our Nordic Allies and friends and our Baltic and other regional partners, including Russia. It is vital that we reinforce ten years of progress in a region of shared values. Opportunities for economic progress, good neighborly relations and democratic institution building are beginning to outweigh the challenges.

OSCE remains a vital element in our engagement with Europe. It is the pre-eminent multilateral institution for upholding democracy, human rights and the rule of law. It undertakes early warning measures, conflict prevention, and post-conflict rehabilitation. OSCE also implements valuable programs to counter corruption and trafficking, and strengthen the rule of law through police training and judicial reform. Its broad membership allows it to operate throughout Europe and Eurasia.

The OSCE has said it will begin to play a role in the war against terrorism. The OSCE can encourage European and Eurasian countries to adhere to the principles of UN Resolution 1373. It will continue to be central to development of pluralistic societies in the Balkans, including solidifying the Framework Agreement in Macedonia. Implementation of CFE [Conventional Armed Forces in Europe] commitments will be an ongoing OSCE oversight responsibility. The organization can offer opportunities for cooperative engagement with Russia and the European Union.

The OSCE plays a critical role in our effort to promote democracy, human rights and rule of law throughout Eurasia. It is working to restore territorial integrity in Moldova. In Belarus, we work with the OSCE and our European partners to urge the Lukashenko regime to adopt OSCE standards of behavior and come out of its self-imposed isolation. Unfortunately, the regime shows no inclination to do so thus far. In Moldova, we work through the OSCE and with key players to resolve the separatist conflict in Transnistria and reincorporate that region into Moldova. Ukrainian involvement is important on this issue and in the region generally. Ukraine's influence is a potential force for regional stability and European integration. Ukrainian success in political and economic reform will fulfill that country's European aspirations and will inspire other post-Soviet states to follow the same path.

In the Caucasus, we are working with Armenia and Azerbaijan to resolve their conflict over Nagorno-Karabakh. We seek a comprehensive settlement through the Minsk Group peace process. Georgian sovereignty is important to the Administration. We are proposing a program to develop Georgia's internal capacity to deal with terrorism now and in the future. We also are working to support the development of democracy and human rights in the Caucasus.

Bilateral U.S.-Russia cooperation is unprecedented. Counterterrorism collaboration is central to this effort, although not the sole focus. The U.S. and Russia are cooperating more closely in intelligence sharing, nuclear weapons reduction, and resolution of Eurasian regional conflicts. We are working together in the fight against HIV-AIDS and other infectious diseases, organized crime and narcotics trafficking. We hope to expand the economic and commercial component of the relationship. While we broaden this new cooperation with the Russians, we have not forgotten the difficult issues. We continue to press our concerns over issues such as the conduct of Russian forces in Chechnya and threats to media freedom in Russia as a whole.

Russia's cooperation with us and our Allies in the war on terrorism also reflects the opportunity to bring Russia closer to NATO. We are working with our Allies on arrangements for a new NATO-Russia body that would focus on concrete, practical projects of mutual benefit. Russia would participate in this "NATO-Russia Council"—which would focus on issues with potential for cooperative initiatives—as an equal. The deepening of the Russia-NATO relationship will not be allowed to undercut NATO's ability to decide and act on its own. Russia would not get a veto over the ability of NATO's 19 Allies to act on their own. The NAC [North Atlantic Council] will continue to meet and make decisions as it always has. The mechanisms and substance of such arrangements are still being worked out. I pledge to keep the Committee apprised of progress. Moreover, I want to reiterate President Bush's and [secretary-general of NATO] Lord [George] Robertson's pledges not to give Russia a veto over NATO operations. This is not a backdoor to membership. This is an opportunity for Russia to develop a new relationship with NATO that would advance not only our interests but also its.

In the spirit of new U.S.-Russia cooperation, we believe it is time to move beyond the Cold War. Russia has made significant progress on religious freedom and emigration. Therefore, the President is pursuing the removal of Russia and eight other Eurasian countries from the application of Jackson-Vanik legislation [1974 Trade Act to Russia amendment that was intended to promote free emigration from the Soviet Union]. We hope that Congress will pass legislation to "graduate" Russia from Jackson-Vanik before the President visits Moscow this spring. The President and Secretary Powell appreciate the support of many Members of this committee in this endeavor.

Success in addressing transnational problems is more important than ever in pursuing America's transatlantic agenda. Stable countries able to withstand terrorist and other threats are based on respect for the rule of law, human rights, religious freedom, and open media. Stable countries have vibrant civil societies. They are committed to the principles of free market economies. The Administration's attention of these values with our European and Eurasian friends is even more critical as we pursue the War on Terrorism with our coalition partners. Enhanced defense and security cooperation and intelligence sharing must be buttressed by societies committed to democratic principles such as those in the Final Act in the Conference on Security and Cooperation in Europe. Moreover, we are continuing efforts with our transatlantic partners to address problems that respect no borders, e.g., HIV/AIDS and infectious disease, narcotics trafficking and environmental degradation.

Critical to the promotion of our policies in Europe and Eurasia is the use of Public Diplomacy. Training programs and exchanges offer an accurate portrayal of American views, values and traditions. Such people-to-people ties will help bind the nations of Europe and Eurasia with the United States, thereby enhancing the transatlantic relationship and American security.

John C. Hulsman  **NO**

Laying Down Clear Markers: Protecting American Interests from a Confusing European Constitution

There is a danger in the European constitutional process for transatlantic relations and American interests because the fine print of European agreements often determines outcomes of which few of its participants are aware. The process, scheduled for approval by the various European governments at the Brussels summit on December 13, 2003, was originally de-signed to make the European Union (EU) political system more transparent and understandable to the average European citizen. Whatever the outcome of the process, however, it is already apparent that this goal has not been met.

Consisting of over 440 articles and written in an incomprehensible style, the constitution will surely not further endear the EU to its citizens. At the same time, though, the importance of the constitutional process cannot be denied. For instance, the constitutional settlement on defense will directly affect the North Atlantic Treaty Organization (NATO), defense issues among European member states, and thus all major transatlantic foreign and defense policy decisions. Therefore, the Bush Administration must inject clarity into the debate on issues of primary concern for U.S.-EU relations, particularly in the area of security policy, which is covered by the European Security and Defense Policy (ESDP).

The U.S. should make it clear to the EU that it will support the ESDP only if NATO remains the primary transatlantic security organization. NATO must continue to have first right of refusal for any proposed military mission before the ESDP is activated, leading to European military action. Further, the ESDP must not lead to either duplication of NATO assets or competition with the alliance for strategic primacy. By laying down markers and making its own policy goals crystal clear, the United States can ensure that areas of potential disagreement lurking in the EU constitution do not poison the already endangered transatlantic alliance.

Common Foreign Policy Simply Not in the Cards

The EU constitutional process has only now begun to outline the future of a Common European Foreign and Security Policy (CFSP). It is highly unlikely that the constitutional process will result in an integrated European foreign policy—a reality that ought to be welcomed by Washington as it allows the U.S. to work with a greater number of committed European partners.

Specifically, there is no evidence that individual European states are willing to abandon their national interests and sign on to a European process that would lead to majority voting on matters of high politics regarding foreign policy. (This would mean that EU member states would be willing to give up their veto in Brussels on foreign policy issues.) The national interests of the individual member states remain the basis for their respective foreign policies and are likely to do so into the indefinite future.

Things become far murkier, however, on the defense side of the equation. The result may well be to paralyze defense policymaking in Europe, making the EU a dragging anchor on U.S. security policy.

The Vague Question of a Common Defense Policy

Like a vampire, European federalist efforts to establish a European defense identity that is separate from and in competition with NATO continue to rise from the dead. Berlin-Plus, the March 17, 2003, agreement between the U.S. and the EU member states, was designed to resolve definitively the questions of compatibility between the ESDP and NATO.

Berlin-Plus sought to avoid the unnecessary duplication of transatlantic resources and had four elements: It assured EU access to NATO operational planning; it made NATO capabilities and assets available to the EU; it made the Deputy Supreme Allied Commander Europe (always a European [the Supreme Allied Commander Europe is always an American]) also commander of any EU-led operations; and it adapted the NATO defense planning system to allow for EU-run operations. This commonsense agreement has now been called into question by Franco-German efforts to set up a wholly separate EU headquarters and planning structure.

Berlin-Plus allowed for greater alliance flexibility and was successfully put to the test in the EU-led mission to Macedonia. While the U.S. should not be forced to participate militarily in every transatlantic mission, America should allow the European allies to act alone using NATO assets if Washington does not object.

As the case of Macedonia amply illustrates, that should have been that. During the height of European opposition to the U.S. stance on Iraq, however, France, Germany, Belgium, and Luxembourg reopened the question, in effect calling Berlin-Plus into doubt. They advocated the establishment of an independent EU military headquarters in Tervuren, Belgium, with an independent planning capacity.

Beyond the obvious operational drawback of such an institutional arrangement, which would lead to unnecessary duplication with NATO, the political ramifications of such an outcome are clear: It is the institutional expression of French political desires to lessen the American role in Europe. Ambassador Nicholas Burns, U.S. representative to NATO, rightly sounded the alarm, calling such an outcome "the greatest threat to the future of the alliance."

It is this threat that the European constitutional process has been designed to resolve. While the EU constitution does not formally address establishment of an independent EU military headquarters, the two issues emerged simultaneously and have since effectively merged. Thanks to the tireless efforts of British Prime Minister Tony Blair, another plan has been put forward to heal the rift threatening the NATO alliance. Under its terms, the EU will establish its own military planning cell, located within NATO headquarters in Brussels.

Accordingly, the French Gaullist [after President Charles De Gaulle (1945–1946, 1959–1969, who sought to restore France's leadership role] initiative has been significantly watered down. Instead of working through the planning cell as a separate entity, EU officers will work from NATO's existing headquarters. Belgium, France, Germany, and Luxembourg had pressed for a full-blown EU military headquarters that would function independently of existing structures.

Critically, the [British Prime Minister Tony] Blair government steered the compromise so that the ESDP will work on projects only when NATO decides not to intervene in a crisis, and NATO retains the first right of refusal in a crisis situation. As with the Berlin-Plus agreement, the EU will continue to be allowed to draw on NATO assets—but only if the alliance as a whole approves, giving the U.S. a de facto veto over the process. What is most important for the United States is that, while there is an EU planning unit, the EU will not possess a separate headquarters for the ESDP process.

The Last, Best Chance for Transatlantic Diplomatic Peace

To safeguard American interests, the [George W.] Bush Administration should:

> 1. *Make it clear that the U.S. views the EU constitutional process as more than merely a tidying-up exercise for existing European agreements, particularly regarding defense issues.*

The President should make it clear not only that the outcome of the convention is entirely up to the EU member states, but also that it will change political relationships, both within the EU and between the EU and the U.S. For example, the United States would view any attempt to make the ESDP a rival to NATO or to do away with the veto of individual states on foreign policy issues as highly detrimental to the transatlantic relationship.

> 2. *Accept the NATO-ESDP compromise worked out during the EU constitutional process, but only if it articulates clear operational lines of control.*

As it retains the key elements of Berlin-Plus, the new European agreement crafted by Prime Minister Blair can be accepted by the United States on three conditions.

First, in the words of Secretary of Defense Donald Rumsfeld, "I am confident that things will sort through in a way that we can have an arrangement that isn't duplicative or competitive of NATO." As NATO Secretary General Lord George Robertson has pointed out, "One thousand operational planners are already available in [Supreme Headquarters Allied Powers Europe] headquarters." What is needed are not further architectural changes in the transatlantic military alliance, but greater capability on the part of the European allies in terms of a greater emphasis on computerized weapons, communications, lift, logistics, and intelligence.

From NATO's 19 member states, excluding the United States, only 55,000 troops are deployed abroad out of a staggering 1.4 million people in uniform. As a senior European defense official recently said, "there are far too many European colonels who have never commanded anyone running Brussels, and far too few deployable troops. Everyone knows this." The duplication of precious resources does nothing to rectify this fundamental problem.

Second, after the failure of Berlin-Plus, the United States must undertake to receive concrete assurances from the European allies that NATO remains the preeminent transatlantic security institution. The operational expression of this political point is that the U.S. should condition its support for the new compromise on NATO's retaining the first right of refusal to act in any European crisis.

Third, the U.S. must clearly express its continued strenuous opposition to the formation of a separate EU headquarters, which would exist only to compete with NATO. As Secretary Rumsfeld stated, "Our policy is very clear: that we strongly support NATO as the primary forum for transatlantic defense. We support ESDP that is NATO friendly." This clear strategic view must guide the overall American reaction to any EU constitutional defense initiative.

3. Make it clear that, with the U.S. acceptance of the ESDP plan as formulated around the EU constitutional settlement, the U.S. regards the issue as closed.

The Bush Administration must be careful not to fall into a pattern of continuing to make concessions on ESDP, only to have subsequent "agreements" nullified by the European partners, who will then ask for further concessions. The United States cannot sit idly by while creeping Gaullist attempts undermine NATO in order to build up the ESDP as a rival to the alliance.

The United States must clearly articulate that other approaches would be highly destructive to the NATO alliance. By laying down such a marker, the Bush Administration can measure subsequent European diplomatic initiatives in a clear and effective manner.

Conclusion

The European experiment has always been driven by the process of functionalism, of obscuring policy goals and not making clear the genuine direction of the EU in order to foster internal political consensus. While such an approach succeeded brilliantly on economic issues, it is impossible to use the same modus operandi when dealing with foreign and security policy. These issues require a degree of clarity that leads to definite policy positions (i.e., one chooses to intervene in a country or not) in a manner entirely different from that applied to economic issues.

The United States must therefore make its position clear on European constitutional foreign and defense policy initiatives; otherwise, the constitution's fine print on foreign and security affairs could weaken NATO, which remains a primary American interest.

While European states have the right to work out whatever internal political relationship they prefer, the U.S. must emphasize that actions have consequences and that a more federal, centralized European defense could lead to a diminution of NATO. This, in turn, would tragically affect the transatlantic relationship.

POSTSCRIPT

Should the United States Continue to Encourage a United Europe?

Jones and Hulsman debate the future of an alliance that has persisted for over a half century. For a review of that history, consult Marc Trachtenberg, ed., *Between Empire and Alliance: America and Europe During the Cold War* (Rowman & Littlefield, 2003). More on current issues is available in Thomas Mowle, *Allies at Odds: The United States and the European Union* (Palgrave Macmillan, 2004).

By some measures, U.S. commitment to working with and strengthening Europe remains. With the support of Washington, NATO has taken up new security roles, such as peacekeeping in Macedonia and waging an air war on Yugoslavia to end its oppression of Kosovo. The alliance has also expanded. Former communist countries, the Czech Republic, Hungary, and Poland joined NATO in 1999. Four years later, seven Eastern European countries (Bulgaria, Estonia, Latvia, Lithuania, Romania, Slovakia, and Slovenia) joined NATO, with the enthusiastic support of Washington.

The European Union is also growing. Its current membership of 15 countries is likely to grow to 28 as 13 applicants for membership are admitted over the next few years. In addition, the "Convention on the Future of Europe" has been established to try to fashion an EU constitution, an eventuality that will be a giant step toward political integration for Europe. Although the United States is wary of any attempt to create a separate European military alliance that might supplant NATO, Washington is at least passively supportive of the latest EU effort to achieve greater unity. For a general view of the future of transatlantic relations, read Tod Lindberg, ed., *Beyond Paradise and Power: Europe, America, and the Future of a Troubled Partnership* (Routledge, 2004). For the view that increasing European unity will dramatically change U.S.-Europe relations, see T. R. Reid, *The United States of Europe: The New Superpower and the End of American Supremacy* (Penguin Press, 2004).

Concerns about general U.S. power in the world and about perceived U.S. unilateralism were arguably part of what caused some European countries—France and Germany in particular—to oppose a U.S.-led war against Iraq in 2003. That policy divide seriously escalated tension in the "Western alliance." Despite the constraints of diplomacy, the two factions were barely able to avoid name-calling at times. At one point, for instance, U.S. defense secretary Donald Rumsfeld described the actions of France as "inexcusable," and at another the German chancellor dismissively referred to the war as President Bush's "adventure" in Iraq.

These tensions led some on both sides of the Atlantic to see the controversy as the death rattle of the Western alliance. Others believe that the alliance will recover its health and remain a key aspect of both U.S. and European foreign policy.

ISSUE 6

Is Russian Foreign Policy Taking an Unsettling Turn?

YES: Ariel Cohen and Yevgeny Volk, from "Recent Changes in Russia and Their Impact on U.S.-Russian Relations," *The Heritage Foundation Backgrounder* (March 9, 2004)

NO: Leon Aron, from Testimony During Hearings on "U.S.-Russia Relations in Putin's Second Term," Committee on International Relations, U.S. House of Representatives (March 18, 2004)

ISSUE SUMMARY

YES: Ariel Cohen, research fellow in Russian and Eurasian studies in the Kathryn and Shelby Cullom Davis Institute for International Studies at The Heritage Foundation, and Yevgeny Volk, The Heritage Foundation's Moscow office director, write that the revival of statism and nationalism has seriously diminished Russia's chances of being regarded as a close and reliable partner that is clearly committed to democratic values.

NO: Leon Aron, director of Russian studies at the American Enterprise Institute, recognizes that there are pressures within Russia to try to take a more confrontational stance but believes that the forces for moderation are stronger.

Russia has experienced two momentous revolutions during the twentieth century. The first began in March 1917. After a brief moment of attempted democracy, that revolution descended into totalitarian government, with the takeover of the Bolshevik Communists in November and the establishment of the Union of the Soviet Socialist Republics.

The second great revolution arguably began in 1985 when reform-minded Mikhail S. Gorbachev assumed leadership in the USSR. The country's economy was faltering because of its overcentralization and because of the extraordinary amount of resources being allocated to Soviet military forces. Gorbachev's reforms, including *perestroika* (restructuring, mostly economic) and *glasnost* (openness, including limited democracy) unleashed strong forces within the USSR. The events of the next six years were complex, but suffice it to say that the result was the collapse of the Soviet Union. What had been the USSR fragmented into

15 newly independent countries. Of these former Soviet republics (FSRs), Russia is by far the largest, has the largest population, and is in reality and potential the most powerful. Russia retained the bulk of Soviet Union's nuclear weapons and their delivery systems.

When Russia reemerged in the aftermath of the collapse of the USSR, its president, Boris Yeltsin, seemed to offer the hope of strong, democratic leadership that would economically rejuvenate and democratize Russia internally and that, externally, would work to make Russia a peaceful and cooperative neighbor.

These prospects soon faded, however, amid Russia's vast problems. The country's economy fell more deeply into shambles, leaving 22 percent of all Russians below the poverty level. Russia's economic turmoil also caused a steep decline in Russia's military capabilities. To make matters even worse, the rekindling of an independence movement by the Chechens, a Muslim nation in the Caucuses Mountains region, let to savage fighting.

Yeltsin's ill-fated presidency ended when he resigned on December 31, 1999. His elected successor and current president, Vladimir Putin, is an individual who spent most of his professional career in the KGB (*Komitet Gosudarstvennoi Bezopasnosti*/Committee for State Security), the Soviet secret police, and who headed its successor, Russia's FSB (*Federal'naya Sluzhba Bezopasnosti*/Federal Security Service). Early on, Putin expressed his determination to regain strong control internally, to reassert Russia's world position, and to rebuild the country's economy.

For good or ill, Putin has brought a level of stability to Russia. Slowly, Russia's economy has steadied itself. Moreover, with a well-educated populace, vast mineral and energy resources, and a large (if antiquated) industrial base, Russia has great economic potential. Similarly, while Russian military forces fell into disarray in the 1990s, the country retains a potent nuclear arsenal. Furthermore, its large population, weapons manufacturing capacity, and huge land mass make it likely that the breakdown of Russia's conventional military capabilities and geostrategic importance will only be temporary.

Putin was reelected by an overwhelming margin to a second term in March 2004, but Russia is less democratic than it once was. Putin has used the various challenges facing his country as a reason to consolidate Moscow's power. Much of the independent new media is gone, local authorities have lost much of their power, and the country's largest company has been seized and its leader jailed on charges of corruption.

There are also numerous issues that divide Russia from the United States. For example, Moscow believes that it is threatened by the the U.S. drive to deploy a ballistic missile defense system and by the expansion of the North Atlantic Treaty Organization to even include some FSRs.

The question, then, is, Wither Russia? What are the chances it will once again become antagonistic toward the United States and its allies? In the first of the following readings, Ariel Cohen and Yevgeny Volk are somewhat pessimistic about the future. They fret that "old think" among Russia's foreign and security policy elites has caused the country to return to a strategic posture that is both prickly and at times anti-United States. Leon Aron is more optimistic in the second reading. He contends that several factors, including Russia's need to have good economic relations with the West, are apt to mean that Putin will maintain a moderate foreign policy.

**Ariel Cohen
and Yevgeny Volk**

 **YES**

Recent Changes in Russia and Their Impact on U.S.-Russian Relations

. . . U.S.-Russian relations are in limbo. The revival of statism and nationalism has seriously diminished Russia's chances of being regarded as a close and reliable partner that is clearly committed to democratic values. . . .

In 2003, the U.S.-Russian relationship was fraught with multiple complications. Russia resisted the U.S. military action in Iraq, continued its military cooperation with Iran, developed multifaceted ties with China, attempted to play the anti-American card in its relations with Western Europe, and stepped up political pressure on the independent nations of the post-Soviet space.

Moreover, a joint war on terrorism and large-scale exploration of natural resources, especially in the energy sector, are yet to become a focal point of the bilateral relationship. At the same time, backtracking on democratic politics could change the nature of the Russian state and challenge America's national interests unless both sides can find a common ground and reconcile their mutual concerns.

In a fundamental difference between Russia and Central Europe, the Russian political establishment underwent little reform after the collapse of communism and has yet to complete the transition from the centuries-old Soviet and czarist worldview. The legacy of Russia's imperial and totalitarian past deeply affects Moscow's foreign policy rhetoric and performance. Anti-Americanism, exaggeration of differences between the United States and Western Europe, heavy-handedness toward smaller Central and Eastern European nations, attempts to recreate a sphere of influence in the former Soviet republics, and continuing relations with Iran, North Korea, and Cuba all continue to frustrate bilateral ties.

The Moscow elite finds America's global leadership overbearing and continues to view its own country as a great power, at times capable of competing with the United States for regional, if not global, dominance. Russian leaders, while recognizing their country's weakness, strive to maximize their freedom of maneuver. They believe in a "multi-polar world" model in which Russia forms part of the great concert of powers, including the U.S., China, and eventually India and a

From *The Heritage Foundation Backgrounder,* March 9, 2004. Copyright © 2004 by The Heritage Foundation. Reprinted by permission.

united Europe. Russia also is anxious to maintain good relations with the Islamic world, as both the September 2003 visit of Saudi Crown Prince Abdullah and [Russia's President Vladimir] Putin's speech at the Organization of the Islamic Conference have demonstrated.

Back to the "Old Think"?

The "old think" among Russia's foreign and security policy elites has caused the country to return to a strategic posture that is both prickly and at times anti-U.S. In military policy, despite low-intensity radical Islamist threats from the South, arms sales to countries that threaten international stability, including Iran and Syria, have been on the rise.

In February, Moscow conducted its largest military exercise in the past 25 years, which culminated in intercontinental ballistic missile firings. Though many missile strikes toward an "unspecified" enemy failed, the exercises were a throwback to the Cold War. So was the surrounding propaganda: Putin announced that the maneuvers were successful, and government TV channels reported only successful launches. By contrast, NTV, which is owned by the Russian gas monopoly Gazprom, also mentioned failures.

The New Authoritarianism

Russia's domestic policy has been marked by the consolidation of President Putin's authoritarian rule, including the control of all TV channels and manipulation of the news media. In addition, there is reason to suspect that the new government appointee at the largest Russian public opinion research organization, known as VTsIOM, will open doors for the manipulation of polling results. The Kremlin has intensified its manipulation of mass media, political parties, and vital financial flows in the economy.

The fairness of the State Duma [Russia's parliament] elections last December, in which pro-Putin parties secured an absolute majority in the Duma, is suspect as the Kremlin exercised its powerful "administrative resources" through which it sways mass media outlets, regional governments, the military, the police, and control over the Central Elections Commission. The outcome of the elections led some in Washington to call for reassessment of the whole paradigm of the U.S.-Russian strategic partnership.

Subsequently, the [President George W.] Bush Administration made an effort to smooth over relations while speaking frankly to Moscow. U.S. Secretary of State Colin Powell's visit to Moscow on January 26 is evidence of those efforts.

Secretary Powell sent clear messages to the Kremlin on issues such as withdrawal of Russian troops from Georgia, securing the independence and territorial integrity of Moldova and Ukraine, and U.S. concerns about backtracking on democratic development in an op-ed published on the front page of Izvestia's January 26 issue, writing that "Russia's democratic system, it seems to us, has yet to find the necessary balance between the executive, legislative and judicial branches of power." Powell also hailed the strength of the bilateral relationship, adding that the two countries should continue developing relations while taking into account their national interests. Powell's op-ed was the shot across the bow, expressing the Bush Administration's concerns with the direction Russia has chosen for Putin's second term.

A Russian Sphere of Influence in the CIS?

The U.S. has expressed concerns about the emerging Russian sphere of influence in the former Soviet Union area. Russia's attempts to entrench its military presence from Moldova to Georgia to Kyrgyzstan, and its efforts to impose a regional free trade zone, all cause insecurity in the capitals of the Commonwealth of Independent States (CIS).

A significant U.S. concern is the future of Georgia and, more broadly, the Caucasus and CIS at large. Continuous Russian pressure on Ukraine, Moldova, and other countries could undermine bilateral U.S.-Russian ties. At the same time, as the U.S. focuses on the war on terrorism, primarily in the Middle East and South Asia, confrontation with Russia is counterproductive. Without clarification of strategies on both sides, and without policies constructed to pursue cooperation and avoid confrontation, Moscow and Washington this year could find themselves—unnecessarily—on a collision course from the Black Sea to the Pamir Mountains.

The tension escalates particularly in relations with Ukraine and Belarus, both of which are ethnically, religiously, and linguistically close to Russia and home to millions of Russian-speakers. Russia hampers their rapprochement with the West. To this end, it has given backing to Alexander Lukashenko's authoritarian regime in Belarus. In Ukraine, Moscow employs political and economic pressure to solidify the pro-Russia forces and weaken the pro-Western, democratic, and nationalist opposition ahead of elections this October.

Tensions are also rampant in Russian-Georgian ties. Georgian President Mikheil Saakashvili made sincere attempts to improve relations with Russia during his February 2004 trip to Moscow, including a suggestion of a trans-Georgian oil pipeline to Russia. These steps were positively received in Moscow.

Washington hopes that Russia will not launch a massive campaign to destabilize Georgia, as Russia should have no interest in turmoil along its southern border, in addition to which it has no alternative candidate to lead the country.

While the U.S. plans to continue to support the Saakashvili administration and to back completion of the Baku-Tbilisi-Ceyhan pipeline, it is likely that internal policy disagreements over Georgia within the Russian establishment will continue [at least for a while]. Ideally, Washington would like to see quick progress for reunification of Georgia, but without Moscow's support, such a development is unlikely.

There is a risk that Russia, which during 2003 has retreated from many global commitments, after this year's presidential elections may focus on its immediate neighborhood, scaling up its involvement in the CIS. This may include further acquisitions of energy, transportation, and other industrial assets; pressures to expand a free trade area; and more military and security cooperation under the umbrella of the CIS Mutual Defense Treaty. . . .

U.S. Interests in Eurasia

As elsewhere, the U.S. has to pursue its national interests in its relationship with Russia and Eurasia. These interests can be divided into two categories: "vital" and "important."

Vital Interest #1: The War on International Terrorism

As the U.S. projects power on a global scale to fight the war on terrorism, the attitude of regional powers, elites, and public opinion toward cooperation in combating terrorism becomes important.

Objectively, the United States and Russia are allies in fighting international terrorism. Forces linked to al-Qaeda are financing acts of terrorism in Russia. The Chechen conflict, which began as resistance to the Russian imperial occupation at the end of the 18th century, has evolved into a separatist movement for national self-determination. Stalin subjected the Chechens to a genocidal deportation in 1944, and they were allowed to return to their homeland only in 1956.

Radical Wahhabi Islam, a recent import into this war, has hijacked the nationalist movement and spread to Daghestan and other regions of the Northern Caucasus. The radical forces aim to build an Islamic state on the doorstep of Europe between the Black Sea and the Caspian, expanding into Tatarstan and Bashkortostan and eventually Islamizing Russia. While Russia could have split the Chechens by conducting talks with non-radical separatists, so far it has chosen not to do so.

Theoretically, Russia and the U.S. should coordinate anti-terrorist policy and work closely to derail the economic foundation of international terrorist networks. . . .

Yet there are other stresses, despite Putin's words. Ongoing Russian-Iranian nuclear cooperation is a highly sensitive issue, especially after supporters of theocratic totalitarianism rigged parliamentary elections in February. Efforts by the International Atomic Energy Agency to prevent Iran from acquiring nuclear weapons may not be sufficient. Iran's acquisition of nuclear weapons can become a major security threat to the U.S. and its allies, and threaten stability in the Persian Gulf. All these developments in the area of terrorism and terrorist-sponsoring states should lead Washington to recognize that partnership in this sphere might have clear-cut limits.

Vital Interest #2: Development of Energy Resources

Since the U.S. relies heavily on imports of foreign oil, the development of energy resources in the Caspian Sea basin and joint exploitation of Russian oil and gas deposits have become an important aspect of U.S.-Russian relations. However, the two countries' interests over these resources may not always coincide. If Moscow pursues an aggressive policy in the South Caucasus and Central Asia, it could derail U.S. plans to establish a reliable pipeline system in these regions. However, a policy of cooperation would benefit both parties.

Western companies are invited to participate in development in Russia only where difficult geological and geographic conditions, such as deep water, permafrost, or extreme climates, necessitate technologies that the Russian companies lack. As long as oil prices remain high, the Russian companies are likely to have access to credit and not to need Western financing, even of larger projects.

The prospects for U.S.-Russian energy cooperation have been endangered by the recent withdrawal of the license previously granted to ExxonMobil and ChevronTexaco to explore and develop the oil and gas fields of the Sakhalin-3 block, as well as by extortionate demands from the energy ministry for a $1 billion fee to pursue the project. The Sakhalin-3 experience could put the future of the total $6 billion-$10 billion U.S. investment in Russian oil at risk. U.S. Ambassador to Russia Alexander Verschbow said that this decision by the Russian government could impede a U.S.-Russian energy dialogue. It is likely that, in the future, the U.S. will react more strongly to hostile Russian actions against American companies. As Russian oil, steel, and software companies increasingly enter the U.S. market, they may become subject to similar hardball tactics.

The situation in the natural-gas industry is even more difficult, principally because the state-controlled Gazprom remains a monopoly. Until that changes, U.S. access to gas fields will remain limited. . . .

Vital Interest #3: Averting a Strategic Threat to Europe, East Asia, and the Persian Gulf

At present, Russia does not pose a genuine military threat to American interests in Europe and Asia. However, it seeks at times to complicate U.S.-European relations. Russia backed the French and German opposition to the U.S. military action in Iraq. For a few years, it waged a harsh but ineffective campaign against NATO enlargement that was designed to weaken the Atlantic alliance—one of the pillars of U.S. security.

Russia opposes the relocation of U.S. military bases eastward. At the international security conference in Munich, Germany, in February 2004, Defense Minister Sergei Ivanov stated that Russia may scrap the CFE Treaty limiting conventional weapons and troop deployments in Europe unless it is changed to include Baltic militaries and rule out NATO forces in the Baltic States.

At the same time, Russia refuses to pull its military out of Georgia and Moldova, even though it vowed to do so in an agreement signed at the 1999 Istanbul summit of the Organization for Security and Cooperation in Europe.

Moscow's efforts to improve relations with the European Union (EU) were rebuffed, and enlargement of the EU is proceeding to Moscow's detriment. Russia's accession to the World Trade Organization (WTO) has been virtually stalled due to EU members' opposition to Russian cheap domestic energy prices, which constitute a hidden subsidy to the Russian economy. . . . This causes the Kremlin's disenchantment with Russia's prospects in Europe and ratchets up the elite's anti-Western attitudes.

Vital Interest #4: Protecting America, Its Borders, and Its Airspace

Intercontinental ballistic missiles armed with nuclear warheads are a major threat to the United States. Russia and China are the only states potentially capable of a massive nuclear attack against the United States. Russia's state-of-the-art intercontinental ballistic missile, Topol-M, is entering service.

Since the end of the Cold War, Russia has pursued a buildup of strategic missile forces, including research and development of new systems allegedly capable of defeating U.S.-built ballistic missile defenses. Putin stated that the new program would not be a threat to the U.S. Yet Russian military doctrine has become increasingly offensive, clearly aimed at repelling the kind of "air-space attack" that only the U.S. and its allies are capable of staging. Russia's doctrine is also allowing pre-emptive use of force, including nuclear weapons, and the development of mini-nukes.

In addition, the Russian military still has vintage ICBMs in service that are armed with multiple, independent re-entry vehicles (MIRVs). These are known as RS-20 "Satan" missiles. As both Russia and the U.S. are likely to abide by the START-III arms control ceilings, the U.S. has called for the destruction of these weapons. [START-III refers to the Moscow Treaty of 2002, which pledges the United States and Russia to reduce their respective arsenals of nuclear warheads and bombs to no more than 2,200 by 2021.] Recently, however, Sergei Ivanov unexpectedly made an announcement that the Satan would remain in service until 2016. This definitely boosts the strength of Russia's strategic nuclear forces.

This challenge demands that the United States and its allies deploy a reliable missile defense system in the near future. The emerging missile defense system, however, would be incapable of defending America from a massive Russian attack.

Important Interest #1: Stability in the Post-Soviet Space

The political pressure that Russia applies to its neighbors to the west and south could impede their development along a democratic and market-oriented model, step up social tensions, endanger territorial disintegration, and instigate armed conflicts. The "big brother" syndrome is ingrained in Russia's dealings with the former Soviet Republics, and the Russian elite continues to look upon the countries of the former Soviet space as its sphere of influence. This leaves open an imperial option or, at least, a scenario of border revisions in the future. Realizing that these nations are truly independent and sovereign is difficult for Moscow.

That is, in part, why Russia concentrates on its military presence in the former Soviet space, including through CIS "peacekeeping missions." Russian military bases and units in the Trans-Dniester (Moldova), Georgia (Abkhazia and Adjara), Kyrgyzstan, and Tajikistan are tools of Russia's political pressure on the governments of these states.

The Russian state is relying too much on its military presence as a political tool in the post-Soviet space. It is also overreacting to U.S. military deployments in the adjacent regions for the purpose of combating international terrorism. Many in the Russian elite are concerned that the Americans have established a permanent presence in the region. "They [the United States] will never go away, we are witnessing a long-term American presence in Central Asia, and possibly, in the Caucasus," says a senior Russia expert who requested anonymity.

Such speculations are broadly used by Russian nationalists to revive the "enemy image" of the U.S. Some experts maintain, though, that "Russia is as yet undecided: should it perceive the United States presence on the broad sweep of the former 'Soviet Motherland' as an ally, partner, rival, or enemy."

The results of the December 2003 parliamentary elections demonstrate that nationalists such as Vladimir Zhirinovsky's Liberal-Democratic Party, Dmitry Rogozin's Motherland Party, Communists, and others have consolidated their position in Russia's political life. They are engendering increased xenophobia. Under the pretext of fighting terrorism, Russian nationalist policymakers call for the deportation of non-Slavic people, primarily Caucasus-born, from Moscow and other large cities.

Russian nationalists are also lobbying for "protecting Russian speakers" and the Russophone population in the post-Soviet space. The selectivity of their complaints exposes a deeper, more sinister agenda, however: While they protest the "violations of Russian speakers' rights" in the Baltic nations, they choose to disregard the infringement of these rights by the Central Asian authoritarian regimes whose anti-democratic worldview they share.

Important Interest #2: Progress of Democracy Abroad

The increasing authoritarian trends in Russia challenge the fundamental U.S. mission to consolidate freedom. In 2003, democracy and the rule of law were declining in Russia. Since 2000, all independent television channels have been shut down under powerful administrative pressure or taken over by the government's allies. Radio stations and print media are also being gradually brought under control. Self-censorship is used across the board: The authorities "guide" journalists on what to report and what to withhold, and are quick to clamp down on dissenters.

Moscow has stepped up its control over regional administrations through the extra-constitutional institution of unelected presidential envoys (four out of seven of whom are former military or security-services generals) and through its power to recall elected governors. This is at odds with the basic principles of federalism and abuses the rights of legitimately elected governors and regional legislatures.

The conduct of last December's State Duma elections provoked discontent among many Russians. Federal, regional, and local administrations have spent vast resources to secure the victory for pro-government parties, primarily United Russia. To back "the party of power," government-run television channels aired elaborate programs on the candidates while denying equal access to the opposition. David Atkinson, head of the Council of Europe Parliamentary Assembly delegation to Russia, described Russia's recent parliamentary elections as "free but unfair."

Only a high level of respect for individual freedom and property rights would guarantee Russia's political stability, economic growth, and integration into a democratic international community. Russia's authoritarian regime is likely to engineer "foreign threats" for domestic consumption, including pursuit of anti-Americanism, to justify its own existence. Authoritarianism and anti-Americanism in Russian public opinion and policies threaten further progress toward the rule of law, civil society, and a market economy. . . .

Conclusion

Russia and the United States are facing a choice: They can build a constructive relationship based on joint repelling of mutual threats and recognition of each other's relative power, capabilities, and limitations or they can revert to Cold War-style confrontation. Put another way, they must choose between respecting each other's national interests and engaging in petty fights over status; developing lucrative economic partnerships or playing power games that benefit third parties; fostering 21st century norms of democracy, human rights, and the rule of law or retreating to heavy-handed authoritarianism and risk international opprobrium.

The U.S. has chosen a path of productive partnership with Russia and should encourage Moscow to choose well. People of both countries want freedom, security, stability, and prosperity. It is up to their leaders to deliver the goods.

Stabilization and a New Consensus

In the next four years, Russian policy toward the United States (as well as Russian domestic politics and economic policy) will be shaped largely by the components of a powerful and complicated social and political trend, which, along with the best economic growth in the past quarter century, is responsible for most of President [Vladimir] Putin's popularity (and for his victory in the [March 2004] presidential election).

This trend, well familiar from the histories of other great revolutions, is a post-revolutionary "stabilization" attendant with a conservative or even reactionary retrenchment, and a drift to the core of the national political and cultural tradition.

This phenomenon consists of two occasionally overlapping but distinct components. First, formerly dominant pre-revolutionary political and economic elites seek to stage a comeback, to regain their power and possessions. In the Russian case, they are the secret police (KGB/FSB), law enforcement functionaries, and the federal bureaucracy—the groups that effectively owned Soviet Russia's politics and economy.

The other part of the "stabilization," well established by many polls and last year's parliamentary elections, is an intense and widespread longing for pre-dictability, security, and continuity—after a decade of political and economic revolutions, the relentless and dizzying onslaught of the new, and the taxing choices and responsibilities of freedom—even at the expense of some (although by no means all) newly-gained liberties.

As in all previous post-revolutionary "restorations," there is a shift in popular sentiment from a near total negation of and shame for the *ancien régime*, to the desire for a partial recovery of traditional polices, institutions, and symbols. Unlike the radical liberal intelligentsia, a plurality of Russians over forty years old is not ready to dismiss the entire Soviet past. While condemning the crimes of Stalinism [under Soviet Leader Josef Stalin, 1924–1953] and the repression and corruption of the [Soviet leader Leonid] Brezhnev era [1964–1982], they continue to take pride in the Soviet Union's role in defeating the Nazis, in its nuclear parity with the United States, and the pioneering achievements in space.

It is to his remarkable "fit" into what amounts to a new national consensus that Vladimir Putin owns a great deal of his extraordinary popularity. Instinctively or by design (or, likely, both), he has come to embody and symbolize to millions

Committee on International Relations, U.S. House of Representative, April 21, 2004.

of Russians a unifying synthesis, a still very precarious balance between the old and the new.

As a result, in the next four years the direction of Russia's foreign and security affairs will be determined largely by the interplay of three sometimes overlapping but distinct and occasionally clashing factors: the bureaucratic reactionary "restoration," a new national consensus on "stability," and President Putin's interpretation of and mediation between them.

The Foreign Policy Consensus

Early in the 1990's, post-Soviet Russia adopted a tri-partite vision of the country's core foreign policy and defense objectives: Russia as nuclear superpower, as the world's great—but no longer super—power, and as the regional superpower. It means that, while insisting on maintaining a nuclear parity with the United States, Russia has given up the Soviet messianic globalism and ideologically-driven worldwide competition with the United States. From the world's leading "revisionist" power (that is, one relentlessly seeking a change in the "balance of forces"), Russia has become a status-quo power.

Secondly, during the same period, there has occurred a startling departure from traditional Russian criteria of national greatness. Asked recently how Russia can best assert its place in the world, 46 percent of the respondents in a national survey named "becoming more competitive economically" and only 21 percent mentioned "maintaining or rebuilding a strong military."

Thirdly, not one reputable poll since 1991 has shown a majority of Russians longing for the re-creation of the unitary Soviet empire in its pre-1991 form. No matter how nostalgic millions of them feel, most reject out of hand a recreation of the empire because of the enormous economic, political and military burden that such a project would entail. The past ten years have demonstrated that barring unlikely sudden threats to its strategic interests, Russia appears to be interested most of all in the preservation of a status-quo in the post-Soviet space.

At the same time Russia's new popular sentiment is strongly in favor of greater assertiveness of national interests. Russians are no longer desperate to be liked by the U.S. (or "the West" in general): they realize that the latter are not going to protect them from Islamic terrorists who have killed over 500 people in Russia in the past 18 months. As a leading Russian expert, Dmitry Trenin, put it recently, Russia wishes "not to belong but to be."

Finding themselves in a very rough neighborhood and sharing thousands of miles of borders with China and North Korea (and with only a string of unstable Central Asian states between them and Iran and Afghanistan) after a decade of unprecedented unilateral disarmament, most Russians support a strong, efficient and modern military.

Enter the "Restorationists"

In foreign and defense policy, the "restorationists" are likely to go outside the consensus and seek to restore Russia as a global superpower counterbalancing the United States. They will go beyond assertiveness and to a tougher, even provocative stance toward the U.S. especially in what they consider Russia's "sphere of influence": the Caucasus, the Central Asia, the Far East, and North Korea.

Another item on the agenda is a massive re-armament and expansion of conventional and nuclear forces. The reactionaries have already succeeded in slowing down and diluting the progressive military reform, which couples modernization with a sharp reduction in the number of soldiers, the abolition of the draft and the creation of all-volunteer armed forces.

Finally, on the territory of the former Soviet Union, the "restorationists" are likely to push beyond the current Russian position of a strongest economic and military power and toward that of an overlord and, perhaps, an imperial master.

Putin

Given the obvious disjoint between the popular and the restorationist versions of foreign and defense policies, Putin's position is critical to policy-making. He may, of course, surprise us, but there is little in his past behavior to indicate that he will adopt an extreme reactionary agenda.

The Russian President is not a man of abrupt changes and risky policies. He is obsessed with and addicted to his popularity. He is thinking of his place in history, and, as far as we can glean from his public statements, he sees his legacy as that of economic revival, restoration of law and order, and the reduction of incompetence, over-bureaucratization and corruption in the Russian state. In the end, Mr. Putin is most likely to stay within the consensus or never deviate too far or for too long.

In addition to such policies' being outside the consensus, an aggressive, Soviet-like anti-Americanism with a global reach would reverse the post-Soviet tradition and directly challenge Mr. Putin's key domestic objectives because of the massive increase in the share of national income devoted to defense that such a policy would necessitate.

After the Yeltsin-Gaidar government [President Boris Yeltsin and Prime Minister Yegor Gaidar] cut military spending by 90 percent in 1992, it was kept at no more than 3 percent of the GDP [gross domestic product] during the 1990's. Putin has generally hewed close to this parameter. Even in the booming economy and state flush with tax receipts and bursting with gold and hard currency (and even with a 19-percent increase in defense appropriation this year, the first such increase in eleven years), Russia spends 2.8–3.7 percent of the GDP on defense (344 billion rubles or an equivalent of slightly over $11 billion in a $300–$400 billion economy). Last year, President Putin rejected calls to use the country's swelling hard-currency reserves for defense because that money "provided the basic foundation for our economic development." In 2004, the spending is set at 411 billion rubles, $14 billion or 3.5–4.6 percent of the GDP—or at least six times smaller than the defense's share during the Soviet era.

In addition to radically skewing national priorities and breaching the consensus, a pro-defense restructuring of the budget would spell the end to Mr. Putin's declared objective of doubling the country's GDP between 2000 and 2010.

"Near Abroad": A Potential Area of Tension

At the same time, there is likely to be a great deal of saber rattling and chest beating on the territory of the former Soviet Union. Like big continental powers, from Babylon, China, Persia and Rome, have done for millennia (and as the U.S. did in Latin America for most of this country's history), Russia will seek to maintain, or enforce, stability by securing friendly policies by friendly regimes on its borders. She will do so by seeing to exerting pressure and control over the "near abroad"—and by continuing to keep some of its impoverished neighbor-states with electricity, oil and gas free of charge or orders of magnitude below the world prices in what amounts to perhaps the world's largest bilateral economic aid program, particularly in Ukraine, Armenia, and Georgia.

Thus, the recent U.S.-Russian tensions over Moldova and Georgia will not be the last. Yet such conflicts are likely to be contained by the overarching mutual strategic agenda, especially war on terrorism.

Conclusion

In developing Russia's strategic posture toward the United States, President Putin is likely to mediate between the national consensus and the "restorationists" agenda. In end, the resultant policies are likely to be closer to the former rather than the latter. The anti-American impulse is likely to be constrained both by the over-arching mutual strategic agenda and by the cost of neo-globalism and massive re-armament that such an impulse would dictate. While increasing Russian assertiveness on the territory of the former Soviet Union, Russia is not likely to undermine the U.S. strategic interests—provided such interests are clearly demarcated and communicated to Russia in no uncertain terms.

POSTSCRIPT

Is Russian Foreign Policy Taking an Unsettling Turn?

The debate over the future of Russia is not a matter of idle speculation. There are two very real policy considerations. The first involves the fact that the direction Russia takes in the future is likely to have important consequences for the world. Both the ultranationalist right and communist/socialist left wings of Russian politics favor a much more aggressive foreign policy. Under President Yeltsin and during the earlier years of President Putin's first term, the Russian government sometimes strongly criticized such U.S.-favored actions as the expansion of NATO's expansion, but Moscow's weakness constrained it from trying to block U.S. preferences.

There is still not a great deal that Russia can do, but it has gotten somewhat more assertive. Russia opposed the U.S. invasion of Iraq in 2003 and helped block Washington's effort to get a supportive resolution passed by the UN Security Council, on which Russia has a permanent seat and a veto. Russia cannot stop the U.S. efforts to build and deploy a ballistic missile defense (BMD) system, but Putin's government has countered by beginning to deploy mobile intercontinental ballistic missiles (ICBM) with improved in-flight maneuverability meant to limit the ability of a BMD system to intercept them. New issues continue to emerge, including the friction between Washington and Moscow over the elections in the Ukraine in 2004. For the view that there are consistencies in the foreign policies of historic Russia, the USSR, and modern Russia, see Robert H. Donaldson and Joseph L. Nogee, *Foreign Policy of Russia: Changing Systems, Enduring Interests* (M. E. Sharp, 2005). Steven Rosefielde, *Russia in the 21st Century: The Prodigal Superpower* (Cambridge University Press, 2004), contends that Russia will try to restore its position as a full-fledged superpower within the next decade.

There are also doubts about whether democracy can survive in a country that is in such poor condition and that has no democratic tradition. The increasing curbs on a free press and other essential democratic elements that have occurred in Russia led U.S. Secretary of State Colin Powell to say publicly soon after Putin's reelection that Washington "was concerned about a level of authoritarianism creeping back in the society" and to encourage Putin to "use the popularity that he has to broaden the political dialogue and not use his popularity to throttle political dialogue and openness in the society." This concern with Russian democracy arguably has important implication for foreign policy. Many scholars contend that democracies generally do not go to war with one another, which means that the collapse of democracy in Russia might increase the potential for clashes between it and the Western democracies. Since much will depend on President Putin, two studies worth reading are

Lilia Shevtsova, *Putin's Russia* (Carnegie Endowment for International Peace, 2004) and Andrew Jack, *Inside Putin's Russia* (Oxford University Press, 2004).

It is too early to accurately predict what course Russia and its foreign relations will take in the decade ahead. What can be said is that Russia has seemed to be down and out financially and military at more than one juncture in its history, and its has always recovered, as it seems to be doing now. If the trend continues, Russia will regain the economic and military muscle necessary to play an important role in global affairs. A number of outstanding policy disputes still divide Moscow from Washington. How those are managed by both sides will be an important determinant of the general tone of future relations. A balanced view of the forces contending to shape the country's future is available in James H. Billington, *Russia in Search of Itself* (Woodrow Wilson Center Press, 2004).

ISSUE 7

Does a Strict "One China" Policy Still Make Sense?

YES: Michael D. Swaine, from Testimony During Hearings on "The Taiwan Relations Act: The Next Twenty-Five Years," Committee on International Relations, U.S. House of Representatives (April 21, 2004)

NO: William Kristol, from Testimony During Hearings on "The Taiwan Relations Act: The Next Twenty-Five Years," Committee on International Relations, U.S. House of Representatives (April 21, 2004)

ISSUE SUMMARY

YES: Michael D. Swaine, senior associate, Carnegie Endowment for International Peace, testifies before Congress that for the foreseeable future, any workable U.S.-China relationship depends on maintaining the long-standing understanding between Beijing and Washington on the status of Taiwan.

NO: William Kristol, editor of *The Weekly Standard* magazine, contends that it is time to question whether U.S. interests and those of Taiwan are served by the long-standing understanding between Beijing and Washington on the status of Taiwan.

The "one China" issue centers on the status of Taiwan. It is an island about the combined size of Maryland and Delaware that is located 100 miles to the east of south-central China and has a population of 23 million. Taiwan was part of China until it was seized by Japan in 1895. China regained the island after World War II, but Taiwan again became politically separated in 1949 when the communists came to power in China and the forces of the defeated nationalist government fled to the island. Both the communists and nationalists maintained that there was only one China and that Taiwan was part of it. What the governments in Beijing and Taipei (Taiwan's capital) disagreed about was which of them was the legitimate government of that one China. For years, the logic of the Cold War meant that the United States and most other governments treated the communists as illegal usurpers and continued to recognize the nationalists in Taiwan as the legal government of all of China.

Over time, more and more countries shifted their legal recognition to Beijing, and in 1972 the mainland government replaced the nationalist government in the United Nations. Washington was one of the last governments to shift its recognition, doing so in 1978.

Despite this change, the United States did not abandon Taiwan. Instead, the Americans reached an understanding with the mainland Chinese that the United States for its part would abide by the legal position that there was just one China. Therefore, the United States would neither recognize Taiwan as a separate country nor support a declaration of independence by Taiwan. For its part, China agreed not to use force to take control of Taiwan. Beijing accepted the U.S. determination to give or sell armaments to Taiwan, but Washington pledged to transfer only modest amounts needed for Taiwan's self-defense. Additionally, the U.S. Congress enacted the Taiwan Relations Act (TRA) in 1979. It reiterated the basic understanding and, in particular, stressed that the United States would react strongly against a military assault on Taiwan and would supply Taiwan with the capability to defend itself.

Over time, Taiwan has prospered. With an annual per capita gross domestic product (GDP) over $13,000, Taiwan is wealthier than even some western European countries, such as Greece ($12,000 per capita GDP). Taiwan is the world's tenth largest economy, and its exports and imports rank eleventh in the world. Its population is equal to that of such countries as Romania. Taiwan also has a substantial and reasonably well-equipped military. In short, it functions in most ways as any independent country would.

Another change over time involves Taiwan's government, its population, and their attitudes. By the late 1990s, Taiwan had become fully democratic. As a result, the ethnic Taiwanese, who make up most of the population, have regained power from the mainland Chinese who dominated the nationalist government. The ethnic Taiwanese have fewer emotional ties to China. These changes along with Taiwan's prosperity have led many there to begin to flirt with the idea of independence.

During the 1996, 2000, and 2004 elections in Taiwan, which styles itself the Republic of China, some of the presidential contenders made statements that overtly or inferentially favored independence. This caused repeated sharp reactions in Beijing, which threatened to invade the island if it were to try to change the status quo. Indeed, some people in China believe that the country's growing power should be used to fully incorporate Taiwan in any case. During each of the three crises, U.S. diplomacy persuaded Taiwan to back away from any declaration of independence and Beijing to pull back from any use of force.

The status of Taiwan remains an anomaly. In fact, it operates as an independent country. Legally, it is a province of China. The gap between its *de facto* (in fact) and *de jure* (legal) status has become increasingly strained. In 2004, the twenty-fifth anniversary of the TRA prompted Congress to hold hearings on whether the act and the understandings behind it still make sense. During those hearings, Michael Swaine argued that the status quo is necessary for maintaining peace and should be preserved. William Kristol disagreed and told Congress that it is time to violate the taboo against discussing a change in the status quo.

Michael D. Swaine　　　　　　　　 **YES**

The Taiwan Relations Act: The Next Twenty-Five Years

The peace, stability, and prosperity of East Asia and the overall advancement of America's security interests depend on the maintenance of stable, workable state-to-state relations between the United States and the People's Republic of China [P.R.C.]. This is still true today even though the original strategic motivation for the normalization of U.S-China relations—the need to balance against the Soviet Union—no longer exists.

At present, and in my view for the foreseeable future, any workable U.S.-P.R.C. relationship depends on the maintenance of an understanding that was reached between Beijing and Washington at the time of normalization. This understanding exchanged a U.S. acknowledgement of the so-called One China position for a P.R.C. commitment to the search for a peaceful means to resolve the Taiwan issue as a first priority.

The Taiwan Relations Act (TRA) codifies, in U.S. domestic law, two of the three central pillars of policy towards both the P.R.C. and Taiwan that derives from this understanding. First, it requires that the Taiwan situation be handled peacefully, and indicates that the U.S. Government will regard "with grave concern" any use of non-peaceful means to resolve Taiwan's status. Second, it requires the U.S. government to maintain (both directly and indirectly via assistance to Taiwan) a credible military capability to counter a Chinese attack, in order to deter Beijing from being tempted to employ force against a diplomatically weakened Taiwan.

These two elements form an essential part of the reason why America's policy toward Taiwan and the P.R.C. has been successful, to date, in preventing conflict, in sustaining a beneficial—if often troubled—Sino-American relationship; and in permitting Taiwan's society and polity to thrive.

However, the TRA is only part of the reason for this policy success. A U.S. commitment to a peaceful solution of the Taiwan problem—and to military deterrence in support of that goal—could not have succeeded in advancing U.S. interests without an equally strong assurance to the P.R.C. that the U.S. will not use its superior military power and its defensive-oriented assistance to Taiwan to encourage Taiwanese independence, or to shield movement by the island toward independence.

Committee on International Relations, U.S. House of Representative, April 21, 2004.

Without such a U.S. assurance, the P.R.C. leadership could not have tolerated what it views as the challenge to China's claim to sovereign authority over Taiwan that the TRA represents. They almost certainly would not have emphasized their desire to pursue a peaceful solution to the Taiwan situation as a top priority.

Therefore, the other important elements of U.S. policy—the agreement not to challenge the P.R.C.'s One China position as well as the expression of the P.R.C.'s priority emphasis on a peaceful resolution to the issue—are essential to the maintenance of stability in the Taiwan Strait. These elements were provided by the three Sino-American communiqués, not the TRA.

Some critics of U.S. policy argue that democratization in Taiwan, the subsequent rejection by the current Taiwan government of the original "One China" notion, and the P.R.C.'s military buildup along the Taiwan Strait require a fundamental change in U.S. policy. They argue that the One China approach should be jettisoned in favor of a policy that recognizes the "reality" of Taiwan's independence and that relies almost exclusively on military deterrence to prevent a rising P.R.C. from reacting forcibly to such a policy shift.

In support of this position, these critics would turn the TRA into a security guarantee to Taiwan (which it is not*), and would provide even greater levels of military assistance to Taipei. They would also negate, by word or deed, much of the three communiqués.

In my view, any effort by the United States to confront the P.R.C. with the so-called "reality" of an independent Taiwan would destroy the foundations of a stable Sino-U.S. relationship, throw Asia into turmoil (especially because no Asian state would support such a policy move), and very possibly result in a war with the Chinese. To those who disagree with this assessment, I would ask, can one be confident enough that the Chinese will not respond with force to make it worth the risk of provoking a confrontation with Beijing by unilaterally rejecting the One China policy—given the likely damage that will ensue if a conflict erupts? On the other hand, to those who accept the likelihood of a Chinese use of force in response to such a U.S. policy shift, I would ask, can one be sure enough that the resulting conflict would be quickly terminated, resolved, or contained in ways that preserve essential U.S. and Taiwanese interests? The danger of escalation in such a confrontation would be very real, once conflict begins.

Ultimately, the U.S. position toward Taiwan must balance two policy objectives:

1. The need to preserve the credibility of America's word, in this case its commitment to a peaceful, non-coerced solution to a potentially volatile international problem, as well as America's support for a long-standing friend; and
2. The need to maintain workable, if not amicable relations with a nuclear power whose long-term stance toward U.S. interests remains unclear, and whose cooperation is essential for the maintenance of many core U.S. interests, including the war on terrorism.

For the United States, both of the above objectives are critical, and, equally important, are not mutually exclusive, as long as the P.R.C. is not resolved to using force against Taiwan.

In fact, China remains committed to a peaceful resolution of the Taiwan imbroglio as a first priority. Chinese military deployments are intended primarily to deter the attainment by Taiwan of *de jure* [legal] independence, not to prepare for an inevitable war. But if deterrence fails, China's leaders will almost certainly fight, to ensure their respect among their colleagues and the Chinese populace, and to defend the legitimacy and stability of their government. Moreover, China's leaders would likely fight even they stood a good chance of losing in the first few rounds. For them, to not fight would mean a certain loss of power; to fight and lose would probably mean that they would survive politically to fight another day.

There is at present no realistic alternative to the One China policy, combined with the TRA, that can provide a more durable basis for stability, for conflict avoidance, and for gaining the time that is required for the two sides to moderate their stance and move toward dialogue and a stability-inducing *modus vivendi*.

However, in order to maintain the credibility of the One China policy and overall stability in the Taiwan Strait, I believe that the United States must consider taking a more active role in influencing calculations in both Taipei and Beijing. Specifically, Washington should, in my view:

1. Reaffirm unambiguously that a danger of conflict with the P.R.C. exists over the issue of Taiwan, i.e., the threat of a use of force and of inadvertent escalation is genuine; thus, provocations by either side are totally unacceptable.
2. State clearly, either publicly or privately, that the United States does not agree with the position of Taiwan President Chen Shui-bian's government that it is engaged in merely "consolidating" a long-term status quo of sovereign independence. This position is dangerously misleading to the Taiwan public. For the U.S. and virtually the entire international community, the sovereign status of Taiwan remains unresolved at best.
3. Communicate clearly that U.S. support for Taiwan is not unconditional—it requires responsibility and restraint—and that efforts to alter the source of sovereignty of the government of Taiwan by unconstitutional means—something that Chen threatens to undertake via a referendum on a new constitution—are potentially dangerous and destabilizing.
4. Pledge that, as long as Taiwan exercises restraint, Washington will undertake greater—albeit still limited—efforts to increase Taiwan's international profile.

With regard to the TRA, I believe that the United States government should consider redefining and delimiting more precisely the type of defense assistance that it will provide to Taiwan in the future. Washington should insist that Taiwan acquire greater capabilities to defend itself against *specific* military actions, especially the possibility of a rapid strike by Beijing designed to achieve success before the U.S. can lend assistance. Such a stance would convey to elites in both Taipei and Washington the fact that some types of military capabilities currently desired by Taipei (and by some in the Pentagon and Congress) are less critical for Taiwan, and that Taiwan must ultimately rely on the U.S. for its defense in key areas, especially its defense against a sustained amphibious and air attack.

Washington should also unambiguously reject the option of "offensive" strike capabilities for Taiwan. Such capabilities are currently under serious consideration by the Chen Shui-bian government. However, in my view, they would not increase Taiwan's security, and would undermine efforts at controlling escalation in a crisis.

Toward Beijing, The United States should assert that the best way for China to lower tensions with Taiwan would be for the People's Liberation Army to reduce its military deployments along the Taiwan Strait, as an indicator of good will, as a first step toward a cross-strait dialogue, and with the clear understanding that such actions would be reciprocated in some manner. To reach such an understanding, Washington should undertake direct discussions with Beijing on reducing the military buildup along the Taiwan Strait, in consultation with Taipei.

Finally, in my view, Washington should undertake more active efforts to advance political reform in China, in order to increase China's attractiveness to the Taiwan public and thereby encourage movement toward a cross-strait dialogue. Concrete initiatives to support the rule of law should be included in this undertaking.

Without a more active effort by the United States to balance deterrence with reassurance, and to counter unilateral efforts by either side to alter the status quo, the chances for an eventual conflict in the Taiwan Strait will in my view increase significantly over the next several years.

William Kristol  **NO**

The Taiwan Relations Act: The Next Twenty-Five Years

As members of Congress consider issues relating to China and Taiwan, they might begin by considering something a "senior administration official" said last week about our policy on Israeli settlements and Palestinian refugees. "Eliminating taboos and saying the truth about the situation is, we think, a contribution toward peace. Getting people to face reality in this situation is going to help, not hurt" (*Washington Post*, April 15, 2004).

This statement applies equally to the Taiwan Strait. America's policy toward Taiwan is ridden with taboos. In fact, one such taboo is the virtual prohibition on questioning whether our interests and those of democratic Taiwan are served by the various communiqués agreed to by Beijing and Washington since 1972. This reluctance to adjust U.S. policy to reflect changes in the strategic and political situation in the region has also meant that the TRA [Taiwan Relations Act] itself—if I am not mistaken—has never been amended.

Today, and in the coming months, we need an honest and public discussion of what we want to happen and not to happen in China and in Taiwan. We have for many years avoided such a discussion. It has been as if Taiwan's survival as a democracy, and, for that matter China's possible evolution into one, are not proper matters of polite conversation. Instead, we have pretended that there can be an unchanging "status quo," that China is not seriously preparing for military action or other forms of coercion against Taiwan, and that Taiwan's people would be amenable to unification if it were handled well.

Congress passed the Taiwan Relations Act to avert the worst consequences of President [Jimmy] Carter's decision in 1978 to break relations with Taipei, withdraw U.S. troops, and abrogate the mutual defense treaty. The TRA established important principles of U.S. policy—chiefly our insistence on a peaceful resolution of Taiwan's fate, our opposition to aggression, including coercive acts such as boycotts or embargoes, and a commitment to Taiwan's defense through the provision of defensive arms and the maintenance of America's own ability to resist Chinese aggression against Taiwan. As you know, Congress also established a role for itself in providing for Taiwan's defense needs and in determining any response to a danger that the president is required to report under the Act. The "one China" policy and the strategic ambiguity that came to govern U.S. policy

Committee on International Relations, U.S. House of Representative, April 21, 2004.

are nowhere to be found in the law's text. Yet, as any observer of U.S. China policy knows, the language of "one China" pervades U.S. policy. It is a mantra that every official must intone on virtually any occasion on which China or Taiwan is discussed.

The "one China" policy began as a way to defer the resolution of Taiwan's fate until better conditions for resolving it prevailed. It purposely left the U.S. neutral about the outcome. Unfortunately, the policy has come to mean denying Taiwanese sovereignty and self-determination. Part of the problem is that the arcane and nuanced language that its advocates believe manages a complicated situation—and deters the non-expert from trying to criticize it—does not reflect the changes that have taken place on both sides of the Strait. It also invites constant pressure for revisions from Beijing. For example, over the past year, Beijing has campaigned to bring about a change in U.S. policy from "not supporting" Taiwan independence to "opposing it." Officially, "not supporting" independence remains U.S. policy. This apparent slight difference is actually important. Not supporting Taiwan's independence is consistent with longstanding policy of not predetermining the outcome of discussions or negotiations between China and Taiwan. Opposing independence appears to settle the matter and might give Beijing reason to believe that the U.S. might not resist China's use of force against Taiwan, or coercive measures designed to bring about a capitulation of sovereignty.

At the same time, independence sentiment on the part of Taiwan's people is neither frivolous nor provocative, but rather the natural manifestation of a process that the U.S. has supported. As my colleague Gary Schmitt wrote recently in the *Wall Street Journal*, "Taiwanese identity has grown in direct relation to the progress of democracy on the island. The people of Taiwan increasingly have come to think of themselves as Taiwanese as they have established themselves over the past decade as a self-governing people." Viewed this way, Taiwan's desirable democratic transformation has an unavoidable implication for U.S. policy on Taiwan—not to tilt *against* independence but toward it.

In short, the "one China" policy expresses neither the situation on the ground in Taiwan, nor U.S. values and interests. No one drafting a new U.S. policy toward Taiwan today would recreate the one the U.S. has pursued since the 1970s and 80s. Ever since its basic premises were set forth, the policy has been under pressure. The reason is obvious: the situation has changed. Taiwan's people have established democracy. More importantly, they no longer claim the mainland or wish to join it. Even the Kuomintang—the Nationalist party—has abandoned its longstanding position regarding unification.

Meanwhile, across the Strait, economic growth has fueled China's military modernization. There are at least 450 missiles pointed at Taiwan, and Beijing is acquiring other capabilities designed to help it take Taiwan, or coerce Taiwan to accept unification on Beijing's terms. China's leaders rely increasingly on nationalism, rather than communism, as the source of legitimacy for the regime. This will become more pronounced if, as predicted, labor unrest, the banking system, and the further collapse of state enterprises become more dire problems.

Future policy on Taiwan should be designed to reflect new realities. In the short term, we can take practical steps that reflect Taiwan's importance as a fellow democracy, maximize its international standing, and improve U.S.-Taiwan defense cooperation.

Bilateral Relations

The U.S. should reduce Taiwan's international isolation by increasing high-level contacts. The number of visits to Taipei and to Washington by senior officials should be increased to the point that it is unremarkable.

The administration must soon decide who will represent the United States at the [2004] inauguration of President Chen [Shui-bian] for his second term. It would be good to send someone of prestige and importance, especially in light of the administration's handling of the congratulations to President Chen on his re-election. This is a perfect opportunity for the administration to signal Beijing that the future of U.S.-Taiwan relations will be more respectful of Taiwan's democratic character. It would help if Washington sent an administration official of high rank from within the [President George W.] Bush administration. Serving cabinet members have visited Taiwan in the past, but none have visited Taiwan since 1998 when Secretary of Energy Bill Richardson was trapped in a high-rise hotel during an earthquake. Why shouldn't the Bush administration send a cabinet officer to represent the U.S. at the May inauguration ceremony?

Washington should also change the way it deals with the president of Taiwan. While the visits of Taiwan's presidents have been increasingly dignified, the ad hoc nature of the policy on visits guarantees intense pressure from China and forces the U.S. to devote unreasonable amounts of effort to placating Beijing. It is frankly absurd that a democratically elected president cannot visit senior U.S. officials or even Washington, but general secretaries of the Chinese Communist Party have been to the White House. Taiwanese officials below the level of the president also need to be able to come to the U.S. and speak freely to the American public and the media. The fact that they do not may not be due to any particular policy directive. However, it is undeniably true that Taiwan's international isolation has created ingrained habits—both here and in Taipei—that are extremely unhealthy and even counterproductive insofar as they prevent a frank sharing of views.

Defense and Regional Security

After the 1995 and 1996 missile volleys, the U.S. realized we were ill-prepared to coordinate defense of Taiwan with Taiwan's own defense forces. Since that time, we have improved our preparations. These efforts should be continued, enhanced and made as public as possible to underscore our commitment to Taiwan's defense. Greater openness about the nature and extent of America's commitment to Taiwan's defense would help deter Beijing and dispel ambiguity. Such openness would also benefit the people of the United States who, far from fearing America's overseas commitments, understand the importance of America defending democratic allies.

Furthermore, Taiwan is more than just a dependent. It also cooperates with America's security objectives. Last August, on receiving a request by the U.S., Taiwan forced a North Korean freighter to unload dual use chemicals. According to an American official, "we provided the intelligence and Taiwan stepped up to the plate." In short, Taiwan is helping the Proliferation Security Initiative [PSI], an effort the Bush administration launched to stop nuclear proliferation. Taiwan should be allowed to join the core group of the PSI, which just recently added three new members. Incidentally, according to the State Department, the PSI is "an activity, not an organization," so the question of statehood for membership is not an issue. By virtue of its democratic character, its strategic location, and its long history of working with the United States, Taiwan's cooperation in regional security is imperative to U.S. interests. There is no reason that Taiwan should not be recognized not only as a participant in PSI, but also in other multilateral discussions, exercises, and operations among democratic countries in Asia.

U.S. efforts to draw Taiwan into the international community should also include a serious initiative to win Taiwan's admission into the World Health Organization [WHO], including sponsoring its nomination for membership. Taiwan's exclusion from the WHO vastly complicated efforts to deal with the spread of SARS [severe acute respiratory syndrome]. No one has forgotten the callous comment of the Chinese ambassador after Taiwan failed to win admission to the WHO as an observer last year: "The bid is rejected. Who cares about your Taiwan?" The Bush administration has expressed its support for Taiwan's WHO membership. China, however, is uniquely talented at using leverage and threats in international fora, and the WHO is no exception. The U.S. and other sympathetic countries need to meet China's ante and raise it.

Finally, a Free Trade Agreement between the U.S. and Taiwan would fit neatly within U.S. policy to build bilateral trade agreements. The Project for the New American Century held a conference on this idea, and found wide acceptance of the idea within the policy and business communities of both our countries. Politically, the impact would be extremely important.

With regard to China, we need to be quite clear that we expect Beijing not to attack or coerce Taiwan in any way, and that the costs to Beijing of attacking Taiwan would be more than it can bear. We also need to be clear that we look forward to China becoming a democratic country, like Taiwan. Then the people on each side of the Strait can decide their relationship and their future.

When Vice President [Dick] Cheney visited China last week [April 2004], he made an impressive speech that spoke about democracy. But the Vice President used one key word that let China know that for now the U.S. does not consider democracy a priority for China. That word is "eventually." Cheney said China's people will "eventually ask why they cannot be trusted with decisions over what to say and what to believe." "Eventually" was used with precision not only in this speech but also in President Bush's widely praised speech establishing democracy as a foreign policy priority to the National Endowment for Democracy last November. America's policy toward China is insufficiently directed toward democratizing China, and so long as that is true it will be more difficult to help Taiwan's democracy survive.

Conclusion

Twenty five years ago [in 1979], Congress checked the Carter administration's policy on Taiwan. At the time, to quote one scholar, Beijing hoped that the U.S. withdrawal of support "would arouse a sufficient sense of vulnerability within the Nationalist government to make it more susceptible to overtures from the mainland." Beijing decided that "[i]f Taiwan would only bow to Beijing's sovereignty, then the Beijing government would promise to concede a very high degree of administrative autonomy to the Taipei authorities." The famous "one country, two systems" formula that China claims to apply in Hong Kong was originally dreamed up with Taiwan in mind.

It didn't work. Congress acted to pass the Taiwan Relations Act and China set its sights on Hong Kong. Since 1997, it has been quite clear that Beijing is not interested in or sincere about respecting autonomy under a "one country, two systems" arrangement.

Don't misunderstand. America's commitment to Taiwan is admirable. No other country could or would do what the United States has done. At the same time, no other country except the U.S. can hurt Taiwan or weaken it as much as the United States can.

The greatest test is still to come. China is very serious about taking Taiwan, and we have not done enough to dissuade it. Taiwan has transformed itself from a dictatorship to a democracy. That momentous change has very likely increased the chances of a conflict in the Taiwan Strait—not because Taiwan is provocative, but because China cannot abide Taiwan's democratic character and the reality that it has become a separate, self-governing people. That is why U.S. clarity and resolve are so important.

A discussion of these and other issues needs to happen now and yield results right away. The Pentagon has estimated that the balance of forces in the Taiwan Strait will begin to tip in Beijing's favor, perhaps as soon as next year. We need above all, therefore to deter any attack or coercion. And we need to rethink policy constraints developed for circumstances decades ago, while confronting greatly changed and still changing conditions in order to develop a new, sustainable policy for security and democracy in Taiwan and China for the present and future.

POSTSCRIPT

Does a Strict "One China" Policy Still Make Sense?

Although it has not dominated the headlines in recent years, the future of Taiwan is as fraught with danger as any potential conflict in the world. To say that the status of Taiwan as a *de facto* country and a *de jure* province of China is an oddity hardly conveys the complex situation that exists.

There are pressures on both sides of the Taiwan Strait to resolve the issue. Taiwan's president since 2000, Chen Shui-ban, is ethnic Taiwanese. He was born in 1951 and is the first member of the post-separation generation to lead Taiwan. In 2003, he persuaded Taiwan's legislature to pass a law allowing for national referendum, a move that Beijing reasonably suspected was setting the stage for a popular vote on independence. Underlining that point, Premier We Jiabao of China charged that the referendum law was an attempt to "use democracy only as a cover to split Taiwan away from China." Such a declaration of independence or any other overt acts tantamount to independence, such as a U.S. diplomatic recognition of Taiwan, would probably lead to war. China's General Peng Guangqian recently told reporters, "Taiwan's independence means war. This is the word of 1.3 billion people, and we will keep our word." A study of the political identity of the people in Taiwan is Melissa J. Brown, *Is Taiwan Chinese?: The Impact of Culture, Power, and Migration on Changing Identities* (University of California Press, 2004).

Even if Taiwan holds to the status quo, there are pressures in China to take control of the island. The transfer of sovereignty of Hong Kong to China by Great Britain in 1997 and of Macao to China by Portugal in 1999 leaves Taiwan as the last remaining "lost territory" of China. Among other indications of the pressures within China to reclaim its last lost territory was a report issued by China's government entitled "The One-China Principle and the Taiwan Issue." The document proclaimed that if Taiwan continued to refuse to reunify peacefully, then China would be "forced to adopt all drastic measures possible, including the use of force." An argument that China's ability to militarily seize Taiwan will decline and that, therefore, China might soon decide to act is Justin Bernier, "China's Closing Window of Opportunity," *Naval War College Review* (Summer 2003). The related issues of China's and Taiwan's respective military capabilities are reviewed by Ivan Eland, in "The China-Taiwan Military Balance: Implications for the United States," *Cato Foreign Policy Briefing No. 74* (February 5, 2003) at http://www.cato.org/pubs.

Whether a military move against Taiwan by China was prompted by a move toward independence by Taiwan or by an unprovoked action by Beijing, the conflict could well bring a U.S. counterintervention. The understandings of the 1970s, including the Taiwan Relations Act, give a virtual security pledge to

Taiwan, and President George W. Bush has declared, "the Chinese must understand" that the United States is committed to "help Taiwan defends herself" if attacked by China. This view is supported by 51 percent of American political, economic, and social leaders according to a 2004 survey, but only 33 percent of the general American public shared their leaders' opinion that the United States should defend Taiwan if China attacks it. For a general overview of this perilous situation, see Nancy Bernkopf Tucker, *Dangerous Strait: New Thoughts on the U.S.-Taiwan-China Crisis* (Columbia University Press, 2005). For China's view, go to the Web site of its embassy in Washington at http://www.china-embassy.org/. The U.S. view is found on the Web site of its embassy in Beijing at http://www.usembassy-china.org.cn/. The site of the *de facto* Taiwanese embassy in the United States, Taipei Economic and Cultural Representative Office in Washington, is http://www.tecro.org/.

ISSUE 8

Should North Korea's Nuclear Arms Program Evoke a Hard-Line Response?

YES: William Norman Grigg, from "Aiding and Abetting the 'Axis,'" *The New American* (February 24, 2003)

NO: Robert J. Einhorn, from "The North Korea Nuclear Issue: The Road Ahead," *Policy Forum Online* (September 14, 2004)

ISSUE SUMMARY

YES: William Norman Grigg, senior editor of *The New American*, argues that North Korea is a dangerous country with an untrustworthy regime and that it is an error for the United States to react to North Korea's nuclear arms program and other provocations by offering it diplomatic and economic incentives to be less confrontational.

NO: Robert J. Einhorn, senior adviser at the Center for Strategic and International Studies and former assistant secretary of state for nonproliferation, maintains that the idea that Pyongyang can be squeezed until it capitulates and surrenders its nuclear weapons capabilities or collapses altogether is wishful thinking.

T he global effort to control the spread of nuclear weapons centers on the Nuclear Non-Proliferation Treaty (NPT). It was first signed in 1968, then renewed and made permanent in 1995. The 85 percent of the world's countries that have agreed to the NPT pledge not to transfer nuclear weapons or to assist in any way a nonnuclear state to manufacture or otherwise acquire nuclear weapons. Under the NPT, nonnuclear countries also agree not to build or accept nuclear weapons and to allow the UN's International Atomic Energy Agency (IAEA) to monitor their nuclear facilities to ensure that they are used exclusively for peaceful purposes.

For all its contributions, the NPT is not an unreserved success. India and Pakistan both tested nuclear weapons in 1998; Israel's possession of nuclear weapons is an open secret; and other countries, such as Iran, have unacknowledged nuclear weapons programs. Then there is North Korea.

The immediate background of this issue dates to early 1993, when North Korea announced its withdrawal from the NPT. Fear that North Korea would try to develop nuclear weapons was heightened by media reports that the CIA believed that North Korea probably already had one or two nuclear weapons.

For a year and a half after North Korea's statement about withdrawing from the NPT, there was diplomatic maneuvering designed to persuade North Korea to continue to abide by the NPT. North Korea eventually agreed to remain a party to the NPT, to suspend work on the nuclear reactors it had under construction, to dismantle its current nuclear energy program over 10 years, and to allow the IAEA inspections to resume. The United States and its allies—principally Japan and South Korea—pledged that they would spend approximately $4 billion to build in North Korea two nuclear reactors that were not capable of producing plutonium for bomb building. The allies also agreed to help meet North Korea's energy needs by supplying it with about 138 million gallons of petroleum annually until the new reactors are on-line.

The issue flared anew to near crisis proportions beginning in late 2002 when North Korea expelled IAEA inspectors, dismantled their monitoring equipment, once again renounced its adherence to the NPT, and moved to restart its Yongbyon nuclear power plant, a facility capable of producing weapons-grade plutonium and uranium.

The alarm over North Korea's actions was even greater than it had been in 1993. The director of the CIA estimated in early 2002 that Pyongyang could reprocess the 8,000 containers of nuclear fuel containers stored at Yongbyon to build a half dozen nuclear weapons within several months and then create another two or more nuclear weapons a year to add to its suspected existing small arsenal. He also warned that the United States could face a "near term" intercontinental ballistic missile (ICBM) threat from North Korea's extensive missile program.

Washington at first responded to North Korea's actions with threats. However, faced with some harsh realities, the Bush administration was forced to retreat from its initial tough stance. One reality was that U.S. options were limited at the time because it was on the edge of war with Iraq. Second, even if it did not use nuclear weapons, North Korea is still capable of inflicting massive damage on South Korea, which has population centers (including its capital, Seoul) within artillery range of the border. Third, almost all other concerned countries, including South Korea and Japan, favored a placating rather than hostile reaction. Fourth, it is possible, given North Korea's alleged nuclear weapons inventory and its missile capabilities, that an overly aggressive response could lead to a nuclear war, including an attack on Japan. As a result, Washington moved to downplay the confrontation, to express assurances that it would not attack North Korea, and to suggest that aid might be resumed or even increased if North Korea abided by the NPT.

There the matter stood when the first of the following selections were written. In it, William Norman Grigg castigates the Bush administration for its willingness to appease North Korea and for its focus on what Grigg considers a much less dangerous Iraq. In the second, Robert Einhorn maintains that North Korea should be offered carrots, not threatened with sticks.

YES

William Norman Grigg

Aiding and Abetting the "Axis"

Saddam Hussein's Iraqi regime may be close to building a nuclear weapon. Kim Jong-Il's North Korean hell state, according to intelligence estimates, currently possesses two nukes, and will shortly develop the capacity to produce an entire arsenal. Under threat of war, Saddam has allowed UN weapons inspectors to canvass Iraq for evidence of weapons of mass destruction. Last December [2002], North Korea summarily evicted UN weapons inspectors from its Yongbyon nuclear plant, and disabled surveillance equipment used to monitor the suspected weapons production facility.

Crippled by the 1991 UN-led Gulf War, intermittent bombings by U.S. and British aircraft, and 12 years of devastating sanctions, Saddam's military poses little threat to Iraq's neighbors, let alone the United States. North Korea, on the other hand, boasts the world's fourth-largest military; it has 37,000 U.S. troops within easy striking range of its artillery. Seoul, the South Korean capital, is 34 miles away from the demilitarized zone and well within striking distance of North Korean artillery tubes. And Kim's regime has successfully tested the Taepo Dong, a missile capable of hitting Japan; the missile's next generation may be able to strike Alaska.

Moreover, North Korea brazenly and unrepentantly sponsors and participates in international terrorism. Adept in using infiltrators and sleeper agents, Pyongyang poses a real threat of nuclear terrorism against the region—and conceivably even the United States.

Of these two members of the "axis of evil," North Korea is—by any rational calculation—a far greater threat than Iraq. Yet in dealing with Pyongyang, the president displays none of the stiff-spined, bellicose rectitude that characterizes his treatment of Baghdad. Crusading for war against a prostrate Iraq, Mr. [George W.] Bush strikes poses of jut-jawed, Churchillian resolution; confronting an insurgent, nuclear-equipped North Korea, he essays a credible Neville Chamberlain impersonation.

Why is this so? How could the same president who identified North Korea as a member of an "axis of evil" now stand ready to lavish that terrorist regime with aid, trade, and technology? Mr. Bush, recall, has condemned not only terrorism but those countries supporting terrorism. "[W]e will pursue nations that provide aid or safe haven to terrorism," he said in a nationally televised address to a joint session of Congress shortly after the 9-11 terrorist attacks. "Every nation, in every region, now has a decision to make. Either you are with us, or you are with the terrorists. From this day forward, any nation that continues to harbor or support terrorism will be regarded by the United States as a hostile regime."

How does Mr. Bush reconcile this tough stance with aiding North Korea, the most militant of the three "axis of evil" regimes he named? And how does he reconcile that stance with counting as allies in the war against terrorism Russia and Communist China, who are the puppet-masters behind the three "axis" nations? Based on Mr. Bush's own definition, would not his policies qualify his own administration as "a hostile regime"?

Power Behind the Axis

Last December 12th, while the attention was focused on Baghdad and Pyongyang, Russian President (and KGB veteran) Vladimir Putin made what the *New York Times* described as "a quick but high-profile visit to Beijing" for a summit with Communist Chinese ruler Jiang Zemin. "China and Russia will be good neighbors, friends and partners forever," proclaimed Jiang during the quickie summit, held to reiterate the Sino-Russian "Good Neighborly Treaty of Friendship and Cooperation" signed in 2001.

One tangible item of business in the December 2002 Beijing meeting was a joint declaration urging the U.S. to normalize relations with North Korea "on the basis of continued observation of earlier reached agreements, including the framework agreement of 1994." Under that agreement, the U.S. and key allies—particularly South Korea and Japan—would pay at least $4 billion to supply North Korea with light-water nuclear reactors (which would be used to produce weapons-grade plutonium) and unspecified amounts to provide Pyongyang with heavy fuel oil and upgrades to its decrepit power grid. In exchange, North Korea supposedly agreed to "freeze" its nuclear program, and submit to international inspections beginning in 1999. In predictable fashion, Kim Jong-Il and his cohorts eagerly accepted these incredible concessions while covertly continuing their "frozen" nuke research.

Incredible as it may seem, the Bush administration allowed oil shipments to North Korea to continue after Pyongyang announced in October 2002 that the 1994 agreement was "nullified." "Can you imagine the uproar if Bill Clinton had let the deliveries to go forward [sic] if he had been told the agreement was nullified?" commented a Democratic congressional aide to the October 23rd [2002] *Washington Post.*

According to the CIA, North Korea attempted to buy equipment for a uranium weapons program from Communist China in 2001. During the same year, Beijing provided crucial missile-related technology to Pyongyang, and Russia concluded a defense agreement setting the stage for arms sales and weapons technology transfers to North Korea. This is curious behavior for powers hailed by President Bush as valued allies in the "war on terror"—and North Korea was hardly the only beneficiary of this treacherous Sino-Russian support. The CIA report, as summarized by *Washington Times* defense affairs analyst Bill Gertz, "identified Russia, China, and North Korea as major suppliers of chemical, biological and nuclear-arms goods and missile systems to rogue states or unstable regions."

A Terror Regime

North Korea is a museum-quality exhibit of Communism in the full flower of its malignancy. In congressional testimony . . . , Norbert Vollertsen, a German physician who lived in North Korea for a year and a half as a humanitarian volunteer, described how that nation's wretched hospitals are filled with people "worn out by compulsory drills, the innumerable parades, the assemblies from 6:00 in the morning and the droning propaganda. They are tired and at the end of their tether. Clinical depression is rampant. Alcoholism is common because of mind numbing rigidities and hopelessness of life."

Mass starvation is a hallmark of Communism, and North Korea has preserved this tragic tradition as well. Since 1992, at least one million North Korean subjects—and perhaps as many as four million, or one-quarter of the population—have died from starvation. And as has been the case in Soviet Russia, Red China, Ethiopia, and Zimbabwe, famine has been used as a weapon of social control. "North Korea is a terror regime," testified Dr. Vollertsen. "They are committing genocide there. . . . They are using food as a weapon against their own people. . . . North Korea [represents] the real killing fields of the 21st Century."

The Bush administration, citing humanitarian concerns, has repeatedly promised to continue providing food shipments to North Korea via the UN's World Food Program. But such aid actually compounds the humanitarian crisis by helping to prop up Kim's regime, which rations the food through the country's Public Distribution System (PDS). A North Korean subject's access to food and other necessities is strictly defined by his loyalty to the regime. According to Sophie DeLaunay of the humanitarian group Doctors Without Borders, "the three class labels—'core,' 'wavering,' and 'hostile'—continue to be used to prioritize access to jobs, region of residence, and entitlement to items distributed through the Public Distribution System. . . ."

"There are two worlds in North Korea," observed Dr. Vollertsen. "The world for the senior military, the members of the [ruling] party and the country's elite. . . . In the world for these ordinary people in a hospital one can see young children, all of them too small for their age, with hollow eyes and skin stretched tight across their faces, wearing blue-and-white striped pajamas like the children in Auschwitz and Dachau in Hitler's Nazi Germany." In September 1995, Kim Jong-Il issued orders to arrest wandering homeless children found outside their home counties and imprison them in the North Korean gulag.

While the common people starve, North Korea's Communist oligarchy lives in royal splendor. Seeking to co-opt Dr. Vollertsen, the Communist government awarded him a "friendship medal" and offered him unprecedented access to the "festivities . . . [of] all those who are in charge of power in the foreign ministry."

In that company, the German physician saw the country's elite "enjoying a nice lifestyle with fancy restaurants, diplomatic shops with European food, nightclubs and even a casino. . . ." The North Korean *nomenklatura* [leadership class] does little to disguise its privileged status. The October 5, 1999 *South China Morning Post* reported that Kim's regime purchased a $20 million fleet of 200 Mercedes-Benz S500 class cars for its leadership.

Gangster State

An unavoidable consequence of Communist central planning, the North Korean famine has been exacerbated by the regime's investment in narco-terrorism. The February 15, 1999 issue of *U.S. News & World Report* observed that up to 17,000 acres of farmland have been locked up by state-mandated opium farming, which began in the mid-1980s under dictator Kim Il-Sung, the present dictator's late father.

Kim Jong-Il has "ordered a major expansion of the drugs-for-export program," noted the magazine, which also reported that "U.S. food aid to the regime—over $77 million worth this year—may be needed in part because farm acreage is used to grow poppies for opium." To that figure can be added millions of additional dollars stolen by the regime from charitable aid sent to North Korea by private and religious relief organizations.

"Interviews with law enforcement officials, intelligence analysts, and North Korean defectors suggest that the regime is now dramatically expanding its narcotics production and that much of the criminal activity is controlled at the highest levels of government," reported *U.S. News*. "[I]t is clear that the worldwide network of North Korean embassies, coupled with the use of diplomatic pouches and immunity, offers the ideal cover for a criminal enterprise. . . ."

"Authorities in at least nine countries have nabbed North Korean diplomats with a virtual pharmacy of illegal drugs: opium, heroin, cocaine, hashish," continued the report. In July 1998, two North Korean diplomats were arrested in Cairo with six suitcases containing 506,000 tablets of Rohypnol, the so-called "date rape drug." During the same month, Japanese authorities intercepted a North Korean methamphetamine shipment worth $170 million.

Pyongyang is also deeply involved in counterfeiting. According to South Korea's National Intelligence Service, North Korea has printed vast quantities of counterfeit bills, including $15 million in "super notes"—bogus bills that are very difficult to detect—using new counterfeiting technology. The South Korean report charges that the counterfeiting operation was authorized at the highest levels of the North Korean government and cites as evidence that, in 1999, an aide to Kim Jong-Il was caught trying to exchange $30,000 in counterfeit notes in Vladivostok.

In typical Communist fashion, the North Korean gangster regime often sends politically suspect subjects to the gulag on spurious criminal charges. This was the case with Sun-Ok Lee, a survivor of Pyongyang's gulag. Lee was convicted of spurious embezzling charges—and eventually escaped from North Korea to bear witness of the regime's unfathomable crimes against its most innocent subjects.

"In the 'reform institute' in Kaechon where I was held, there were 200 women housewives as prisoners," recalled Lee in congressional testimony. "In the case of these women, if any is pregnant, the baby would be killed. If the baby's mom was a political criminal, inside her the baby is the same political criminal. So the seed of a political criminal should not be allowed to be born."

Lee personally witnessed instances in which gulag officers would murder newborn infants by "stepping on the baby's neck with his boots once he or she was born. If the mom would cry for help to save her child, it was an expression of dissatisfaction against the party. So such a woman would be dragged out of the building and put to public execution by firing squad."

True Face of Evil

Such is the nature of the regime directly supported by our "allies" Russia and Communist China—and which the Bush administration is courting with humanitarian aid and promises of economic and technical assistance.

The Bush administration's treatment of North Korea exemplifies the utter phoniness of the "war on terrorism." Of the three "axis of evil" states, North Korea is undoubtedly the most oppressive and aggressive, and it poses the most immediate threat to U.S. citizens. Yet the administration has chosen to temporize in its dealing with Pyongyang in order to focus on the unnecessary, UN-authorized confrontation with Iraq.

And indeed, the North Korean hell state is a direct product of our nation's tragic entanglement with the UN. . . . [I]n the early stages of the Korean War the U.S.-led coalition liberated the entire peninsula from Communist hands—only to see the UN reverse this victory. That betrayal, and its tragic consequences, serves as a compelling illustration of the utter foolishness of fighting a "war on terrorism" through the UN.

Robert J. Einhorn

 **NO**

The North Korea Nuclear Issue: The Road Ahead

From the time it took office [in 2001], the [President George W.] Bush administration has been deeply divided on North Korea. One camp has assumed that North Korea would never voluntarily give up nuclear weapons, believed that the North Koreans would cheat on any new agreement, and feared that such an agreement would prop up a tyrannical regime. This group has supported regime change as the most reliable way of disarming North Korea. The other camp, skeptical about Pyongyang's [the capital of North Korea] willingness to give up nuclear arms but dubious about prospects for regime change in the near term, has favored exploring a negotiated solution. For much of the past three years, differences between the two camps have blocked a coherent approach toward North Korea's nuclear program.

However, after Colonel [Muammar] Gaddafi [Lybia's head of state] agreed in December 2003 to give up his nuclear program, a compromise (actually, more of a truce) was reached on the basis of what the administration started calling the "Libya model"—according to which an autocratic regime, looking to end its international isolation, makes a strategic decision to give up its nuclear program quickly, completely, and transparently without the U.S. having to make concessions up front. Consistent with that model, if [North Korea's leader] Kim Jong-il were prepared to follow Gaddafi's example and disarm on U.S. terms, the Bush administration would be willing to support the DPRK's [Democratic People's Republic of Korea, North Korea] integration into the world community, provide it assistance, and not seek to topple its regime.

The Libya model was the basis for the U.S. proposal tabled in late June 2004 at the third round of the six-party talks [North Korea, South Korea, China, Russia, Japan, and the United States]. It calls on North Korea first to make a clear commitment to dismantle all of its nuclear programs. Once that commitment is made, a three-month "preparatory period" begins during which North Korea makes a full declaration about its programs (including uranium enrichment), all nuclear activities are verifiably halted, any nuclear weapons are disabled, and preparations are made for the elimination (including by shipment out of North Korea) of all nuclear facilities, equipment, and materials. As Pyongyang takes

From *Policy Forum Online* by Robert J. Einhorn, September 14, 2004. Copyright © 2004 by Robert J. Einhorn. Reprinted by permission of the author.

credible steps during this preparatory period, the other parties reciprocate in various ways. Non-U.S. parties provide heavy fuel oil. A "provisional" multilateral security assurance is provided. The U.S. begins a "discussion" with the DPRK about its non-nuclear energy requirements, the lifting of remaining U.S. economic sanctions, and the removal of North Korea from the U.S. list of state sponsors of terrorism. At the end of the three-month period, a relatively short, but as yet unspecified, elimination period begins.

The U.S. proposal was welcomed by the other participants. South Korea and China, which had repeatedly asked the U.S. to show more flexibility, were relieved that a detailed U.S. offer had been made. Even North Korea reportedly deemed the proposal constructive. Meanwhile, [U.S.] administration officials stressed that their fundamental position had not changed. North Korea still needed to make a strategic decision to abandon nuclear weapons, after which total elimination would have to be achieved quickly—not via a prolonged series of steps that would give the North opportunities to stall, renege, or extort further concessions. Moreover, the Bush administration would not pay the DPRK to live up to existing commitments. It would "discuss" future benefits and allow others to provide them early in the process, but the U.S. would not provide its own tangible rewards until dismantlement was essentially complete.

The DPRK's approach is very different. It calls on the U.S., as a means of showing that it no longer has hostile intent, to provide inducements from the start. It resists dismantling its programs quickly, claiming that would forfeit its leverage to get the U.S. to live up to the deal. To preserve its deterrent for as long as possible, it presumably will seek to prolong the process and structure it in such a way as to enable it to opt out if it judges that getting rid of its capability entirely would put it at the mercy of a still-hostile U.S. Indeed, North Korea's position so far in the talks reinforces doubts that it is prepared to give up its capability. While saying it supports de-nuclearization, it still denies having a uranium enrichment program, and its freeze proposal seems to exclude plutonium produced before its January 2003 NPT withdrawal—enough for eight or nine bombs.

There is little likelihood North Korea will accept the current U.S. proposal. It sees itself in a strong bargaining position. With American forces tied down in Iraq and stretched thin worldwide (and some U.S. troops even shifting from South Korea to Iraq), Pyongyang must calculate that a military threat from the U.S. is remote. Indeed, even before Iraq, the DPRK's ability to devastate Seoul with its massive, forward-deployed conventional artillery and rocket forces was a strong inhibitor of U.S. military action. Economic factors do not compel DPRK flexibility either. Kim Jong-il appears to be serious about pursuing market reforms and probably realizes that such reforms cannot get very far if his country remains economically isolated because of the nuclear issue. But given the choice between maintaining his "powerful deterrent" and promoting the success of the reform effort, it is clear that he will give priority to security.

Pressures facing North Korea have also been reduced by Pyongyang's successful strategy of seeking separate accommodations with its neighbors. By adopting a more conciliatory approach to North-South interactions in the economic, humanitarian, and even military areas (e.g., military confidence-building measures in the West Sea), it has reinforced the inclination of [South Korea's]

Roh Moo-hyun's government to address the nuclear issue by offering carrots rather than threatening sticks. By taking steps to resolve the issue of Japanese abductees, it has increased prospects for resumed bilateral normalization talks and opened some daylight between U.S. and Japanese positions on North Korea. Its charm offensive is also paying dividends with China and Russia, both of which gave considerable support to North Korea's position at the last six-party round.

There may still be one more round of talks before the November [2004] election but very little chance of making progress this year. North Korea knows the U.S. administration won't turn up the heat in an election year and, in any event, Pyongyang will await the election results before making changes in its own position. Meanwhile, the Bush administration, believing that its recent proposal has undercut Kerry campaign criticism of its North Korea policy, will be content to run out the clock for 2004.

In 2005, the next administration could move in one of two very different directions. After a decent interval of trying unsuccessfully to get the DPRK to accept the U.S. offer, some in a second Bush administration may push for blaming North Korea for the deadlock, declaring the talks a failure, and seeking to ratchet up multilateral pressure against Pyongyang—including by calling for UN Security Council sanctions, stepping up Proliferation Security Initiative interdiction operations, and urging Japan to curtail trade and remittances. The goal would be to force North Korea to accept disarmament on U.S. terms or, if that does not prove possible, to contain, deter, and stifle the regime until it eventually collapses.

Such a strategy has little chance of succeeding. The idea that Pyongyang can be squeezed until it capitulates or collapses is wishful thinking. The regime has been surprisingly resilient, defying repeated predictions of its imminent demise. Moreover, neither China nor South Korea wants a sudden, destabilizing collapse in the North. Especially in the absence of a U.S. negotiating position that Beijing and Seoul consider reasonable—and since the June round, they have already begun calling for more U.S. flexibility—both can be expected to resist U.S. appeals for squeezing the North and to continue providing the assistance needed to keep it afloat.

Reliance on pressure alone to disarm North Korea could result in the worst of all worlds—Kim Jong-il shores up his grip on power by resisting U.S. coercion; the DPRK continues augmenting its nuclear arsenal; the U.S. strains relations with the ROK [Republic of Korea, South Korea] and China in a futile effort to pressure them to coerce North Korea; and the South Korean people and government continue to realign themselves toward China and away from the U.S., harming long-term American interests in East Asia.

A preferable approach, one likely to be supported by some in [the] second Bush administration . . . , would be to explore whether a sound agreement is possible. It would adopt elements of the Bush proposal (e.g., clear commitment to complete elimination and full disclosure of all programs from the outset), but it would provide for a phased elimination in a longer timeframe. At the same time, the U.S. would join the others in offering incentives in each of the phases, including from the beginning.

To be sure, this approach has downsides, including the risk that the North Koreans sooner or later would try to cheat or pull out of the process before dismantlement is complete. These risks can be minimized but not avoided, reflecting the reality that has faced the last three U.S. Presidents: there are no good options in dealing with North Korea. An imperfect agreement is the least bad option.

Of course, if the North Koreans have decided they must have a substantial nuclear weapons capability whatever we may do (hardly a remote possibility), they would likely reject a reasonable offer. In that event, the next U.S. administration would have little choice but to turn to a longer-term strategy of pressure, containment, and eventual rollback. But having made a proposal that North Korea's neighbors considered fair and balanced, we would be in a stronger position to gain multilateral support for that strategy.

POSTSCRIPT

Should North Korea's Nuclear Arms Program Evoke a Hard-Line Response?

The simmering crisis over North Korea's nuclear program was on the back burner during 2004 amid the U.S. travails in Iraq and the U.S. presidential election. But the issue is almost certain to reintensify. The confrontation with North Korea over its nuclear weapons program raises a host of important, long-term issues that extend far beyond the immediate concern. One is that fashioning an appropriate response to another country's action is often difficult because one cannot be sure what is motivating the other country. North Korea is perhaps the most secretive country on Earth. During the 1993–1995 crisis, one U.S. official commented, "The fact of the matter is that we don't really understand what they are doing." Knowledge of what caused North Korea in late 2002 to ignite the crisis once again could provide better direction for policymakers. For example, a hard-line response would arguably be appropriate if, as some people think, North Korea was merely taking advantage of the fact that the United States was snarled up in the Iraq crisis, using that opening to renew its efforts to acquire nuclear weapons in order to increase its power, and using that augmented power to cow South Korea and Japan and perhaps to force a U.S. military withdrawal from the region.

By contrast, a soft-line approach might be better if North Korea was trying to play a "nuclear chip" to garner more aid to ease the horrendous economic conditions (including widespread starvation) in the country. Arguably, a very soft approach would be best if North Korea was restarting its nuclear weapons program out of a defensive fear that it would be attacked by the United States. Americans are hesitant to see their country as the aggressor, but that reluctance is not shared worldwide. After all, President George W. Bush had labeled North Korea a member of an "axis of evil," and Washington was threatening and soon did attack Iraq, one of the other "evil axis" countries. It would not be unreasonable, then, for North Korea to think it was next on the American "hit list." To gain further insight into the politics of the Korean peninsula, including the U.S. involvement, read Bruce Cumings, *North Korea: Another Country* (New Press, 2003). Two perspective on U.S. policy in the region are Jihwan Hwang, "Realism and U.S. Foreign Policy Toward North Korea: The Clinton and Bush Administrations in Comparative Perspective," *World Affairs* (June 2004) and Fred Kaplan, "Rolling Blunder: How the Bush Administration Let North Korea Get Nukes," *Washington Monthly* (May 2004).

Another long-term issue is how to contain the spread of nuclear weapons. At least some countries ask why it is acceptable for some countries to have nuclear weapons and unacceptable for others. A particularly sore point in some regions of the world is the U.S. silence on Israel's all-but-official nuclear arsenal. When the NPT was being renewed in the mid-1990s, many nonnuclear countries were reluctant to do so. They wanted countries that had nuclear weapons to set a timetable for dismantling their arsenals because, as Malaysia's delegate to the conference noted, without such a pledge renewing the treaty would be "justifying nuclear states for eternity" to maintain their monopoly. These objections were partly overcome when the states with nuclear weapons pledged to conclude a treaty to ban all nuclear testing. The United States reneged on that pledge when the Senate refused to ratify the resulting Comprehensive Test Ban Treaty in 1999.

A third extended issue is what effect the North Korean nuclear program will have on nuclear proliferation in general and, more particularly, in Northeast Asia. China and Russia already have such weapons, as does another major player in the area—the United States. If North Korea also has an arsenal, can defense planners in South Korea and Japan reasonably be asked not to build their own capabilities? As George Tenet, the director of the CIA, has commented, "The desire for nuclear weapons is on the upsurge. Additional countries may decide to seek nuclear weapons as it becomes clear their neighbors and regional rivals are already doing so. The 'domino theory' of the 21st century may well be nuclear." The importance of the issue has led to the publication of several worthy, commentaries, including Victor D. Cha and David C. Kang, *Nuclear North Korea: A Debate on Engagement Strategies* (Columbia University Press, 2005); Ted Galen Carpenter and Doug Bandow, *The Korean Conundrum: America's Troubled Relations with North and South Korea* (Palgrave Macmillan, 2004); and Michael O'Hanlon and Mike M. Mochizuki, *Crisis on the Korean Peninsula: How to Deal With a Nuclear North Korea* (McGraw-Hill, 2003).

ISSUE 9

Would It Be an Error to Establish a Palestinian State?

YES: P. J. Berlyn, from "Twelve Bad Arguments for a State of Palestine," *A Time to Speak,* http://www.israel.net/timetospeak/bad.htm (December 2002)

NO: Rosemary E. Shinko, from "Why a Palestinian State," An Original Essay Written for This Volume (2004)

ISSUE SUMMARY

YES: P. J. Berlyn, an author of studies on Israel, primarily its ancient history and culture, refutes 12 arguments supporting the creation of an independent state of Palestine, maintaining that such a state would not be wise, just, or desirable.

NO: Rosemary E. Shinko, who teaches in the department of political science at the University of Connecticut, contends that a lasting peace between Israelis and Palestinians must be founded on a secure and sovereign homeland for both nations.

T he history of Israel/Palestine dates to biblical times when there were both Hebrew and Arab kingdoms in the area. In later centuries, the area was conquered by many others; from 640 to 1917 it was almost continually controlled by Muslim rulers. In 1917 the British captured the area, Palestine, from Turkey.

Concurrently, a Zionist movement for a Jewish homeland arose. In 1917 the Balfour Declaration promised increased Jewish immigration to Palestine. The Jewish population in the region began to increase slowly, then it expanded dramatically because of refugees from the Holocaust. Soon after World War II, the Jewish population in Palestine stood at 650,000; the Arab population was 1,350,000. Zionists increasingly agitated for an independent Jewish state. When the British withdrew in 1947, war immediately broke out between Jewish forces and the region's Arabs. The Jews won, establishing Israel in 1948 and doubling their territory. Most Palestinian Arabs fled (or were driven) from Israel to refugee camps in Gaza and the West Bank (of the Jordan River), two areas that had been part of Palestine but were captured in the war by Egypt and Jordan, respectively. As a result of the 1967 Six Day War between Israel and Egypt, Jordan, and Syria, the Israelis again expanded their territory by capturing several areas, including the Sinai Peninsula, Gaza, the Golan Heights, and the West Bank. Also in this period the Palestine Liberation Organization (PLO) became the major representative of Palestinian Arabs. True peace was not possible because the PLO and the Arab states would not recognize Israel's legitimacy and because Israel refused to give up some of the captured territory.

Since then, however, continuing violence, including another war in 1973, has persuaded many war-exhausted Arabs and Israelis that there has to be mutual compromise to achieve peace. Perhaps the most serious remaining sore point between the Arabs and Israelis is the fate of the Palestinians, who live primarily in the West Bank and Gaza.

In 1991 Israelis and Palestinians met in Spain and held public talks for the first time. Israeli elections brought Prime Minister Yitzhak Rabin's liberal coalition to power in 1992. This coalition was more willing to compromise with the Arabs than had been its more conservative predecessor. Secret peace talks occurred between the Israelis and Palestinians in Norway and led to the Oslo Agreement in 1993. Palestinians gained limited control over Gaza and parts of the West Bank and established a quasi-government, the Palestinian authority led by Yasser Arafat.

The peace process was halted, perhaps even reversed, when in 1995 Prime Minister Rabin was assassinated by a Jewish fanatic opposed to Rabin's policy of trying to compromise with the Palestinians. Soon thereafter, the conservative coalition headed by Prime Minister Benjamin Netanyahu came to power. He dismissed any possibility of an independent Palestine, made tougher demands on the PLO, and moved to expand Jewish settlements in the West Bank. With some 200,000 Jews already in the West Bank and East Jerusalem, these actions compounded the difficult issue of the fate of those people in a potentially Palestinian-controlled area.

Pressure from a number of quarters, including the United States, kept the Israelis and Palestinians talking. Meeting in 1997 at the Wye River Plantation in Maryland under the watchful eye of President Bill Clinton, Israel agreed to give the Palestinians control over additional areas of the West Bank, and the Palestinians agreed to work to protect Israel from Arab terrorist attacks and to remove language in the PLO charter that called for the destruction of the Jewish state. The immediate impact of the Wye River Agreement was negligible.

A liberal government under Prime Minister Ehud Barak failed to move the peace process forward and was replaced by a conservative government led by Ariel Sharon, who favors a very stern approach to the Palestinians. Under Sharon, the Israeli military has responded to terror attacks by conducting extensive military operations in Gaza and the West Bank. Whatever the justice or wisdom of that approach by Israel, it has not stemmed the suicide bombings and other attacks on Israelis by the Palestinians. In the following selections, P. J. Berlyn argues that creating an independent Palestine would be a grave error, while Rosemary E. Shinko maintains that there will be no end of the violence until Palestinians have their own independent homeland.

P. J. Berlyn

 **YES**

Twelve Bad Arguments for a State of Palestine

In 1991, the United States, during the administration of President George H. W. Bush, sponsored the Madrid Conference at which Israel is invited to meet with Jordan and other Arab States to negotiate peace. In a letter to the Government of Israel, the Government of the United States pledges:

> "In accordance with the United States, traditional policy, we do not support the creation of an independent Palestinian state. [. . . .] Moreover, it is not the United States' aim to bring the PLO [Palestine Liberation Organization] into the process or to make Israel enter a dialogue or negotiations with the PLO."

This pledge was indeed consonant with history, strategy, justice and common sense. It was not, however, to be honored. In 2001, President George W. Bush announced that it was a "vision of long standing" of U.S. policy to create a Palestinian State west of the Jordan River. Such a state would be under the rule of the PLO and must be recognized by Israel. The United States proceeded swiftly to have this newly-discovered long-standing vision ensconced in a resolution of the United Nations Security Council.

It is now conventional to suppose that the invention of a PLO State in the Land of Israel is wise, just and desirable, even inevitable. Among the platitudes strung into a mantra:

1. It will rectify an historic injustice to the Arabs.

On the contrary: Of all that Arabs have demanded for themselves since the end of World War I, they have been given 99.5 percent.

In 1921 the League of Nations defined Mandate Palestine, as The Jewish National Home, to be "open to close Jewish Settlement." In 1922, the British Mandatory Government subtracted the entire region east of the Jordan River, more than 75 percent of the Jewish National Home, to create the Arab Kingdom of [Trans]-Jordan.

Great Britain then progressively restricted or banned Jewish immigration and settlement even west of the Jordan River, rigidly blockading the Land of their fathers to Jews trying to escape the gas chambers of Europe. At the same time, the British authorities permitted massive immigration of the Arabs into Mandatory Palestine, whose families now number among those who claim it as their ancient ancestral homeland.

In 1947, the United Nations attempted to whittle away the remnant of the Jewish National Home with a second partition to create a second Arab state in Palestine, this one west of the Jordan River where most of the Arab population were but recently arrived, with no roots and no history in the land. Had the Arabs accepted that offer, they would have had 83 percent of the Land of Israel-Jewish National Home. Instead, they went to war to get 100 percent of it.

The real injustice is depriving Israel of its historic homeland, in order to invent a 23rd Arab state where none ever existed.

2. It will end Israel's occupation of Palestinian territory.

On the contrary: There is no such thing as "Palestinian territory" and there is no "occupation" of what never belonged to any Arab nation. Furthermore, of the Arab residents of Judea, Samaria [biblical terms used for the West Bank] and Gaza, 98 percent now live under the rule of the PLO.

3. Israel must comply with United Nations resolutions.

On the contrary: The Arab attack on Israel in June 1967 left Israel in possession of Judea and Samaria, that Jordan had seized in the attack of 1947, Gaza, that Egypt had seized in the war of 1947, the Sinai and the Golan Heights. The UN Security Council, that had done nothing to prevent or even deplore the Arab attack, took it upon itself to pass a resolution to guide a future settlement. That was Resolution 242, that calls on Israel, in the context of a peace settlement, to withdraw from unspecified "territories" to "agreed and secure borders."

This was very specifically *not* a demand for a return to the borders of June 4, 1967 [prior to Israel's victory in the 1967 war]—which were themselves merely the ceasefire lines of the War of Independence, a war launched by the Arabs to destroy Israel in 1948.

The author of Resolution 242 was Lord Caradon, representative of the United Kingdom. He explained to the British Parliament: "It would be wrong to demand that Israel return to its positions of June 4, 1967, because those positions were undesirable and artificial."

The United States was a co-sponsor of Resolution 242, and its representative stated: "The notable omissions—which were not accidental—in regard to withdrawal are the words 'the' or 'all' and 'the June 5, 1967 lines'."

It is now widely and repeatedly alleged that this resolution demands that Israel withdraw from *all* the territories. *That is a lie.* It is alleged that this withdrawal is unconditional. *That is a lie.* It is even alleged that it calls for an independent Palestine-Arab state in the territories. *That is a lie.*

In accordance with the Israel-Egypt treaty of 1978, Israel withdrew from the entire Sinai peninsula, 91 percent of the "territories." That may well be considered more than sufficient to satisfy the terms of Resolution 242.

In more recent resolutions, the sponsors of No. 242 have reneged on their own positions. The United States has reversed all previous policy statements and promises to Israel and called for a PLO state. Great Britain has turned its own resolution upside down by demanding full withdrawal as well as a PLO state.

These fickle flip-flops show that the authors and sponsors of the resolution cannot be trusted to stand by their own creation. Israel thus betrayed should not be expected to bow to every new whim of the moment. Add to that betrayal the role of the United Nations as the world epicenter for hatred of Israel and Jews, and it is absurd to argue that Israel has any obligation to submit to its demands.

4. It will bring peace and stability to the Middle East.

On the contrary: It will establish a Middle Eastern national base for terror, that will spread incitement, bloodshed, and mortal danger not only to Israel but also to Arab regimes in the neighborhood.

The citizens supposed to build this peaceful and stable state will be the ones that the PLO regime has programmed to hatred and contempt, to yearn to earn martyrdom by murdering Jews. They will be the hysterics who run through the streets, some in costumes to rival the Ku Klux Klan, brandishing weapons and shrieking curses and threats.

5. It will satisfy the demands of the Palestinian Arabs, who will give up terrorism and war and settle down to building a society.

On the contrary: The PLO Charter of 1964 defines its sole purpose as the destruction of Israel. (That was three years *before* 1967, when there were no "occupied territories" to liberate.) Despite flimflam to the contrary, that Charter still stands unamended, and so does the goal.

The PLO openly and repeatedly proclaims that it will never settle for less than every inch between the Jordan River and the Mediterranean Sea—including all of what is now the State of Israel. That is how "Palestine" is shown on its maps and emblems. If it condescends to accept a smaller state it will be only as an interim measure, to facilitate the future destruction of Israel.

6. A State of Palestine will honor a pledge to respect Israel's right to exist.

On the contrary: The PLO has already made at least six formal agreements with Israel and has pledged itself to at least seven ceasefires. Not a single term or clause of any of them has been kept for a single day. To expect any other behavior in the future defies basic common sense.

Repeated statements by officials of the Fatah, Hamas, and other member bodies of the PLO declare over and over again that their goal is the end of Israel, the expulsion of the Jews, and an Arab Palestinian State from the Jordan River to the Mediterranean.

7. A State of Palestine will be demilitarized and thus no danger to Israel.

On the contrary: There were limitations in the Oslo Accords—a police force of no more than 8,000 and no heavy weapons. Today the PLO has a military force of at least 50,000 with heavy weapons. More are smuggled in constantly, from Egypt, Syria and elsewhere.

If Israel is deprived of the strategic defense line of the Jordan valley and the highlands to the west, then its width will be reduced to *nine miles.* The PLO state can become militarized in mere hours (as was the "demilitarized" Sinai in 1967), and forces from Iraq and Syria can sweep in without hindrance.

8. It will secure the human rights of the Palestinian Arabs.

On the contrary: The PLO regime in the areas it controls in Judea, Samaria and Gaza has nothing at all to its credit in human rights, and everything to its discredit. (This should have been expected from the example of its rule over southern Lebanon in the 1980s.) Overseers of human rights who used to keep captious watch on these areas when they were under Israel's administration have been on an extended vacation since September 1993.

9. It will solve the Arab refugee problem.

On the contrary: The PLO insists that it will not absorb these exploited people, but will demand their "return" to Israel—meaning the destruction of Israel. The residents of the UN refugee camps also insist they will settle for nothing less than "return", and the UN that runs the camps makes sure the residents do not budge from this determination.

10. It will encourage civic and economic development, raise the standard of living and bring contentment to the people.

On the contrary: In areas administered by the PLO, the standard of living drops and hardship increases. Economic development is strangled by graft and corruption, and revenues are squandered. The United States and the European Union have given large donations for development, but the bulk of the money melts away or ends up in private foreign bank accounts.

11. It will win the respect of world opinion for Israel.

On the contrary: "World opinion" is a jigsaw of many mismatched pieces. It includes a professional contempt for Israel, conspicuous especially in journalism and academia. It includes taste-makers of a Europe that have been trying to crush the Jews for two millennia and have not yet given up the habit. It has ancient roots in religious convictions, and in history, and new shoots of resentment and envy.

There are even a few examples here and there of good sense and fair play, but those are not found in the portion of "world opinion" that makes relentless demands on Israel—sometimes masquerading as "friendly advice in your own best interest." Israel is not obliged to satisfy those demands by making itself shrunken, demoralized, discredited, and vulnerable, nor would it be any better liked if it did.

That is not to say "world opinion" never approves of anything Israel does. It did welcome with delight the self-demeaning and self-destructive Oslo Accords [agreement between Israel and the PLO signed in 1993].

> 12. *If a State of Palestine commits aggression against Israel, then Israel can fight its military forces and win back what it gave away.*

On the contrary: The supporters of the Oslo Accords in 1993 also said "If they [the PLO] do not keep their commitment to peace, we will just take the land back." But those who said it took their words back—or ignored that they ever said them.

Now these areas are used as bases for terrorism against Israel. When the IDF [Israel Defense Force] goes in even briefly, to close down terror bases and weapons factories and dumps, the world—including even the United States—howls for Israel to "get out of Palestinian territory immediately." If those areas were to become territory of an Arab "State of Palestine" any defensive actions by Israel would be branded aggression against a sovereign state. It would be condemned and threatened even more harshly than when it moved against terrorism when these areas were held by Jordan and Egypt—without sovereignty.

If these areas of the Land of Israel became a PLO state, Israel would lose even minimal control. It could not restrict import of heavy weapons, destroy weapons factories and depots, intercept terrorist activities or arrest terrorists. It could not even prevent the entry of foreign troops from other Muslim countries.

Israel, drastically restricted geographically, will be exposed and vulnerable. When the PLO and its allies launch all-out war, the cost to Israel will be horrendous.

If Israel wins a battle for survival, it still will not be able to regain what it gave away. Even if a PLO State is defeated in battle, it will not cease to exist. In all of the Arab wars against Israel, outside powers have intervened to save them from total defeat. A PLO state can thus survive defeats and repeat its aggressions.

Giving up Judea-Samaria would also mean that Jews would be cut off from the cities and sites that are the heart of their historic homeland. Israel could not prevent Arabs from destroying ancient Jewish sites and relics. There would be no chance to make new discoveries that shed new light on the history of Israel and its people.

Some who are made aware of all these circumstances nevertheless say: "It is useless to oppose a State of Palestine—it is inevitable." Such passive submission, such moral indolence, is tacit consent to an act inimical to the Jewish people and the Land of Israel. It is a limp surrender of both the past and the future.

For 2000 years, the Jewish people did not despair of restoration to their Land. When the restoration has at last come, those who toss it away betray both their ancestors and their descendants.

NO

Rosemary E. Shinko

Why a Palestinian State

On July 8, 1937 the Palestine Royal Commission (Peel Commission) offered its recommendations to the British government regarding the disposition of the Palestinian question. The commission expressed serious reservations about the possibility of reconciliation between Arabs and Jews and thus they concluded, "only the 'surgical operation of partition' offers a chance of ultimate peace" (www.guardiancentury.co.uk). They proposed the establishment of two separate states—a sovereign Arab State and a sovereign Jewish State. United States President, Bill Clinton, reiterated these same sentiments in a speech he delivered on January 7, 2001. "I think there can be no genuine resolution to the conflict without a sovereign, viable, Palestinian state that accommodates Israel's security requirements and the demographic realities." Any settlement must ultimately be "based on sovereign homelands, security, peace and dignity for both Israelis and Palestinians."

Why is it then that P. J. Berlyn argues that the "invention" of a 23rd Arab state would be unwise, unjust and undesirable? Her arguments revolve around the following five main assertions: there is no such thing as "Palestinian territory" nor do the Arabs constitute a "Palestinian people," if such a state were "invented" the Arabs would be unable to fulfill the rights and duties associated with statehood, it would not be in the self-interest of the State of Israel to allow such a state to be created in its midst, and finally it would betray the sacrifices of the past and the promises of the future of the Jewish people.

What is a "state" and why does the possibility of the creation of a Palestinian State, in particular, provoke such a strong, emotional response from Ms. Berlyn? What does the term "state" signify? According to Hegel, "only those peoples that form states can come to our notice" because it is the state that provides the foundation for "national life, art, law, morality, religion, [and] science" (*Reason In History*). The political identity of most peoples is inextricably bound up with the notion of statehood (Rourke, 2001: 189). According to an international relations text, *International Politics on the World Stage,* written by John Rourke and Mark Boyer, "States are territorially defined political units that exercise ultimate internal authority and that recognize no legitimate external authority over them." The political implications of legitimacy that would flow from the establishment and recognition of a Palestinian state are extremely significant in this particular instance. As scholar Malcom Shaw notes in his 1999 Cambridge University Press book *International Law*, a state is recognized as having a "legal personality," which includes the capacity to possess and exercise certain rights and to perform specific duties. These rights and duties encompass the attributes of independence, legal equality, and peaceful coexistence. Thus a Palestinian State would claim the right to exercise jurisdiction over its population and territory, as well as the right to self-defense. Such a state would also have a concomitant duty not to intervene in the internal affairs of another state and a duty to respect the territorial integrity and sovereignty of other states.

The establishment and recognition of a Palestinian State would confirm the political legitimacy and legal equality of the Palestinian people. Historically they have been denied recognition as a people and the legitimacy of their claims to the territory of Palestine have been dismissed. Ms. Berlyn's arguments are designed to foster the sense of Palestinian illegitimacy with her bold assertion that the Arabs have "no roots and no history" in Palestine, and that there is "no such thing as a Palestinian territory." To round out her argument, she employs the term "invent" when referring to the establishment of a Palestinian State in order to further deny and delegitimate Palestinian claims to their homeland. All states are man made creations, all states are reflective of the political, legal, social, economic and historical conditions which led to their rise. All states, even the State of Israel, are a political creation of men and women. Ms. Berlyn's arguments, which attempt to dismiss a Palestinian presence and history, are an extension of the earlier Zionist attempts to portray Palestine as a "land without people for a people without land." A perception was fostered that the territory of Palestine was "empty" and that its only inhabitants were uncivilized, backward nomads.

Demographic realities, however, proved otherwise. "There were always real, live Palestinians there; there were census figures, land-holding records, newspaper and radio accounts, eyewitness reports and the sheer physical traces of Arab life in Palestine before and after 1948," according to Edward W. Said, and Christopher Hitchens in their 1988 study, *Blaming the Victims, Spurious Scholarship and the Palestinian Question*. In 1947 when the United Nations Committee on Palestine (UNSCOP) made its recommendation that Palestine be portioned into two separate states, there were 1.2 million Arabs as compared to 570,000 Jews living in the territory. Clearly the Arabs formed the majority of the population in Palestine. On what basis then can it be maintained that the Palestinians had no history, had no roots, and had no presence in Palestine? "The fact is that [when] the people of Israel . . . came home, the land was not all vacant," President Clinton commented in 2001. Statehood confirms presence, establishes legitimacy, confers recognition, and provides a focal point for a peoples' identity.

David Shipler in an October 15, 2000 article in the *New York Times* commented astutely that "Recognizing the authenticity of the other in that land comes hard in the midst of the conflict. Yet the conflict cannot end without that recognition." Ultimately legitimacy and recognition are the keys to the end of conflict and the establishment of peace in the Middle East. Peace cannot occur without the recognition of the Palestinian people's right to self-determination and without their consent to the government that exercises authority over them. "The six wars with the Arabs created a situation in which 3 million Jews came to control territories that contained nearly 2 million Arabs," scholar John G. Stoessinger observed in his 2001 book, *Why Nations Go to War.* The United Nations General Assembly also concluded that without "full respect for and the realization of these inalienable rights of the Palestinian people," namely the right to self-determination and the right to national independence and sovereignty, there would be no resolution of the question of Palestine (http:// domino.un.org).

Statehood implies the capacity to maintain certain rights and the performance of specific duties. Ms. Berlyn maintains that even if the Palestinians were granted their own state they would not live up to the duties of a state because a Palestinian State would be committed to the destruction of Israel and merely serve as a base for further acts of terror. In her estimation a Palestinian State would not respect the territorial integrity and sovereignty of the State of Israel. Furthermore, she even questions the ability of such a state . . . to fulfill the requirements of statehood, including civic and economic development and the promotion of human rights. Fundamentally such a negative assessment rests on conjecture and an underlying sense of distrust born of conflicting claims of legitimacy to the same parcel of land. No one can claim to profess the future, and not even Ms. Berlyn can with any certainty predict the actions of a State of Palestine. One thing does however appear to be foreseeable, and that is the continued agitation of the Palestinians for recognition, self-determination and legitimacy. As Professor Stoessinger put it in *Why Nations Go to War,* "The shock inflicted on the Arab consciousness by the establishment of Israel and the resulting homelessness of a million native Palestinians grew more, rather than less, acute as Arab nationalism gathered momentum." The Arabs perceive Israel as the ever-expanding and ever-growing threat to their survival, thus a state is the only way to insure their continued existence as a people. Declaring that a Palestinian State would be unable to fulfill the requirements of statehood is merely a thinly veiled ethnocentric critique, which smacks of patronization and cultural superiority. What precisely do the Palestinians lack that would deem them ill-suited to exercise self-rule and incapable of founding a government that rests on consent which would secure their rights to life, liberty and property? The Lockean assertion that the only legitimate form of government is that which rests on consent is as true for the Jews as the Arabs of Palestine.

Israel is exposed and vulnerable as a result of the Palestinians' unrelenting quest for legitimacy and recognition. Israel's self-interest may ultimately rest with the establishment of a separate Palestinian State in order to diffuse the longstanding animosities and hatreds that have arisen between the two peoples. We have seen where the denial of legitimacy has taken us, and it has not nurtured the seeds of peace. Peace can only be established in the wake of the recognition of legitimacy and equality between the two peoples. In order to secure the national character and the cultural identity of the Israelis, the national character and the cultural identity of the Palestinian Arabs must likewise be secured. This can only occur with the establishment of a separate, sovereign Palestinian State.

POSTSCRIPT

Would It Be an Error to Establish a Palestinian State?

The Middle East's torment is one of the most intractable problems facing the world. In addition to the ancient territorial claims of Jews and Palestinian Arabs, complexities include long-standing rivalries among various religious and ethnic groups and countries in the region. To learn more about the history of the current conflict, consult Bernard Wasserstein, *Israelis and Palestinians: Why Do They Fight? Can They Stop?* (Yale University Press, 2004).

Complicating matters for Israel is the fact that the country is divided between relatively secular Jews, who tend to be moderate in their attitudes toward the Palestinians, and Orthodox Jews, who regard the areas in dispute as land given by God to the Jewish nation and who regard giving up the West Bank and, especially, any part of Jerusalem as sacrilege. Furthermore, there are some 200,000 Israelis living in the West Bank, and removing them would be traumatic for Israel. The issue is also a matter of grave security concern. The Jews have suffered mightily throughout history; repeated Arab terrorism represents the latest of their travails. It is arguable that the Jews can be secure only in their own country and that the West Bank (which cuts Israel almost in two) is crucial to Israeli security. If an independent Palestine centered in the West Bank is created, Israel will face a defense nightmare, especially if new hostilities with the Palestinians occur. Additonal material on a prospective Palestinian state is available in David Gompert, *Building a Successful Palestinian State* (Rand, 2004).

Thus, for the Israelis the "land for peace" choice is a difficult one. Some Israelis are unwilling to cede any of what they consider the land of ancient Israel. Other Israelis would be willing to swap land for peace, but they doubt that the Palestinians would be assuaged. Still other Israelis think that the risk is worth the potential prize: peace.

Palestinians do not march in political lockstep any more than Israelis, and a key factor in the approach of the Palestinians to peace and conflict centers on changes in their leadership following the death of their longtime leader, Yasser Arafat, in November 2004. His successor, Mahmoud Abbas, was elected as president of the Palestinian National Authority in January 2004. Abbas has called for an end to the violence, but whether he can control violence-prone Palestinian factions, such as Hamas, remains to seen. If he can, then there is a better chance that the U.S.-outlined "roadmap" to peace can progress. For a study that criticizes Israel approach to the Palestinians and the U.S. support of it, read Noam Chomsky, *Middle East Illusions* (Roman & Littlefield, 2004). For the opposite perspective, see Alan Dershowitz, *The Case for Israel* (John Wiley & Sons , 2003). For the first-hand views of the two sides, see the Web sites of the Palestinian National Authority at http://www.pna.gov.ps/ and of Israel at http://147.237.72.16/eng/mainpage.asp.

ISSUE 10

Was War with Iraq Justified?

YES: Richard Cheney, from "Meeting the Challenge of the War on Terrorism," Address at the Heritage Foundation (October 17, 2003)

NO: Robert Byrd, from "Invasion of Iraq," Remarks in the U.S. Senate, *Congressional Record* (November 25, 2003)

ISSUE SUMMARY

YES: Vice President Richard Cheney argues that Saddam Hussein's drive to acquire weapons of mass destruction, links with terrorists, and brutal dictatorship warranted U.S. action to topple his regime.

NO: West Virginia Senator Robert Byrd criticizes the decision to invade Iraq in the first place as ill-founded and further contends that the consequences have been too costly.

In August 1990, Iraq overran Kuwait. President George H. W. Bush responded by sending 250,000 troops to Saudi Arabia to protect that country and the other oil states to the south from any possible further move by Iraq. Working through the United Nations, the United States also built a coalition to apply economic, then military, pressure on Iraq to withdraw from Kuwait. The UN passed several resolutions denouncing Iraq's actions and imposing sanctions on Iraq. Then in November 1990, the UN Security Council passed Resolution 678. It demanded that Iraq withdraw from Kuwait by January 15, 1991, and authorized UN members to use "all necessary means" after that date to expel Iraq from Kuwait.

Iraqi intransigence was followed on January 17, 1991, by a U.S.-led attack by a coalition of more than a dozen countries that defeated Saddam Hussein's forces after a six-week air campaign followed by a four-day ground operation involving more than 500,000 coalition troops. In the aftermath, huge stocks of Iraqi chemical weapons were uncovered. Security Council Resolution 687 (1991) spelled out the terms of peace for Iraq. Among these were that Iraq was barred from possessing, producing, or seeking to acquire any weapons of mass destruction (WMDs) and that UN arms inspectors would have unhindered access anywhere in the country to ensure that Iraq was complying. An economic embargo would remain in place until Iraq fully complied with the WMD and inspection clauses of Resolution 687. A cat-and-mouse game ensued. Iraq claimed to be complying with Resolution 687, but it often blocked or delayed UN inspections. In 1998 it even expelled the inspectors. In response, economic sanctions continued, and there were periodic attacks by U.S. warplanes and cruise missiles.

The U.S. attitude hardened even more once George W. Bush became president. Many of his advisors had long advocated toppling Saddam Hussein, and that belief was intensified by the terrorist attacks of September 11, 2001. Still, the pattern maneuver between the United States and Iraq continued. In September 2002, Bush told the UN General Assembly that Iraq constituted a "grave and gathering danger" and that Iraqi lack of cooperation was making conflict "unavoidable." Days later, inspectors were readmitted into Iraq. They found no hard evidence of Iraqi WMDs, but they also reported repeated barriers to their inspections. In this tense atmosphere, the U.S. Congress authorized the use of U.S. force against Iraq if necessary in October 2002. In November 2002, Washington successfully sponsored UN Security Council Resolution 1441 demanding full Iraqi compliance. Barriers to inspection continued, however, leading the Bush administration, the British government, and some others to push for a UN resolution declaring Iraq in "material breach" of numerous resolutions dating back to 1991 and authorizing member states to take action against Baghdad.

Among other causes of war, Bush and British Prime Minister Tony Blair claimed that intelligence reports indicated that Iraq was seeking to acquire nuclear weapons. The president also connected the Iraqi government with the support of terrorism. France, Russia, and China blocked the American and British efforts in the Security Council. Each had a veto, and they believed that more time should be given to the UN inspectors. Most other council members, and indeed, most other countries, seemed to agree. Faced with a defeat in the Security Council, Washington abandoned the effort to win UN authorization and chose to act without one. In March 2003, Bush, supported by Great Britain and a number of countries dubbed the "coalition of the willing," issued an ultimatum to Iraq that demanded that Saddam Hussein and his sons leave the country and that Iraqi forces surrender their arms. Iraq refused, and another U.S.-led invasion ensued. Iraq's forces were soon vanquished, and Saddam Hussein was later captured.

Securing the peace proved much more difficult. Terrorist-style operations began against occupying U.S. and other coalition troops, civilians, and anyone, including Iraqis, who cooperated with them. Soon coalition casualties exceeded the number killed and wounded during invasion. It was at this point, six to eight months after the invasion, that the two speeches in this debate occurred. Vice President Cheney argued that despite the travails, bringing down Saddam Hussein was the right thing to do. Senator Robert Byrd differed, finding fault with both the rationale for invasion and fretting about the damage that he claimed the action had done and would continue to do to the United States.

YES

Richard Cheney

Meeting the Challenge of the War on Terrorism

I've come here this morning to discuss the war on terror, the choices America has made in that war, and the choices still before us.

For most of this year, the attention of the world has centered on Iraq. From the final ultimatum to Saddam Hussein last March [2003], to the removal of his regime, and on up to the present, as we continue to battle with Saddam loyalists and foreign terrorists. Iraq has become the central front in the war on terror. It was crucial that we enforced the U.N. Security Council resolutions. Now, having liberated that country, it is crucial that we keep our word to the Iraqi people, helping them to build a secure country and a democratic government. And we will do so.

Our mission in Iraq is a great undertaking and part of a larger mission that the United States accepted now more than two years ago. September 11, 2001, changed everything for this country. We came to recognize our vulnerability to the threats of the new era. We saw the harm that 19 evil men could do, armed with little more than airline tickets and box cutters and driven by a philosophy of hatred. We lost some 3,000 innocent lives that morning, in scarcely two hours' time.

Since 9/11, we've learned much more about what these enemies intend for us. One member of al-Qaeda said 9/11 was the "beginning of the end of America." And we know to a certainty that terrorists are doing everything they can to gain even deadlier means of striking us. From the training manuals we found in the caves of Afghanistan to the interrogations of terrorists that we've captured, we have learned of their ambitions to develop or acquire chemical, biological, or nuclear weapons. And if terrorists ever do acquire that capability—on their own or with help from a terror regime—they will use it without the slightest constraint of reason or morality.

That possibility, the ultimate nightmare, could bring devastation to our country on a scale we have never experienced. Instead of losing thousands of lives, we might lose tens of thousands, or even hundreds of thousands of lives in a single day of war. Remembering what we saw on the morning of 9/11, and knowing the nature of these enemies, we have as clear a responsibility as

Address by Richard Chaney at the Heritage Foundation, October 17, 2003.

could ever fall to government: We must do everything in our power to keep terrorists from ever acquiring weapons of mass destruction.

This great and urgent responsibility has required a shift in national security policy. The strategy of deterrence, which served us so well during the decades of the Cold War, will no longer do. Our terrorist enemy has no country to defend, no assets to destroy in order to discourage an attack. Strategies of containment will not assure our security, either. There is no containing terrorists who will commit suicide for the purposes of mass slaughter. There is also no containing a terror state that secretly passes along deadly weapons to a terrorist network. There is only one way to protect ourselves against catastrophic terrorist violence, and that is to destroy the terrorists before they can launch further attacks against the United States.

Sustained Campaign of Terrorism

For many years prior to 9/11, it was the terrorists who were on the offensive. We treated their repeated attacks against Americans as isolated incidents and answered, if at all, on an ad hoc basis, and rarely in a systematic way. There was the attack on the Marine barracks in Beirut in 1983, killing 241 men; the bombing of the World Trade Center, in 1993; five more murders when the Saudi National Guard Training Center in Riyadh was struck in 1995; the killings at Khobar Towers in 1996; the East Africa Embassy bombings in 1998; and in 2000, the attack on the USS *Cole*.

There was a tendency to treat incidents like these as individual criminal acts to be handled primarily through law enforcement. . . . [As a result,] leads were not successfully followed, the dots were not adequately connected, the threat was not recognized for what it was. For al-Qaeda, the World Trade Center attack in 1993 was part of a sustained campaign. . . . [by a] as a growing network with operatives inside and outside the United States, waging war against our country. For us, that war started on 9/11. For them, it started years ago, when Osama bin Laden declared war on the United States. . . . Since September 11th, the terrorists have continued their attacks in Riyadh, Casablanca, Mombasa, Bali, Jakarta, Najaf, and Baghdad.

New, Global U.S. Strategy

Against this kind of determined, organized, ruthless enemy, America requires a new strategy—not merely to prosecute a series of crimes, but to conduct a global campaign against the terror network. . . . [Among U.S. initiatives] we are applying the Bush doctrine: Any person or government that supports, protects or harbors terrorists is complicit in the murder of the innocent and will be held to account.

The first to see this doctrine in application were the Taliban, who ruled Afghanistan by violence, while turning the country into a training camp for terrorists. With fine allies at our side, we took down the regime and shut down the al-Qaeda camps. . . .

In Iraq, we took another essential step in the war on terror. The United States and our allies rid the Iraqi people of a murderous dictator, and rid the world of a menace to our future peace and security. Saddam Hussein had a lengthy history of reckless and sudden aggression. He cultivated ties to terror—hosting the Abu Nidal organization [a largely Palestinian terrorist group in existence since 1974], supporting terrorists, making payments to the families of suicide bombers in Israel. He also had an established relationship with al-Qaeda, providing training to al-Qaeda members in the areas of poisons, gases, making conventional bombs. Saddam built, possessed and used weapons of mass destruction. He refused or evaded all international demands to account for those weapons.

Twelve years of diplomacy, more than a dozen Security Council resolutions, hundreds of U.N. weapons inspectors, thousands of flights to enforce the no-fly zones, and even strikes against military targets in Iraq—all of these measures were tried to compel Saddam Hussein's compliance with the terms of the 1991 Gulf War cease-fire. All of these measures failed.

Last October [2002], the United States Congress voted overwhelmingly to authorize the use of force in Iraq. Last November, the U.N. Security Council passed a unanimous resolution finding Iraq in material breach of its obligations, and vowing serious consequences in the event Saddam Hussein did not fully and immediately comply. When Saddam Hussein failed even then to comply, our coalition acted to deliver those serious consequences. In that effort, the American military acted with speed and precision and skill. Once again, our men and women in uniform have served with honor, reflecting great credit on themselves and on the United States of America.

Unacceptable Risks

In the post-9/11 era, certain risks are unacceptable. The United States made our position clear: We could not accept the grave danger of Saddam Hussein and his terrorist allies turning weapons of mass destruction against us or our friends and allies. And, gradually, we are learning the details of his hidden weapons programs. This work is being carried out under the direction of Dr. David Kay, a respected scientist and former U.N. inspector who is leading the weapons search in Iraq.

Dr. Kay's team faces an enormous task. They have yet to examine more than a hundred large conventional weapons arsenals—some of which cover areas larger than 50 square miles. Finding comparatively small volumes of extremely deadly materials hidden in these vast stockpiles will be time-consuming and difficult. Yet, Dr. Kay and his team are making progress, and have compiled an interim report, portions of which were declassified last week. Let me read to you a couple of passages from Dr. Kay's testimony to Congress, which deserve closer attention.

He notes:

> Iraq's WMD programs spanned more than two decades, involved thousands of people, billions of dollars and were elaborately shielded by security and deception operations that continued even beyond the end of Operation Iraqi Freedom.

Dr. Kay further stated,

> We have discovered dozens of WMD-related program activities and significant amounts of equipment that Iraq concealed from the United Nations

during the inspections that began in late 2002. The discovery of these deliberate concealment efforts has come about both through the admissions of Iraqi scientists and officials concerning information they deliberately withheld, as well as through physical evidence of equipment and activities that the Iraq survey group has discovered [that] should have been declared to the United Nations.

Among the items Dr. Kay and his team have already identified are the following:

- a clandestine network of laboratories and safe houses within the Iraqi intelligence service that contained equipment suitable for continuing chemical and biological weapons research;
- a prison laboratory complex, possibly used in human testing of biological weapons agents, that Iraqi officials were explicitly ordered not to declare to the United Nations;
- reference strains of biological organisms, concealed in a scientist's home, one of which can be used to produce biological weapons;
- new research on BW [biological weapon]-applicable agents, Brucella and Congo Crimean Hemorrhagic Fever, and continuing work on ricin and aflatoxin, which has not been declared to the United Nations;
- documents and equipment hidden in scientists' homes that would have been useful in resuming uranium enrichment by centrifuge and electro-magnetic isotope separation;
- a line of unmanned aerial vehicles, not fully declared, and an admission that they had been tested out to a range of 500 kilometers—350 kilometers beyond the legal limit imposed by the U.N. after the Gulf War;
- plans and advanced design work for new long-range ballistic and cruise missiles with ranges capable of striking targets throughout the Middle East, which were prohibited by the U.N. and which Saddam sought to conceal from the U.N. weapons inspectors;
- clandestine attempts between late 1999 and 2002 to obtain from North Korea technology related to 1,300-kilometer range ballistic missiles, 300-kilometer range anti-ship cruise missiles and other prohibited military equipment.

Ladies and gentlemen, each and every one of these finding confirms a material breach by the former Iraqi regime of U.N. Security Council Resolution 1441. Taken together, they constitute a massive breach of that unanimously passed resolution and provide a compelling case for the use of force against Saddam Hussein.

Critics' Arguments

Even as more evidence is found of Saddam's weapons programs, critics of our action in Iraq continue to voice other objections. And the arguments they make are helping to frame the most important debate of the post-9/11 era.

Some claim we should not have acted because the threat from Saddam Hussein was not imminent. Yet, as the President has said, "Since when have terrorists and tyrants announced their intentions, politely putting us on notice before they strike?" I would remind the critics of the fundamental case the President has made since September 11th. Terrorist enemies of our country hope to strike us with the most lethal weapons known to man. And it would be reckless in the extreme to rule out action, and save our worries, until the day they strike. As the President told Congress earlier this year, if threats from terrorists and terror states are permitted to fully emerge, "all actions, all words and all recriminations would come too late."

That is the debate, that is the choice set before the American people. And as long as George W. Bush is President of the United States, this country will not permit gathering threats to become certain tragedies.

Critics of our national security policy have also argued that to confront a gathering threat is simply to stir up hostility. In the case of Saddam Hussein, his hostility to our country long predates 9/11, and America's war on terror. In the case of the al-Qaeda terrorists, their hostility has long been evidenced. And year after year, the terrorists only grew bolder in the absence of forceful response from America and other nations. Weakness and drift and vacillation in the face of danger invite attacks. Strength and resolve and decisive action defeat attacks before they can arrive on our soil.

Another criticism we hear is that the United States, when its security is threatened, may not act without unanimous international consent. Under this view, even in the face of a specific, stated, agreed-upon danger, the mere objection of even one foreign government would be sufficient to prevent us from acting. This view reflects a deep confusion about the requirements of our national security. Though often couched in high-sounding terms of unity and cooperation, it is a prescription for perpetual disunity and obstructionism. In practice, it would prevent our own country from acting with friends and allies, even in the most urgent circumstance.

To accept the view that action by America and our allies can be stopped by the objection of foreign governments that may not feel threatened, is to confer undue power on them, while leaving the rest of us powerless to act in our own defense. Yet we continue to hear this attitude in arguments in our own country. So often, and so conveniently, it amounts to a policy of doing exactly nothing.

In Afghanistan, in Iraq, on every front in the war on terror, the United States has cooperated with friends and allies, and with others who recognize the common threat we face. More than 50 countries are contributing to peace and stability in Iraq today—including most of the world's democracies—and more than 70 are with us in Afghanistan. The United States is committed to multilateral action wherever possible. Yet this commitment does not require us to stop everything, and neglect our own defense merely on the say-so of a single foreign government. Ultimately, America must be in charge of her own national security.

Choosing Action

This is the debate before the American people, and it is of more than academic interest. It comes down to a choice between action that assures our security and inaction that allows dangers to grow. And we can see the consequences of these choices in real events. The contrast is greatest on the ground in Iraq. Had the United States been constrained by the objections of some, the regime of Saddam Hussein would still rule Iraq, his statues would still stand, and his sons would still be running the secret police. Dissidents would still be in prison, the apparatus of torture and rape would still be in place, and the mass graves would be undiscovered. We must never forget the kind of man who ran that country, and the depravity of his regime.

Last month, Bernard Kerik, the former police commissioner of New York, returned from Iraq after spending four months helping to activate and stand up a new national police force. Bernie Kerik tells of many things he saw, including the videos of interrogations in which the victim is blown apart by a hand grenade. Another video, as he describes it shows: "Saddam sitting in an office, allowing two Doberman Pinschers to eat alive a general because he did not trust his loyalty."

Those who declined to support the liberation of Iraq would not deny the evil of Saddam Hussein's regime. They must concede, however, that had their own advice been followed, that regime would rule Iraq today.

President Bush declined the course of inaction, and the results are there for all to see. The torture chambers are empty, the prisons for children are closed, the murderers of innocents have been exposed, and their mass graves have been uncovered. The regime is gone, never to return. And despite difficulties we knew would occur, the Iraqi people prefer liberty and hope to tyranny and fear.

Building an Example of Freedom

Our coalition is helping them to build a secure, hopeful and self-governing nation which will stand as an example of freedom to all the Middle East. We are rebuilding more than a thousand schools, supplying and reopening hospitals, rehabilitating power plants, water and sanitation facilities, bridges and airports. We are training Iraqi police, border guards and a new army, so that the Iraqi people can assume full responsibility for their own security. Iraq now has its own Governing Council, has appointed interim government ministers, and is moving toward the drafting of a new constitution and free elections.

The contrast of visions is evident as well throughout the region. Had we followed the counsel of inaction, the Iraqi regime would still be a menace to its neighbors and a destabilizing force in the Middle East. Today, because we acted, Iraq stands to be a force for good in the Middle East.

Comparing both sides of the debate, we can see certain consequences for the world beyond the Middle East, consequences with direct implications for our own security. If Saddam Hussein were in power today there would still be active terror camps in Iraq, the regime would still be allowing terrorist leaders into the country, and this ally of terrorists would still have a hidden biological weapons program capable of producing deadly agents on short notice. There would be today, as there was six months ago, the prospect of the Iraqi dictator providing weapons of mass destruction, or the means to make them, to terrorists for the purpose of attacking America.

Today we do not face this prospect. There are terrorists in Iraq, yet there is no dictator to protect them, and we are dealing with them one by one. Terrorists have gathered in that country and there they will be defeated. We are fighting this evil in Iraq so we do not have to fight it on the streets of our own cities.

The current debate over America's national security policy is the most consequential since the early days of the Cold War and the emergence of a bipartisan commitment to face the evils of communism. All of us now look back with respect and gratitude on the great decisions that set America on the path to victory in the Cold War and kept us on that path through nine presidencies. I believe that one day, scholars and historians will look back on our time and pay tribute to our 43rd President, who has both called upon and exemplified the courage and perseverance of the American people. In this period of extraordinary danger, President Bush has made clear America's purposes in the world, and our determination to overcome the threats to our liberty and our lives.

Sometimes history presents clear and stark choices. We have come to such a moment. Those who bear the responsibility for making those choices for America must understand that while action will always carry cost, measured in effort and sacrifice, inaction carries heavy costs of its own. As in the years of the Cold War, much is asked of us and much rides on our actions. A watching world is depending on the United States of America. Only America has the might and the will to lead the world through a time of peril, toward greater security and peace. And as we've done before, we accept the great mission that history has given us.

 **NO**

Invasion of Iraq

It was the prophet Hosea who lamented of the ancient Israelites, "For they have sown the wind, and they shall reap the whirlwind."

I wonder if it will come to pass that the President's [George W. Bush's] flawed and dangerous doctrine of preemption on which the United States predicated its invasion of Iraq will some day come to be seen as a modern-day parable of Hosea's lament. Could it be that the Bush administration, in its disdain for the rest of the world, elected to sow the wind, and is now reaping the whirlwind?

I ponder this as the casualties in Iraq continue to mount, long past the end of major conflict, and as the vicious attacks against American troops, humanitarian workers, and coalition partners increase in both intensity and sophistication. I ponder this as the number of terrorists attacks bearing the hallmarks of al-Qaida appear to be increasing, not just in Iraq but elsewhere, including Saudi Arabia and, most recently, Turkey. I cannot help but wonder, as I view these developments with a sorrowful heart, what the President has wrought. By failing to win international support for the war in Iraq and by failing to plan effectively for an orderly post-war transition of power, has the President managed to create in Iraq the very situation he was trying to preempt?

The deaths of three more American soldiers in Iraq over the weekend, and the vicious mob attack on the bodies of two of them, are but the latest evidence of a plan gone tragically awry. The death toll of American military personnel in Iraq since the beginning of the war has now reached 427, and it continues to climb on a near-daily basis. Most troubling of all is the fact that more than two-thirds of those soldiers who have died in Iraq have been killed since the end of major combat operations. At that time, 138 American fighting men and women had died in Iraq, at the time major combat operations had ended. Instead of making headway in the effort to stabilize and democratize post-war Iraq, the administration seems to be losing ground. If the current violence cannot be curbed, if Iraq is allowed to descend unchecked into a holy hell of chaos and anarchy, the implications could be catastrophic for the region and the world.

An article earlier this month in the *Los Angeles Times*, entitled "Iraq Seen As Al Qaeda's Top Battlefield," raises the alarming specter that Iraq already is replacing Afghanistan as the global center of Islamic jihad. According to the

U.S. Congressional Record, November 25, 2003.

article, as many as 2,000 Muslim fighters from a number of countries, including Sudan, Algeria and Afghanistan, may now be operating in Iraq. No one knows the numbers for certain, but foreign Islamic terrorists are suspected in some of the deadliest attacks in Iraq, including the bombing of the United Nations headquarters and the Red Cross offices in Baghdad.

It seems only yesterday that the President and his advisers were warning the United Nations that Saddam Hussein must be disarmed at once, forcibly if necessary, to preempt Iraq from becoming the next front in the war on terrorism. On May 1, when the President announced the end of major combat operations in Iraq as he basked in the glow of a banner that was waving overhead proclaiming "Mission Accomplished," he described the liberation of Iraq as "a crucial advance in the campaign against terror."

What a difference a few months makes. Before the war, it was Afghanistan and al-Qaida, not Iraq, that constituted the central front in the war on terror. It was Osama bin Laden, not Saddam Hussein, who orchestrated the September 11 attacks on the United States, and it was Osama bin Laden, not Saddam Hussein, who orchestrated earlier attacks on the USS *Cole* and on the American embassies in Kenya and Tanzania. It is Osama bin Laden who continues to taunt the United States and who continues to plot against us, and it is Osama bin Laden who has exhorted his followers to gather in Iraq to avenge the U.S. invasion.

Today, while the Taliban appears to be regrouping in Afghanistan, it is Iraq that has become the most powerful magnet for Islamic terrorists. It is Iraq where these forces have coalesced with Saddam Hussein loyalists to create an increasingly sophisticated and deadly insurgency that has paralyzed U.S. efforts to establish postwar stability. Ironically, Saddam Hussein and his henchmen are more of a threat to the United States today than they were before the war began.

Could it be that the war on Iraq, while succeeding in chasing one monster into hiding, has created another, equally vicious, monster in his stead, a hydra-headed monster that is spewing terrorism against both the Iraqi people and their would-be liberators? Could it be that the convergence of Islamic jihadists and Baathist loyalists constitutes a more potent adversary than we ever imagined possible in Iraq?

Could it be, that instead of providing a "crucial advance" in the war on terrorism, as the President suggested, the war on Iraq has provided crucial new resources—money, weapons, and manpower, as well as motivation—for the terrorists themselves? Could it be that instead of curbing terrorism, the war on Iraq has served to fan the flames of terrorism?

If only the President had listened more closely to his father, and his father's advisers. In the 1998 book that he co-authored with former National Security Adviser Brent Scowcroft, *A World Transformed*, the first President Bush said of his decision to end the 1991 Gulf War without attempting to remove Saddam Hussein from power, "We would have been forced to occupy Baghdad and, in effect, rule Iraq. . . . there was no viable 'exit strategy' we could see, violating another of our principles."

The former President Bush and his national security adviser further cautioned that, "Going in and occupying Iraq, thus unilaterally exceeding the United Nations' mandate, would have destroyed the precedent of international response to aggression that we hoped to establish. Had we gone the invasion route, the United States could conceivably still be an occupying power in a bitterly hostile land. It would have been a dramatically different—and perhaps barren—outcome."

Clearly the situation in Iraq today is far more difficult and dangerous than the administration ever envisioned or prepared for before the war. Although the President declared an end to major combat operations more than six months ago, U.S. forces in Iraq have recently been forced to resort to a new bombing campaign in and around Baghdad—the most intense aerial offensive since active combat ended—in an effort to stem the insurgency. More than 6 months after the end of major combat operations, the situation in Iraq appears to be deteriorating, not improving.

While the President and his military advisers remain upbeat about Iraq, the top CIA [Central Intelligence Agency] official in Baghdad appears to have reached a far bleaker assessment of the situation on the ground. According to news reports, a top secret CIA analysis from Baghdad has concluded that growing numbers of Iraqi citizens are turning against the American occupation and supporting the insurgents. It may well have been this report that prompted the President to recall the U.S. administrator of the Coalition Provisional Authority to Washington two weeks ago for a hastily arranged round of meetings on accelerating the transition of power to an Iraqi provisional government.

Nothing could do more to spotlight the Administration's abysmal failure to rally international support for the stabilization and rebuilding of Iraq than this frantic scramble to arrange a Hail Mary pass of power from the United States to a provisional government in Iraq that does not yet exist. The Administration has slapped a new deadline on the democratization of Iraq—an Iraqi "transitional assembly" is to be in place by June 1—but it has come up with no blueprint as to how that assembly is to function or how it can be expected to stem the violence in Iraq.

Once again, the administration is ignoring the obvious—the United States cannot go it alone in Iraq. The United Nations and NATO need to be brought on board as full partners with a personal stake in the governance, the stabilization, and the future of Iraq.

Every day that the administration continues to spurn the United Nations is another day that the insurgents have to choreograph their attacks in Iraq and further isolate the United States from the rest of the world. The pattern is becoming chillingly clear. Systematic attacks, including those against the United Nations and the Red Cross headquarters in Baghdad and the Italian military police headquarters in Nasiriyah, have succeeded in driving most humanitarian workers from Iraq and have rocked the resolve of U.S. allies to support the Iraq operation. In the wake of the attack on the Italian troops, Japan is reconsidering its offer to send troops to Iraq, and South Korea continues to procrastinate. Help from other countries on which the United States had pinned its hopes, including Turkey and Pakistan, has evaporated.

Even in the streets of London, the seat of government of America's strongest ally, tens of thousands of demonstrators marched on Trafalgar Square last week to protest President Bush's state visit and his policies in Iraq.

Because of the administration's arrogance and impatience, the United States, for better or worse, is the make-or-break force in Iraq. Could it be that the President, in his haste to impose his will on the rest of the world, has inadvertently sown the wind and must now confront the whirlwind?

Mr. President, in a short time—perhaps the next day or so—the Senate will adjourn for the year. We are privileged and blessed to return to the comfort of our families for the holidays. Not all families in America will share in our blessings.

Many families will wait out the holidays in fear and tension as they worry about their loved ones in Iraq and Afghanistan.

We in the Senate will not be here to absorb the news from the battle fronts in Iraq and Afghanistan or to voice our response to these developments. I pray that all will be calm, that "Silent Night, Holy Night" will be more than the strain of a familiar carol. But I worry it will not be so, that reality will be harsher than sentimentality.

The war in Iraq is far from over. When we will ultimately be able to declare victory, I do not know and I dare not venture a guess. I only hope that the President will be able to put the good of the Nation over the pride of his administration and accept a helping hand from the United Nations to turn the tide of anarchy in Iraq. Perhaps he may finally be ready to do so. Senior administration officials have been quoted as suggesting that the United States is preparing to seek another U.N. resolution endorsing a new plan for the transition of power in Iraq. I urge the President to do so without delay. This time around, the effort must be genuine, and the resolution must be meaningful.

The facts are stark and hard to accept. If not outright losing, the United States is far from winning the peace in Iraq. Only a significant turnabout in the handling of the security and reconstruction effort, centered on giving the United Nations a leading role in the transition of power, holds any hope for a constructive course change in Iraq. It is a course change that is desperately needed.

As the crisis in Iraq deepens, leadership and statesmanship are urgently needed. I pray that the President, in his desperate quest for a new solution to the chaos in Iraq, will demonstrate those qualities, abandon the U.S. stranglehold on Baghdad, and forge a meaningful partnership with other nations of the world, a partnership with the United Nations so that a swift, orderly, and effective transition of power in Iraq can be achieved and American fighting men and women can come home.

POSTSCRIPT

Was War with Iraq Justified?

Most Americans supported the war in March 2003. Just days before the war, a poll found 67 percent of Americans approved "of the United States taking military action against Iraq to try to remove Saddam Hussein from power," while only 29 percent disapproved, and 4 percent were unsure. Another poll found 54 percent of Americans favoring action even if the UN refused to authorize it, with 40 percent disapproving such a move, and 6 percent unsure. Of course, at that time, most Americans believed that Iraq had weapons of mass destruction (WMDs) and that it was supporting terrorism.

Yet no evidence of WMDs was subsequently uncovered, and there was little or no indication that Iraq played an important role in aiding and abetting terrorism. This reality and the ongoing cost in lives and treasure undermined American opinion against the war, but only marginally so. A December 2004 poll that asked, "Do you approve or disapprove of the United States' decision to go to war with Iraq in March 2003?" recorded 51 percent disapproving, but 48 percent still approved, and 1 percent unsure. This finding, and, to a degree, President Bush's reelection in 2004 are indications that while American support of the war had softened considerably, it had not collapsed.

The question of whether the war with Iraq was justifiable or not contains a host of important issues that relate to the road to war. That path is detailed in Todd S. Purdum, *A Time of Our Choosing: America's War in Iraq* (Times Books, 2004).

One issue relates to the use of force without UN consent. This topic is taken up in Issue 14 of this volume. A second issue, which is at the heart of this debate, is whether the U.S. invasion was justified even if some of the beliefs in March 2003 were wrong. There is no conclusive proof either way about whether President Bush and Prime Minister Blair knew then that Iraq did not have WMDs. Various intelligence failures and disputes about what subordinates told the leaders may forever cloud that question. If the two leaders did know that there were no WMDs and deliberately misled the public, then they disserve severe blame. But what if they believed there were WMDs, even though that turned out to be wrong? Certainly, Saddam Hussein was a brutal leader and had waged aggressive war against Kuwait. His forces had earlier possessed chemical weapons and had used them against Kurdish rebels and against Iranian troops externally during the two countries' war (1980–1988). Additionally, Iraq in 2003 had a 12-year record of obstructing UN inspections and had even expelled the inspectors completely for four years (1998–2002). Thus, it was not far-fetched to believe that Saddam Hussein had or was trying to obtain WMDs and might one day use them again. Some have also argued that toppling Saddam Hussein was the right war for the wrong reason—that the war was justified by the removal of a savage dictator and the freeing of Iraqis from often gruesome oppression, even if no WMDs or direct links to aiding terrorism were uncovered. One side of this debate is advanced by Paul Savoy in "The Moral Case Against the Iraq War," *The Nation* (May 31, 2004). Taking the opposite view, the editors of the *The New Republic* argue in "Were We Wrong?" (June 28, 2004) that while the "strategic rationale for war has collapsed," the "moral one has not." Similarly, the faulty U.S. performance in organizing and conducting the occupation and transition back to Iraqi control of the country is not necessarily an indictment of the decision to act. For more on the post-war problems, see Larry Diamond, "What Went Wrong in Iraq," *Foreign Affairs* (September 2004).

ISSUE 11

Are Strict Sanctions on Cuba Warranted?

YES: Commission for Assistance to a Free Cuba, from "Hastening Cuba's Transition," *Report to the President: 2004* (May 6, 2004)

NO: William Ratliff, from "The U.S. Embargo Against Cuba Is an Abysmal Failure. Let's End It," *Hoover Digest* (Winter 2004)

ISSUE SUMMARY

YES: The Commission for Assistance to a Free Cuba, which President George W. Bush established on October 10, 2003, and charged with making recommendations about how to hasten a transition to democracy in Cuba, argues in its report to the president that the U.S. government should take stronger measures to undermine the Castro regime and to promote conditions that will help the Cuban people hasten the end of President Fidel Castro's dictatorial regime.

NO: William Ratliff, a research fellow at the Hoover Institution, argues that sanctions on Cuba only hurt the Cuban people because nothing the United States is doing today contributes significantly to the achievement of any change in the Castro regime.

U.S. concern for Cuba goes back to the origins of the Monroe Doctrine of 1823. President James Monroe declared that the Western Hemisphere was not subject to "further colonization" and that any attempt by a power outside the hemisphere aimed at "oppressing . . . or controlling by any other manner" part of the so-called New World would be viewed as "the manifestation of an unfriendly disposition toward the United States." The Monroe Doctrine created in American minds the notion that they have the right to exercise some special authority over the hemisphere, especially Central America and the Caribbean. From early on, Cuba was a central focus of this self-declared sphere of influence. For example, President James K. Polk considered trying to purchase Cuba from Spain for $100 million in 1848.

The tensions over slavery and the ensuring Civil War precluded any attempt to acquire Cuba, but the island remained an issue. The outbreak of an independence movement in Cuba in 1870 spurred more American interest. Americans sympathized with the Cuban revolutionaries. Lurid stories about the treatment of Cubans by Spanish authorities filled the American press, and hostility toward Spain reached a fever pitch after the U.S. battleship *Maine* blew up in Havana's Harbor and the press accused (incorrectly) the Spanish of attaching a mine to it. The ensuing Spanish-American War of 1898 was a lopsided U.S. victory, and Cuba and Puerto Rico in the Caribbean, as well as the Philippines and Guam in the Pacific, came under U.S. control. In 1904, President Theodore Roosevelt declared the Roosevelt Corollary to the Monroe Doctrine, which asserted a U.S. right to intervene in the affairs of other countries in the Western Hemisphere to stop actions that Washington deemed unacceptable. American intervened repeatedly in the region, including occupying Cuba (1895–1922), among other countries.

Cuba was frequently controlled by dictators, and they were supported or tolerated by Washington as long as they did not contravene U.S. interests. This situation changed in 1959 when rebels led by Fidel Castro toppled right-wing dictator Fulgencio Batista. In the atmosphere of the Cold War, Americans were alarmed by Castro's leftist leanings, and tensions grew worse when Castro aligned Cuba with the Soviet Union. A U.S.-sponsored but ill-supported attempt of expatriate Cubans to land at the Bay of Pigs in 1961 and topple Castro was an utter failure. The following year, the U.S.-U.S.S.R. confrontation over the placement of Soviet nuclear weapon–armed missiles in Cuba was resolved when the Soviets withdrew the missiles and Washington secretly pledged not to use military force to try to overthrow Castro.

This left the United States unable to try to remove Castro by force, so it continued a policy of stringent economic sanctions against Cuba. During the 1990s, several factors worked to ease tension. The end of the Cold War did away with images of Cuba as part of a global communist threat. Also, President Bill Clinton was less antagonistic toward the Castro government than had been his immediate predecessors. Castro further helped relations by pledging in 1992 to no longer provide weapons or advisors to leftist revolutionary movements in the hemisphere.

Yet these moderating influences were offset by other factors. The Castro government remained staunchly communist and undemocratic. Periodic floods of refugees from Cuba brought protests from Americans, particularly those in Florida. Given the state's importance in the electoral college, U.S. presidents were wary of appearing "soft" on Castro, and this sentiment was furthered by the voting importance of the ardently anti-Castro Cuban-American community in Florida. An incident in 1998 in which Cuban fighters shot down a small American plane from which Cuban Americans were dropping anti-Castro leaflets over Cuba sparked Congress to pass the Helms-Burton Act that strengthened sanctions even further.

During the first years of his presidency, George W. Bush resisted pressures to ease the sanctions on Cuba by various groups, including business groups that want to trade with and invest in Cuba. Indeed, Bush appointed the Commission for Assistance to a Free Cuba, whose report is featured in the first reading to investigate how to speed the end of the Castro regime. The commission recommended even stronger sanctions, an approach opposed by William Ratliff in the second reading.

 **YES**

Hastening Cuba's Transition

Fidel Castro continues to maintain one of the world's most repressive regimes. As a result of Castro's 45-year strategy of co-opting or crushing independent actors, Cuban civil society is weak and divided, its development impeded by the comprehensive and continuous repression of the Castro regime.

Yet despite decades of suppression, degradation, and deprivation, the aspiration for change is gathering momentum and growing in visibility on the island. Brave Cubans continue to defy the regime and insist that it recognize their fundamental rights, as guaranteed by the Universal Declaration on Human Rights, which Cuba signed, but to which Cuba now outlaws any reference. The March-April 2003 crackdown on peaceful opposition activists was only the most recent and brutal high-profile effort by the regime to eliminate democratic civil society. While these actions set back the consolidation of that movement, they did not end the Cuban people's quest for freedom.

The Castro regime continues to be a threat not only to its own people, but also to regional stability, the consolidation of democracy and market economies in the Western Hemisphere, and the people of the United States. The Castro regime harbors dozens of fugitives from U.S. justice, including those convicted of killing law enforcement officials. It aggressively conducts espionage against the United States, including having operated a spy network, one of whose members was convicted of conspiring to kill U.S. citizens. The Castro regime also has engaged in other hostile acts against its neighbors and other democracies in the Hemisphere. On several occasions, Castro has threatened and orchestrated mass sea-borne migrations to Florida of tens of thousands of Cubans in an effort to intimidate and harm the United States.

This dictatorship has every intention of continuing its stranglehold on power in Cuba and is pursuing every means at its disposal to survive and perpetuate itself, regardless of the cost to the Cuban people. In furtherance of this goal, the regime ruthlessly implements a strategy to maintain the core elements of the existing political and repressive structure to ensure that leadership passes from Fidel Castro to his selected successor, Raul Castro. Under this "succession strategy," the core governmental and Communist Party elite would survive the departure of Fidel Castro and would seek to effect a new relationship with the United States without undergoing fundamental political and economic

Report to the President: 2004, May 6, 2004.

reform. An element that is critical to the success of the regime's strategy is its repressive security apparatus, which instills fear in the Cuban people and uses their impoverishment as a means of control.

This strategy cannot succeed without the continued flow of resources to the regime from outside Cuba. To this end, the Castro regime has built an economic structure on the island designed specifically to exploit all outside engagement with Cuba. One of the regime's central goals is to obtain additional sources of income from the United States, especially through tourism receipts. Overall, these efforts annually subsidize the regime in the amount of more than $3 billion in gross revenues. Specifically, the tourism sector has been developed to generate hard currency as well as to contribute to an image of "normalcy" on the island and to promote international acceptance of the regime.

The Castro regime also cynically exploits U.S. humanitarian and immigration policies, primarily remittances and "family visits," to generate millions in hard currency flows from its victims: those seeking freedom and the Cuban diaspora. Further, Cuba maintains a beneficial arrangement with the sympathetic government of Hugo Chavez in Venezuela, whereby Castro receives up to 82,000 barrels of oil per day on preferential terms; this arrangement nets more than $800 million in annual savings to Cuba (mirrored by an identical amount of lost revenues to Venezuela). Cuba continues to exploit joint economic ventures with third-country investors, who enter these arrangements despite the absence of the rule of law or neutral dispute resolution mechanisms and despite Cuba's lack of respect for basic labor rights. Under these ventures, international employers pay hard currency to the Castro regime for each Cuban worker, who is in turn paid in worthless Cuban pesos.

Another facet of the regime's survival strategy is to control information entering, circulating within, and coming from the island. The regime seeks to minimize the information available to the Cuban people, as well as to manipulate what the outside world knows about the Castro dictatorship and the plight of the average Cuban citizen.

Cuba presents itself internationally as a prime tourist destination, as a center for bio-technological innovation, as a successful socialist state that has improved the standard of living of its people, and as a model for the world in terms of health, education, and race relations. This image belies the true state of Cuba's political, economic and social conditions and the increasingly erratic behavior of its leadership.

Despite the aggressive internal and international propaganda effort by the regime, there is a growing international consensus on the need for change in Cuba. This consensus has been strengthened by the regime's March-April 2003 suppression of peaceful pro-democracy activists, the summary executions of three Afro-Cubans attempting to flee the island, and the courageous effort by many peaceful activists to continue to reach out to the Cuban people and the international community. This flagrant repression, along with the continued work of pro-democracy groups in Cuba, has caused the international community to again take stock of the Castro regime and condemn its methods.

This re-evaluation provides an opportune moment to strengthen an evolving international consensus for democratic change in Cuba. America's commitment to support the Cuban people against Castro's tyranny is part of our larger commitment to the expansion of freedom. In furtherance of this commitment, President George W. Bush mandated that the Commission for Assistance to a Free Cuba identify additional means by which the United States can help the Cuban people bring about an expeditious end to the Castro dictatorship.

This comprehensive framework is composed of six inter-related tasks considered central to hastening change. . .

Empower Cuban Civil Society

The Castro dictatorship has been able to maintain its repressive grip on the Cuban people by intimidating civil society and preventing the emergence of a credible alternative to its failed policies. . . . Now, the tide of public opinion has turned and Castro's loyalists must constantly work to restrain the Cuban people from organizing and expressing demands for change and freedom. . . . Cuban civil society is not lacking spirit, desire, or determination; it is hampered by a lack of materials and support needed to bring about these changes. [The Commission report goes on to recommend a series of efforts to support dissident individuals, groups and organizations in Cuba and of Cuban expatriates]. . . .

Break the Information Blockade

The Castro regime controls all formal means of mass media and communication on the island. Strict editorial control over newspapers, television, and radio by the regime's repressive apparatus prevents the Cuban people from obtaining accurate information on such issues as the Cuban economy and wide-scale and systematic violations of human rights and abridgement of fundamental freedoms. It also limits the ability of pro-democracy groups and civil society to effectively communicate their message to the Cuban people. . . . This blockade on information must be broken in order to increase the availability to the Cuban people of reliable information on events in Cuba. [The Commission report goes on to recommend a series of efforts to facilitate the flow of outside information and viewpoints into Cuba]. . . .

Deny Revenues to the Cuban Dictatorship

A. Undermine Regime-Sustaining Tourism

"Flooding the island with tourists" is part of the Castro regime's strategy for survival. Since 1992, it has been aggressively developing and marketing a tourism infrastructure, including a cynically-orchestrated campaign to make Cuba attractive to U.S. travelers. The estimated annual total number of international travelers is in the 1.8 to 2 million range. Of this global figure, some 160,000 to 200,000 legal and illegal travelers have come from the United States on an annual basis over the past decade. Since the October 10, 2003 implementation of increased U.S. enforcement efforts, there has been a decrease in the number of U.S. travelers, reducing the total to about 160,000.

The regime has a target of hosting 7.5 million international tourists by 2010 and 10 million by 2025. Currently, tourism is Cuba's largest single source of revenue, generating some $1.8–$2.2 billion in annual gross revenues. Of this amount, it is estimated that the regime nets 20 percent, although its take may be greater given the Cuban regime's routine failure to pay creditors or honor contracts with foreign investors. . . .

Tourism & travel-related exports   Central to the marketing of Cuba as a tourist destination, including to U.S. nationals traveling for licensed activities, is its "sand, sun, rum, and cigars" image. Closely related to tourism is the marketing and export of alcohol and tobacco products, a significant revenue-generating activity. Under current U.S. Department of Treasury regulations, licensed U.S. travelers to Cuba can import up to $100 worth of Cuban goods as accompanied baggage. In practice, these goods are almost exclusively Cuban rum and tobacco. In 2003, such imports by licensed travelers could have generated as much as $20 million in revenues to the regime through sales. . . . Revenue gained from the sale of Cuban state-controlled commodities and products strengthens the regime. . . .

Educational travel   Under current regulations, accredited academic institutions receive specific licenses, usually valid for up to two years, to permit students to travel to Cuba for certain educational activities. . . . In practice, while there are well-meaning participants who use this license category as intended, other travelers and academic institutions regularly abuse this license category and engage in a form of disguised tourism. . . . Moreover, the regime has often used the visits by U.S. education groups to cultivate the appearance of international legitimacy and openness to the exchange of ideas. Requiring that educational licenses be granted only to programs engaged in full-semester study in Cuba would support U.S. goals of promoting the exchange of U.S. values and norms in Cuba, would foster genuine academic study in Cuba, and would be less prone to abuse than the current regulations. . . .

Recommendations   Continue to strengthen enforcement of travel restrictions to ensure that permitted travel is not abused and used as cover for tourism, illegal business travel, or to evade restrictions on carrying cash into Cuba. . . [Among other changes:]

- Eliminate the regulatory provision allowing for the import of $100 worth of Cuban goods produced by Cuban state entities, including cigars and rum, as accompanied baggage. . . .
- Eliminate abuses of educational travel by limiting educational travel to only undergraduate or graduate degree granting institutions and only for full-semester study programs, or for shorter duration only when the program directly supports U.S. policy goals. . . .
- Eliminate the general license provision for amateur or semi-professional athletic teams to travel to Cuba to engage in competitions and require that all such travel be specifically licensed.
- Eliminate the specific license provision for travel related to clinics and workshops in Cuba. . . .

B. Limit the Regime's Manipulation of Humanitarian U.S. Policies

To alleviate the hardships of a portion of the Cuban population, the United States has implemented various measures by which those with family members in Cuba can send cash remittances to them; travel to Cuba carrying gifts; and ship "gift parcels." Castro has exploited these policies by effectively shifting burdens that ought to be assumed by the Cuban state and by profiting enormously from these transactions. Not only has he benefited from the pacifying effects of these humanitarian outreaches within the population—relying on the exile community to provide the Cuban people what he refuses to—but he attaches high fees to the various transactions involved. Whether sending remittances or care packages or traveling to the island, the costs to the exile community far exceed market rates and translate into a significant cash windfall to the regime. The Commission found that more than $1 billion annually in funds and goods are sent to Cuba from those living outside the island. . . .

Recognizing the humanitarian need in Cuba as a basis for U.S. policies on remittances, gift parcels, and family travel, the Commission recommends a tightening of current policies to decrease the flow of resources to the regime. [Among recommended actions:]

- Permit individuals to send remittances only to immediate family (grandparents, grandchildren, parents, siblings, spouses, and children) in Cuba.
- Limit gift parcels to medicines, medical supplies and devices, receive-only radios, and batteries, not to exceed $200 total value, and food (unlimited in dollar amount)
- Limit gift parcels to one per month per household, except for gift parcels exclusively containing food, rather than the current policy of allowing one gift parcel per month per individual recipient.
- Limit family visits to Cuba to one trip every three years; . . . [l]imit the definition of "family" for the purposes of family visits to immediate family (including grandparents, grandchildren, parents, siblings, spouses, and children); [and l]imit the length of stay in Cuba for family visitation to 14 days. . .

C. Deny Other Sources of Revenue to the Regime Foreign Investment in Cuba

Starting in the early 1990s as part of its effort to replace lost Soviet subsidies, the Castro regime has pursued an aggressive effort to attract third-country investors for joint ventures. A number of these ventures involve properties expropriated by the regime without adequate and effective compensation. The Castro regime continues to promote foreign investment opportunities in Cuba, including in confiscated properties, claims to which are owned by U.S. nationals. An unfavorable investment climate, a hostile Cuban bureaucracy, and unwieldy and frequently changing laws have limited the levels and types of foreign investment in recent years. In 2003, for example, the number of foreign joint ventures in Cuba dropped by 15 percent, the most notable decline in recent years and a sign of the Castro regime's failed economic and investment policies. However, it continues to actively seek foreign investment, especially from Europe, Canada, and Latin America, in its drive to reap more hard currency.

The Cuban Liberty and Democratic Solidarity (LIBERTAD) Act provides measures to discourage foreign investments in Cuba that involve confiscated property, claims to which are owned by U.S. nationals. Implementation of these laws must address the legitimate desire of U.S. citizens to seek redress for the confiscation of their property, an objective consistent with efforts to implement a comprehensive strategy to deny hard currency to the Castro regime. For those U.S. nationals who hold an ownership claim to property wrongfully confiscated by the Castro government, . . . the LIBERTAD Act provides that U.S. national the right to bring an action in U.S. federal court against foreign nationals benefiting from that property. Furthermore, . . . [the act] provides authority to impose visa sanctions against foreign nationals who benefit from properties wrongfully confiscated when a claim to the property is owned by a U.S. national. The U.S. Government should seek to deter investment in Cuba by devoting additional personnel and resources to implementation and enforcement.

Cuban government front companies   The Cuban government is believed to operate a number of front companies in the United States, Latin America, and Europe that are used to circumvent travel and trade restrictions and to generate additional hard currency. These front companies are believed to be involved in efforts to encourage illegal tourism and the sending of illegal remittances and gift parcels to Cuba, as well as helping the regime acquire high-end computer equipment and other sensitive technologies. These front companies provide another source of currency and technological information for the regime and function as a base for economic espionage.

Venezuelan oil   Cuba maintains an extremely favorable oil arrangement with Venezuelan President Hugo Chavez, whereby up to 82,000 barrels of oil per day is received on preferential terms, a portion of which is then sold on the spot market. This arrangement nets more than $800 million in annual savings to Cuba, mirrored by an identical amount of lost revenues to Venezuela. In exchange, according to his own accounts, Castro has provided Chavez with an army of up to 12,000 Cubans, including doctors, medical personnel, and other technicians to bolster Chavez's popularity with the poorer segments of Venezuelan society, as well as more senior political and military advisors to help Chavez strengthen his authoritarian grip on the nation. Reports from Venezuela also indicate that Cuban doctors are engaging in overt political activities to boost Chavez's popularity. Cheap Venezuelan oil is vital to keeping the Cuban economy functioning, generates additional hard currency, and enables Cuba to postpone much needed economic reforms.

Recommendations   [Among the commissions recommendations are:]

- To deter foreign investment in Cuba in confiscated properties, claims to which are owned by U.S. nationals, aggressively pursue sanctions against those foreign nationals trafficking in (e.g., using or benefiting from) such property. . .

- Neutralize Cuban government front companies by establishing a Cuban Asset Targeting Group. . . . [to] work to identify and close Cuban government front companies.

Illuminate the Reality of Castro's Cuba

The current survival of the regime is in part dependent upon its carefully crafted international image. Cuba presents itself internationally as a prime tourist destination, as a center for bio-technological innovation, and as a successful socialist state that has improved the standard of living of its people. This image belies the true state of Cuba's economic and social conditions and the increasingly erratic behavior of its leadership. [The Commission goes on to recommend a series of efforts to counter Cuba's public diplomacy effort.]

Encourage International Diplomatic Efforts to Support Cuban Civil Society and to Challenge the Castro Regime

There is a growing international consensus on the nature of the Castro regime and the need for change. This consensus was brought about, in part, by the March–April 2003 crackdown on peaceful pro-democracy activists, the valiant effort by these same activists to continue to reach out to the Cuban people and the international community, and Castro's political attacks against the European Union and other nations. Many of those who once stood by Castro have now begun to speak out publicly against the regime's abuses. This same international consensus has limits. All too frequently, moral outrage and international condemnation have not translated into real actions that directly assist the Cuban people in their quest for freedom and basic human rights. On the positive side, the European Union and its member states have denounced the Cuban regime, curtailed assistance and, in some cases, stepped up contacts with Cuban dissidents. However, much more needs to be done. Encouraging international solidarity, challenging the regime in international organizations where appropriate, and strengthening international proactive support for pro-democracy groups in Cuba form a cornerstone of our policy to hasten an end to the Cuban regime and the transition to a free Cuba. [The Commission goes on to recommend a series of efforts to support this goal.]

Undermine the Regime's Succession Strategy

Approaching his 78th birthday, Fidel Castro has conspicuously deteriorated physically and probably mentally as well. His decline over the last few years has been apparent on several occasions at public events in Cuba and abroad. The quality of Castro's decision-making has declined along with his physical condition. . . . The senior Cuban leadership is now faced with the reality that Fidel Castro's physical end is at hand and is making preparations to manage a "succession" of the regime that will keep the senior leadership in power. The regime's survival strategy is to maintain the core elements of the existing political structure in passing eventual leadership of the country from Fidel to Raul Castro and others currently in the senior leadership. U.S. policy must be targeted at undermining this succession strategy by stripping away layers of support within the regime, creating uncertainty regarding the political and legal future of those in leadership positions, and encouraging more of those within the ruling elite to shift their allegiance to those pro-democracy forces working for a transition to a free and democratic Cuba. To these ends, attention and pressure must be focused on the ruling elite so that succession by this elite or any individual is seen as what it would be: an impediment to a democratic and free Cuba. Targeting current regime officials for U.S. visa denials is one instrument available to the United States to hold them accountable for human rights abuses against the Cuban people and others, including the torture by Castro regime officials of American POWs in South East Asia, or for providing assistance to fugitives in Cuba from U.S. justice. [The Commission goes on to make a series of efforts to undermine any Castro-connected succession regime.]

 **NO**

The U.S. Embargo Against Cuba Is an Abysmal Failure. Let's End It

More and more Americans are asking how the United States can come up with a realistic policy toward Cuba during the remainder of Fidel Castro's seemingly interminable lifetime. [Castro has led Cuba since 1959.] The underlying problem with our policy today is that while its objectives are desirable, they cannot be realized with the resources we are willing to commit. That means that no matter how much we and the Cuban people want Castro out, and no matter how much we support democracy, human rights, market reform, and a peaceful transition, in reality nothing we are doing today contributes significantly to the achievement of any of these outcomes. In fact, since the end of the Cold War our policy has been about as effective as our earlier comical efforts to make the dictator's beard fall out. Our policy is a plodding example of how a legitimate lobby (militant Cuban-Americans, especially in Miami) and its political target (U.S. political parties seeking money and votes, especially in Florida) can to some degree reach their own goals at the expense of the general populations of two countries.

Decades of Soviet and U.S. policies toward Cuba have shown that Castro is his own player and will not change his game more than temporarily to suit even a big foreign bidder. Therefore, unless we are willing to force regime change violently, we will just have to endure Castro until he dies. That is what the vast majority of the Cuban people have decided to do. No one likes such a passive policy, but the alternatives, such as what we have been doing or greater intervention, are worse. With this in mind, rather than seeking to heighten tensions across the strait, as we do now, we should try to reduce conflict in order to make life a little easier for the Cuban people in the interim and actually improve the prospects of a more peaceful transition in the post-Castro period. The easiest way to move in this direction is to increase personal and other contacts between the two countries. At very long last, in late 2003 a substantial majority of legislators from both parties in both houses of Congress voted to do just that by easing travel restrictions. But, alas, a clique of Republican legislators in conference committee, contemptuous of the popular will, struck out the provision, which the president had vowed to veto in any event. This is the kind of political skullduggery that gives the political profession its bad image. The good news, however, is that the majority of the American people and their legislators now support change, and in time they undoubtedly will rout the minority.

From the beginning, we must recognize three things: (1) We still do have legitimate concerns about what Castro is up to; (2) our misguided policy is bipartisan; and (3) the embargo causes us and the Cuban people more problems than it does Castro. Beginning with number one, the issues that warrant attention include whether Castro is obstructing the war on terrorism, involved in the trans-shipment of drugs and money-laundering, developing a germ warfare capacity, and nurturing anti-Americanism across Latin America. To the extent that he is doing some or all of these things, we would be able to monitor and counter them better with more rather than fewer people of all sorts on the island. An embargo targeting the entire country is a blunt instrument against the possible activities of a tiny elite. We should instead bomb Havana with Big Macs, that is, increase the American presence there in every possible way.

Second, our current policy is bipartisan. The most militant spokesman for the embargo today is President George W. Bush, but during the 1990s it was President Bill Clinton. In 1992 it was presidential candidate Clinton who got our post–Cold War policy going in the wrong direction by supporting the legislation of former New Jersey senator Robert Torricelli that tightened the embargo, a policy then-president George H.W. Bush had wisely opposed. And in 1998–99, toward the end of his time in office, it was Clinton again who killed the proposed Presidential Bipartisan Commission on Cuba. The idea of a commission to review U.S. policy was conceived by former undersecretary William Rogers and endorsed by former secretaries of state Henry Kissinger, George Shultz, and Lawrence Eagleburger, along with many members of Congress and others.

Finally, since the end of the Cold War the embargo has been largely a vote-getting, feel-good stunt rather than a serious international policy. It has not caused Cuba's main political, economic, or other problems. Cuba's problems are the entirely predictable consequences of Castro's own flawed and self-serving decisions. Today almost every country in the world has political and economic relations with Cuba, but Cuba produces almost nothing anyone abroad wants to buy and has very little money to buy what others produce. Ironically, according to polls conducted in the Miami area by Florida International University (FIU), only slightly more than 25 percent of Cuban-Americans themselves think the embargo really works, though a majority wants to continue it nonetheless. A character in Cuban novelist Pedro Juan Gutierrez's *Dirty Havana Trilogy* remarks that she is "pained to witness so much poverty and so much political posturing to disguise it." One should only add that the embargo is also sustained by massive moral posturing and that although Gutierrez was speaking of posturing in Cuba, this same quality pervades support for sanctions in the United States. To the degree that the embargo does have a real impact, however, it is mainly negative on the lives of ordinary Cuban people and on our international stature.

Why the Embargo Doesn't Work

In the post–Cold War period, the embargo has become a strategic liability. Even though most embargo supporters say the sanctions are intended to promote a peaceful transition, in fact they heighten tensions and hardships for the Cuban people and exacerbate differences among Cubans in Cuba and with those in exile. Our policy encourages conflict and instability and, to the degree that it succeeds, could promote a violent uprising or even civil war. (A few embargo supporters quietly admit they want war to clean out the flotsam and jetsam of the Castro period.) But if the democratic forces do rise up against Castro, which is unlikely in the foreseeable future, and are in danger of being crushed by Castro's repressive apparatus, which is highly probable, the U.S. government would be under heavy pressure to intervene militarily to rescue the reformers. FIU polls show 61 percent of Cuban-Americans want U.S. military intervention, as do a few militant gringo politicians. Most Americans, however, emphatically including officers of the U.S. armed forces I have talked to at many levels, wisely oppose this course of action.

The Helms-Burton Law, the heart of our current policy, is destructive of U.S. interests because it makes demands on the current *and future* governments of Cuba that are imperialistic, logically inconsistent, and counterproductive. [Helms-Burton is the common name for the Cuban Liberty and Democratic Solidarity Act of 1996, which imposed stringent limits on interactions with Cuba and which was sponsored by Senators Jesse Helms (R-NC) and Representative Dan Burton (R-IN).] Helms-Burton is a new Platt Amendment, 100 years after the first. Mark Falcoff of the American Enterprise Institute has noted correctly that the original Platt Amendment—the 1901 agreement that in important respects made Cuba a U.S. dependency—planted "seeds of a long smoldering resentment" in Cuba and that Helms-Burton is doing the same.

Cuba is no military or economic threat to the United States, but our policy there is destructive of what Joseph Nye has called our "soft power" worldwide, that is, the international goodwill that persuades other nations to cooperate with us in matters of real importance, such as the war on terrorism. International polls, confirmed by my recent personal experiences on four continents, tell us that America's image around the world today is low and that hostility toward us is rising. Much of the hostility results from what foreigners consider our bullying of other nations, exemplified in particular by Bill Clinton's bombing of Yugoslavia in 1999 and George W. Bush's invasion of Iraq in 2003. Thus although Cuba is not a major cause of our current reputation, it fits what critics consider to be our objectionable and dangerous style. A unilateral lifting of the embargo should relax tensions on the island and improve our image, especially in Latin America, where regard for the United States is plummeting.

Some embargo supporters say lifting the embargo now would reward Castro for his stubbornness and brutal repression in early 2003. No. We must announce that Castro simply has been dumped from our "most wanted" to our "least relevant" list. His international admirers would shout *"Viva Fidel! Hasta la Victoria Siempre,"* ["to victory always"] but the rest, or at least those few who notice, would just say, "Good Lord, it's about time."

On the other hand, the embargo feeds Castro's soft power. It gives him a scapegoat for his repression and economic failures, and it also refurbishes his image as the enduring scourge of American imperialism, the central drive and purpose of his entire career. If we take this scapegoat from him, it will be a major step toward killing him while he is still alive. One of the tragic and bitter ironies of the embargo is that those who most hate Castro are the ones who most relentlessly and effectively feed his insatiable ego and sustain his international image.

Many embargo supporters note that the dollars Cuba draws from investments, trade, and tourism support repression and therefore should be kept from the island. Yes, some do, but many also pass through the hands and lives of millions of Cubans, giving them hope for survival in a destitute economy. If that isn't so, why do so many Cubans dig so hard to find any kind of freelance work to earn a few greenbacks? Even Cuban-Americans know this is true when they set their passions aside for a moment, which is why they are the main ones who shatter the spirit of the embargo by using a legal loophole that allows them to send remittances to family and friends on the island. Cuban-Americans now send an estimated $1 billion per year to Cuba. All of these dollars, the island's main source of foreign currency, also funnel into Castro's hands, and many support the repressive apparatus. Therefore a really honest embargo would forbid the sending of *any* dollars to Cuba. Cutting *all* dollars to the island would reduce almost everyone to serious poverty and, more than anything we can do short of an invasion, might indeed spark an uprising against Castro. In my judgment, the remittances are humanitarian acts that should continue, but they shatter the logic of the sanctions regime. One important result of the general tourism industry is that it makes at least some dollars available to many Cubans who have no family abroad, thus reducing somewhat the increasingly explosive divide between classes of Cubans, namely those who have dollars and those who do not.

Finally, has the embargo improved human rights? Embargo supporters say they champion human rights, thus trying to seize the moral high ground in the debate over the embargo. In fact there is more evidence to the contrary. Embargo supporters say the arrests, trials, and executions of March and April 2003, which I witnessed firsthand while leading educational tours (now banned) of the island, prove that the moderate engagement line of the Europeans has failed. But if anything the repression proves that *our* policy of assisting the dissidents has backfired. The arrests and imprisonments were a severe blow to the pro-democracy movement. For almost 45 years Castro has shown that neither threats nor enticements, from the former Soviet Union in the past or from Europe or us today, will move him more than temporarily closer to policies he does not support. Castro is like T. S. Eliot's Rum Tum Tugger, if we paraphrase thusly: "Fidel Castro *will* do, as Fidel Castro *do* do, and there's no doing *anything* about it." So unless the United States is prepared to go in and take Castro out, Cubans must just find a way to survive his rule, as they have already done for decades, and so must we.

Constructive Engagement

Two years ago I spent a long evening with Cuban dissidents Elizardo Sanchez, Oscar Espinosa Chepe, Mirian Leyva, and Hector Palacios. (Chepe and Palacios are now serving decades-long prison terms.) Palacios, who had recently polled dissidents on the island, found that 90 percent wanted the embargo lifted. Why? Because, as Sanchez said, "isolation is oxygen to totalitarians." And yet the United States is not only maintaining the embargo but cutting back on the number of Americans who can visit the island legally, thus increasing the isolation that is the oxygen to Castro's totalitarian regime.

One might ask, who knows better how to deal with Fidel Castro? A few Cuban-American legislators who have lived in the United States for decades and represent militantly anti-Castro constituencies with their own understandable (but misguided) agendas and their declining supporters around the country? Or the people of Cuba, including the dissidents on the front lines and in jail? The latter, of course, and Cubans generally and the dissidents in particular want to see more Americans, not fewer, on the island. Embargo supporters, however, promise fewer visitors but more moral and material support for the dissidents. Two years ago, Palacios told me that dissidents are very uncomfortable with aid from abroad, particularly from Washington, even when they take it, because it "burns." It certainly "burned" the 75 dissidents arrested in early 2003 because the Castro regime used contacts with the U.S. government and receipt of support from Americans to "prove" democratic advocates were agents of American imperialism.

Embargo supporters are quick to proclaim support for dissidents, or democratic reformers, when they condemn Castro's repression or call for peaceful reform. The dissident most lauded by Americans today is Oswaldo Paya, who in 2002 won the European Union's top human rights award (the Sakharov Prize) and in December 2003 published a detailed peaceful transition program for Cuba. Paya approved the United Nations General Assembly's condemnation of the embargo in November 2003 (while fruitlessly calling upon the world body to simultaneously condemn Castro's repression). If we Americans really admire courageous men and women like Paya, Palacios, and many others who stand up to Castro to demand a little more freedom, why don't we show them the ultimate respect of doing what they ask of us? Why don't we open the doors wider to Americans who want to visit Cuba or, better yet, lift the embargo altogether?

POSTSCRIPT

Are Strict Sanctions on Cuba Warranted?

T he Commission for Assistance to a Free Cuba reported to President Bush on May 6, 2004. The following day, he imposed most of the increased sanctions on Cuba that were proposed by the commission. Bush praised the commission's recommendations as "strategy that says 'we're not waiting for the day of Cuban freedom, we are working for the day of freedom in Cuba'." Differing sharply, Wayne Smith of the Center for International Policy called Bush's policy "a farce, pure political theater," and predicted, "It will be a nuisance, but it's not going to have a significant effect." A general overview of current Cuba-U.S. relations is found in Chris McGillion and Morris H. Morley, eds., *Cuba, the United States, and the Post–Cold War World: The International Dimensions of the Washington-Havana Relationship* (University Press of Florida, 2004). A good site for the U.S. government view is the Web site of the unofficial embassy in Cuba, the U.S. Interests Section, Havana, http://usembassy.state.gov/havana/wwwhmain.html. The English version of the Web site of the government of Cuba is at http://www.cubagob.cu/ingles/default.htm.

Of the proposed measures, the most important one that Bush did not follow was the recommendation to reduce the funds that Cuban Americans could send to relatives still in Cuba. Reportedly, the White House felt that doing so might anger Cuban Americans. This would have been politically unwise given the impending presidential election, the fact that Cuban Americans make up 5 percent of Florida's population, and that the state casts 27 electoral college votes—the fourth largest total in the country. These political realities were evident in a quip published in the *Miami Herald*: "What irony! Cubans can't elect their president in Cuba, but they elect one in the United States." For a generally pro-sanction Cuban-American group, visit the Web site of the Cuban American National Foundation, http://www.canf.org/2005/principal-ingles.htm. A Cuban American group that views sanctions as harmful is the Cuban American Alliance, http://www.cubamer.org/index.asp.

While supporters of President Bush pointed to his well-established aversion to leftist governments and his repeated pronouncements in favor of encouraging democracy as reasons for imposing the sanctions, critics of the sanctions were more apt to condemn them as motivated by partisan politics. Representing that view, Montana's Senator Max Baucus, a Democrat, criticized the increased sanctions as "absurd" and expressed the hope that "we could move beyond a policy toward Cuba that is held hostage by the politics of the Electoral College."

Perhaps, but to any extent that the U.S. policy toward Cuba is driven by electoral considerations, the same electoral college calculus influences Democrats as well as Republicans. Indeed, the 2004 Democratic nominee for the presidency, Senator John Kerry, said that he would continue sanctions on Cuba as long as it remained a "Stalinist dictatorial state."

Whatever the U.S. government does or does not do, the most likely end to the Castro regime will come with his death. Born in 1926 and in power since 1959, Castro is approaching 80 years old and is in increasingly poor health. Despite his remarkable longevity, he cannot last forever. A step toward understanding the future is knowing the past, so a good place to begin is with a review of Cuba's history, available in Richard Gott, *Cuba: A New History* (Yale University Press, 2004). For a view of the possibilities in a post-Castro Cuba, see Edward Gonzalez and Kevin F. McCarthy, *Cuba After Castro: Legacies, Challenges, and Impediments* (National Defense Research Institute, 2004).

International Development Exchange (IDEX)

This is the Web site of the International Development Exchange (IDEX), an organization that works to build partnerships to overcome economic and social injustice. The IDEX helps people gain greater control over their resources, political structures, and the economic processes that affect their lives.

http://www.idex.org

United Nations Development Programme (UNDP)

This United Nations Development Programme (UNDP) site offers publications and current information on world poverty, the UNDP's mission statement, information on the UN Development Fund for Women, and more.

http://www.undp.org

Office of the U.S. Trade Representative

The Office of the U.S. Trade Representative (USTR) is responsible for developing and coordinating U.S. international trade, commodity, and direct investment policy and leading or directing negotiations with other countries on such matters. The U.S. trade representative is a cabinet member who acts as the principal trade adviser, negotiator, and spokesperson for the president on trade and related investment matters.

http://www.ustr.gov

The U.S. Agency for International Development (USAID)

This is the home page of the U.S. Agency for International Development (USAID), which is the independent government agency that provides economic development and humanitarian assistance to advance U.S. economic and political interests overseas.

http://www.usaid.gov/

World Trade Organization (WTO)

The World Trade Organization (WTO) is the only international organization dealing with the global rules of trade between nations. Its main function is to ensure that trade flows as smoothly, predictably, and freely as possible. This site provides extensive information about the organization and international trade today.

http://www.wto.org

Third World Network

The Third World Network (TWN) is an independent, nonprofit international network of organizations and individuals involved in economic, social, and environmental issues relating to development, the developing countries of the world, and the North-South divide. At the network's Web site you will find recent news, TWN position papers, action alerts, and other resources on a variety of topics, including economics, trade, and health.

http://www.twnside.org.sg

PART 3

Economic Issues

*I*nternational economic and trade issues have an immediate and personal effect on individuals in ways that few other international issues do. They influence the jobs we hold and the prices of the products we buy in short, our lifestyles. In the worldwide competition for resources and markets, tensions arise between allies and adversaries alike. This section examines some of the prevailing economic tensions.

- Is Capitalism the Best Model for the Global Economy?

- Should the Rich Countries Forgive All the Debt Owed by the Poor Countries?

ISSUE 12

Is Capitalism the Best Model for the Global Economy?

YES: Johan Norberg, from "Three Cheers for Global Capitalism," *American Enterprise Online* (June 2004)

NO: Walden Bello, from "Justice, Equity and Peace Are the Thrust of Our Movement," Acceptance Speech at the Right Livelihood Award Ceremonies (December 8, 2003)

ISSUE SUMMARY

YES: Johan Norberg, a fellow at the Swedish think tank Timbro, portrays capitalism as the path to global economic prosperity and argues further that free markets and free trade mean free choices for individuals that transfer power to them at the expense of political institutions.

NO: Walden Bello, executive director of Focus on the Global South, the Bangkok, Thailand–based project of Chulalongkorn University's Social Research Institute, and professor of sociology and public administration at the University of the Philippines, contends that global capitalism is the source of societal and environmental destruction.

T he intersection of politics and economics is called the study of political economy or, at the global level, international political economy (IPE). Whether the focus is a national economy or international economic relations, it is possible to divide economic approaches into three main categories. One, capitalism, favors little if any political manipulation of the economy. Capitalism, which Johan Norberg supports in the first reading, is also sometimes referred to as economic liberalism, free enterprise, or laissez-faire economics.

The other two economic approaches each feature considerable political control of the economy. One of these favors manipulating the economy to improve a country's power and to promote its political goals. This idea is variously called economic nationalism, mercantilism, and economic statecraft. The other theory of political intervention in the economy favors manipulating to achieve social ends, such as ensuring that wealth is fairly (that is, at least somewhat evenly) distributed. One name for this approach is economic structuralism, because its adherents believe that the very economic structure must be dramatically changed to improve the system. Socialism and Marxism, and at the IPE level, dependency theory and neoimperialism theory all also represent variations of this third approach, which is advanced by Walden Bello in the second reading.

In one of the classic works on capitalist theory, Adam Smith wrote in *The Wealth of Nations* (1776) that, "It is not from the benevolence of the butcher, the brewer, or the baker, that we expect our dinner, but from their regard to their own interest." He argued that the self-interest of individuals collectively constitutes an "invisible hand" of competition that creates the most efficient economies and that political interference lessens economic efficiency and, therefore prosperity. On the international level, for instance, Smith would oppose regulating multinational corporations (MNCs) or having tariffs to protect jobs from international competition.

It is important to note that at the national or international level, none of these approaches is or perhaps has ever been utilized in a pure sense. For example, the United States is basically a capitalist country, but it still has many policies, such as a minimum wage, that manipulate the economy to improve equity. Other U.S. policies are mercantilist, including the sanctions on Cuba debated in Issue 11, in that they manipulate the economy to promote U.S. strength and foreign policy.

Thus, most modern capitalist thinkers advocate a modified capitalism that has political restraints to guard against the worst potential aspects of capitalism, such as the formation of monopolies, and to otherwise prevent the competition and unequal distribution of wealth that is inherent in capitalism from becoming too skewed.

The current international system has aspects of all three economic approaches, but capitalism, at least in its modified form, is the most prominent. At the center of the capitalist approach to IPE is free economic interchange, the notion that trade, investment, and currencies should flow across international borders with as few restrictions as possible. Whatever the relative advantages of the three approaches, an important reason that free economic interchange is such a prominent approach is because the United States has been the world's most powerful country for nearly three-quarters of a century. Based on their philosophical beliefs and U.S. interests, American diplomats were the driving force behind the conclusion of the General Agreement on Tariffs and Trade (GATT), the establishment of the International Monetary Fund, and other treaties and organizations that came into being during or after World War II to promote free trade.

More than any other international organization, the World Trade Organization (which administers the GATT) represents and promotes this approach to IPE. Johan Norberg praises capitalism for not only advancing prosperity but also freedom. Walden Bello speaks for the movement, including the organization Focus on the Global South, which he heads, that believes that global capitalism brings about crises, widens inequalities within and across countries, and increases global poverty. Bello's speech was given during his acceptance of the Right Livelihood Award, which has been called the "Alternative Nobel Prize" by those friendly to the ideals it represents.

Three Cheers for Global Capitalism

Under what is rather barrenly termed "globalization"—the process by which people, information, trade, investments, democracy, and the market economy tend more and more to cross national borders—our options and opportunities have multiplied. We don't have to shop at the big local company; we can turn to a foreign competitor. We don't have to work for the village's one and only employer; we can seek alternative opportunities. We don't have to make do with local cultural amenities; the world's culture is at our disposal. Companies, politicians, and associations have to exert themselves to elicit interest from people who have a whole world of options. Our ability to control our own lives is growing, and prosperity is growing with it.

Free markets and free trade and free choices transfer power to individuals at the expense of political institutions. Because there is no central control booth, it seems unchecked, chaotic. Political theorist Benjamin Barber speaks for many critics when he bemoans the absence of "viable powers of opposing, subduing, and civilizing the anarchic forces of the global economy." "Globalization" conjures up the image of an anonymous, enigmatic, elusive force, but it is actually just the sum of billions of people in thousands of places making decentralized decisions about their own lives. No one is in the driver's seat precisely because all of us are steering.

No company would import goods from abroad if we didn't buy them. If we did not send e-mails, order books, and download music every day, the Internet would wither and die. We eat bananas from Ecuador, order magazines from Britain, work for export companies selling to Germany and Russia, vacation in Thailand, and save money for retirement by investing in South America and Asia. These things are carried out by businesses only because we as individuals want them to. Globalization takes place from the bottom up.

A recent book about the nineteenth-century Swedish historian Erik Geijer notes that he was able to keep himself up to date just by sitting in Uppsala reading the *Edinburgh Review* and the *Quarterly Review*. That is how simple and intelligible the world can be when only a tiny elite in the capitals of Europe makes any difference to the course of world events. How much more complex and confusing everything is now, with ordinary people having a say over their own lives. Elites may mourn that they have lost power, but everyday life has vastly improved now that inexpensive goods and outside information and different employment opportunities are no longer blocked by political barriers.

To those of us in rich countries, more economic liberty to pick and choose may sound like a trivial luxury, even an annoyance—but it isn't. Fresh options are invaluable for all of us. And the existence from which globalization delivers people in the Third World—poverty, filth, ignorance, and powerlessness—really is intolerable. When global capitalism knocks at the door of Bhagant, an elderly agricultural worker and "untouchable" in the Indian village of Saijani, it leads to his house being built of brick instead of mud, to shoes on his feet, and clean clothes—not rags—on his back. Outside Bhagant's house, the streets now have drains, and the fragrance of tilled earth has replaced the stench of refuse. Thirty years ago Bhagant didn't know he was living in India. Today he watches world news on television. The stand that we in the privileged world take on the burning issue of globalization can determine whether or not more people will experience the development that has taken place in Bhagant's village.

Critics of globalization often paint a picture of capitalist marauders secretly plotting for world mastery, but this notion is completely off the mark. It has mostly been pragmatic, previously socialist, politicians who fanned globalization in China, Latin America, and East Asia—after realizing that government control-freakery had ruined their societies. Any allegation of runaway capitalism has to be tempered by the observation that today we have the largest public sectors and highest taxes the world has ever known. The economic liberalization measures of the last quarter century may have abolished some of the recent past's centralist excesses, but they have hardly ushered in a system of laissez-faire.

What defenders of global capitalism believe in, first and foremost, is man's capacity for achieving great things by means of the combined force of market exchanges. It is not their intention to put a price tag on everything. The important things—love, family, friendship, one's own way of life—cannot be assigned a monetary value. Principled advocates of global economic liberty plead for a more open world because that setting unleashes individual creativity as none other can. At its core, the belief in capitalist freedom among nations is a belief in mankind.

People have a natural tendency to believe that everything is growing worse, and that things were better in the "old days." In 1014, Archbishop Wulfstan of York declared, "The world is in a rush and is getting close to its end."

Today, we hear that life is increasingly unfair amidst the market economy: "The rich are getting richer, and the poor are getting poorer." But if we look beyond the catchy slogans, we find that while many of the rich have indeed grown richer, so have most of the poor. Absolute poverty has diminished, and where it was greatest 20 years ago—in Asia—hundreds of millions of people have achieved a secure existence, even affluence, previously undreamed of. Global misery has diminished, and great injustices have started to unravel.

One of the most interesting books published in recent years is *On Asian Time: India, China, Japan 1966–1999*, a travelogue in which Swedish author Lasse Berg and photographer Stig Karlsson describe their visits to Asian countries during the 1960s. They saw poverty, abject misery, and imminent disaster. Like many, they could not bring themselves to hold out much hope for the future, and they thought that socialist revolution might be the only way out. Returning to India and China in the 1990s, they could not help seeing how wrong they were. More and more people have extricated themselves from poverty; the problem of hunger is steadily diminishing; the streets are cleaner. Squalid huts have given way to buildings wired for electricity with TV antennas on their roofs.

When Berg and Karlsson first visited Calcutta, fully one tenth of its inhabitants were homeless, and every morning trucks were sent by the public authorities or missionary societies to collect the bodies of those who had died in the night. Thirty years later, setting out to photograph people living on the streets, they had difficulty in finding such people. The rickshaw was disappearing from the urban scene, with people traveling by car, motorcycle, and subway instead.

When Berg and Karlsson showed young Indians photographs from the 1960s, they refused to believe it was even the same place. Could things really have been so dreadful? A striking illustration of the change is provided by a pair of photographs in their book. In the old picture, taken in 1976, a 12-year-old Indian girl named Satto holds up her hands. They are already furrowed and worn, prematurely aged by many years' hard work. The new picture shows Satto's 13-year-old daughter Seema, also holding up her hands. They are smooth and soft, the hands of a girl whose childhood has not been taken away from her.

The biggest change of all is in people's thoughts and dreams. Television and newspapers bring ideas and images from the other side of the globe, widening people's notions of the possible. Why make do with this kind of government when there are alternative political systems available?

Lasse Berg writes, self-critically:

> Reading what we observers, foreigners as well as Indians, wrote in the '60s and '70s, nowhere in these analyses do I see anything of present-day India. Often nightmare scenarios—overpopulation, tumult, upheaval or stagnation—but not this calm and steady forward-jogging, and least of all this modernization of thoughts and dreams. Who foresaw that consumerism would penetrate so deeply in and among the villages? Who foresaw that both the economy and general standard of living would do so well? Looking back, what the descriptions have in common is an overstatement of the extraordinary, frightening uncertainty, and an understatement of the force of normality.

Note that all of the dramatic development described by Berg has resulted from sharp moves over the past few decades toward international capitalism and trade.

This progress is all very well, many critics of globalization will argue, but even if the majority are better off, gaps have widened and wealthy people and countries have improved their lot more rapidly than others. The critics point out that 40 years ago the combined per capita GDP of the 20 richest countries was 15 times greater than that of the 20 poorest, and is now 30 times greater.

There are two reasons why this objection to globalization does not hold up. First, if everyone is better off, what does it matter that the improvement comes faster for some than for others? Only those who consider wealth a greater problem than poverty can find irritation in middle-class citizens becoming millionaires while the previously poverty-stricken become middle class.

Second, the allegation of increased inequality is simply wrong. The notion that global inequality has increased is largely based on figures from the U.N.'s 1999 *Human Development Report*. The problem with these figures is that they don't take into account what people can actually buy with their money. Without that "purchasing power" adjustment, the figures only show what a currency is worth on the international market, and nothing about local conditions. Poor people's actual living standards hinge on the cost of their food, their clothing, their housing—not what their money would get them while vacationing in Europe. That's why the U.N. uses purchasing-power-adjusted figures in other measures of living standards. It only resorts to the unadjusted figures, oddly, in order to present a theory of inequality.

A report from the Norwegian Institute for Foreign Affairs investigated global inequality by means of figures adjusted for purchasing power. Their data show that, contrary to conventional wisdom, inequality between countries has continuously declined ever since the end of the 1970s. This decline has been especially rapid since 1993, when globalization really gathered speed.

More recently, similar research by Columbia University development economist Xavier Sala-i-Martin has confirmed those findings. He found that when U.N. figures are adjusted for purchasing power, they point to a sharp decline in world inequality. Sala-i-Martin and co-author Surjit Bhalla also found independently that if we focus on inequality between persons, rather than inequality between countries, global inequality at the end of 2000 was at its lowest point since the end of World War II.

Estimates that compare countries rather than individuals, both authors note, grossly overestimate real inequality because they allow gains for huge numbers of people to be outweighed by losses for far fewer. For instance, country aggregates treat China and Grenada as data points of equal weight, even though China's population is 12,000 times Grenada's. Once we shift our focus to people rather than nations, the evidence is overwhelming that the past 30 years have witnessed a strong shift toward global equalization.

One myth about trade is the notion that exports to other countries are a good thing, but that imports are somehow a bad thing. Many believe that a country grows powerful by selling much and buying little. The truth is that our standard of living will not rise until we use our money to buy more and cheaper things. One of the first trade theorists, James Mill, rightly noted in 1821 that "The benefit which derives from exchanging one commodity for another arises in all cases from the commodity received, not the commodity given." The only point of exports, in other words, is to enable us to get imports in return.

The absurdity of the idea that we must avoid cheap imports becomes clear if we imagine it applied to non-national boundaries. If imports really were economically harmful, it would make sense for one city or state to prevent its inhabitants from buying from another. According to this logic, Californians would lose out if they bought goods from Texas, Brooklyn would gain by refusing to buy from Manhattan, and it would be better for a family to make everything itself instead of trading with its neighbors. It's obvious that such thinking would lead to a tremendous loss of welfare: The self-sufficient family would be hard pressed just to keep food on the table. When you go to the store, you "import" food—being able to do so cheaply is a benefit, not a loss. You "export" when you go to work and create goods and services. Most of us would prefer to "import" so cheaply that we could afford to "export" a little less.

Trade is not a zero-sum game in which one party loses what the other party gains. There would be no exchange if both parties did not feel that they benefited. The really interesting yardstick is not the "balance of trade" (where a "surplus" means that we are exporting more than we are importing) but the quantity of trade, since both exports and imports are gains. Imports are often feared as a potential cause of unemployment: If we import cheap toys and clothing from China, then toy and garment manufacturers here will have to scale down. But by obtaining cheaper goods from abroad, we save resources in the United States and can therefore invest in new industries and occupations. Free trade brings freedom: freedom for people to buy and sell what they want. As an added benefit, this leads to the efficient use of resources. A company, or country, specializes where it can generate the greatest value.

Economic openness also leads to an enduring effort to improve production, because foreign competition forces firms to be as good and cheap as possible. As production in established industries becomes ever more efficient, resources are freed up for investment in new methods, inventions, and products. Foreign competition brings the same benefits that we recognize in economic competition generally; it simply extends competition to a broader field.

One of the most important but hard to measure benefits of free trade is that a country trading a great deal with the rest of the world imports new ideas and new techniques in the bargain. If the United States pursues free trade, our companies are exposed to the world's best ideas. They can then borrow those ideas, buy leading technology from elsewhere, and hire the best available manpower. This compels the companies to be more dynamic themselves.

The world's output today is six times what it was 50 years ago, and world trade is 16 times greater. There is reason to believe that the trade growth drove much of the production growth. One comprehensive study of the effects of trade was conducted by Harvard economists Jeffrey Sachs and Andrew Warner. They examined the trade policies between 1970 and 1989 of 117 countries. The study reveals a statistically significant connection between free trade and economic growth. Growth was between three and six times higher in free-trade countries than in protectionist ones. Factors like improved education turned out to be vastly less important than trade in increasing economic progress.

Over those two decades, developing countries that practiced free trade had an average annual growth rate of 4.5 percent, while developing countries that practiced protectionism grew by only 0.7 percent. Among industrial countries, the free traders experienced annual growth of 2.3 percent, versus only 0.7 percent among the protectionists. It must be emphasized that this is not a matter of countries earning more because others opened to their exports. Rather, these countries earned more by keeping their own markets open.

If free trade is constantly making production more efficient, won't that result in the disappearance of job opportunities? When Asians manufacture our cars and South Americans produce our meat, auto workers and farmers in the United States lose their jobs and unemployment rises. Foreigners and developing countries will increasingly produce the things we need, until we don't have any jobs left. If increasing automation means everything we consume today will be able to be made by half the U.S. labor force in 20 years, doesn't that mean that the other half will be out of work? Such are the horror scenarios depicted in many anti-globalization writings.

The notion that a colossal unemployment crisis is looming began to grow popular in the mid 1970s. Since then, production has been streamlined and internationalized more than ever. Yet far more jobs have been created than have disappeared. We have more efficient production than ever before, but also more people at work. Between 1975 and 1998, employment in countries like the United States, Canada, and Australia rose by 50 percent.

And it is in the most internationalized economies, making the most use of modern technology, that employment has grown fastest. Between 1983 and 1995 in the United States, 24 million more job opportunities were created than disappeared. And those were not low-paid, unskilled jobs, as is often alleged. On the contrary, 70 percent of the new jobs carried a wage above the American median level. Nearly half the new jobs belonged to the most highly skilled, a figure which has risen even more rapidly since 1995.

So allegations of progressively fewer people being needed in production have no empirical foundation. And no wonder, for they are wrong in theory too. Imagine a pre-industrial economy where most everyone is laboring to feed himself. Then food production is improved by new technologies, new machines, foreign competition, and imports. That results in a lot of people being forced to leave the agricultural sector. Does that mean there is nothing for them to do, that consumption is constant? Of course not; the manpower which used to be required to feed the population shifts to clothing it, and providing better housing. Then improved transport, and entertainment. Then newspapers, telephones, and computers.

The notion that the quantity of employment is constant, that a job gained by one person is always a job taken from someone else, has provoked a variety of foolish responses. Some advocate that jobs must be shared. Others smash machinery. Many advocate raising tariffs and excluding immigrants. But the whole notion is wrong. The very process of a task being done more efficiently, thus allowing jobs to be shed, enables new industries to grow, providing people with new and better jobs.

Efficiency does, of course, have a flip side. Economist Joseph Schumpeter famously described a dynamic market as a process of "creative destruction," because it destroys old solutions and industries, with a creative end in view. As the word "destruction" suggests, not everyone benefits from every market transformation in the short term. The process is painful for those who have invested in or are employed by less-efficient industries. Drivers of horse-drawn cabs lost out with the spread of automobiles, as did producers of paraffin lamps when electric light was introduced. In more modern times, manufacturers of typewriters were put out of business by the computer, and LP records were superseded by CDs.

Painful changes of this kind happen all the time as a result of new inventions and methods of production. Unquestionably, such changes can cause trauma for those affected. But the most foolish way to counter such problems is to try to prevent them. It is generally fruitless; mere spitting into the wind. Besides, without "creative destruction," we would all be stuck with a lower standard of living.

The idea that we should halt change now is as misguided as the idea that we should have obstructed agricultural advances two centuries ago to protect the 80 percent of the population then employed on the land. A far better idea is to use the economic gains that flow from transformation to alleviate the consequences for those adversely affected. As a Chinese proverb has it, "When the wind of change begins to blow, some people build windbreaks while others build windmills."

But the troubles of change are seldom as widespread as a scan of the newspaper headlines might suggest. It is easy to report that 300 people lost their jobs in a car factory due to Japanese competition. It is less easy and less dramatic to report on the thousands of jobs that have been created because we were able to use old resources more efficiently. Costs affecting a small group on an isolated occasion are simple to spotlight, while benefits that gradually accrue to nearly everyone creep up on us without our giving them a thought. Hardly any of the world's consumers have been informed by their news sources that wider selection, better quality, and lower prices spurred by competition following the Uruguay Round of trade liberalization have gained them between $100 billion and $200 billion annually. The difference is visible, though, in our refrigerators, our home electronics, and our wallets.

A review of more than 50 surveys of adjustments after trade liberalization in different countries shows clearly that adjustment problems are far milder than the conventional debate suggests. For every dollar of trade adjustment costs, roughly $30 is harvested in the form of welfare gains. A study of trade liberalization in 13 different countries showed that in all but one, industrial employment had already increased just one year after the liberalization. The process turns out to be far more creative than destructive.

If there are problems resulting from unshackled capitalism, they ought to be greatest in the United States, with its constant swirling economic transformations. But our job market is a bit like the Hydra in the legend of Hercules. Every time Hercules cuts off one of the beast's heads, two new ones appear. The danger of having to continue changing jobs all one's life is exaggerated: The average length of time an American stays in a particular job actually increased between 1983 and 1995, from 3.5 years to 3.8. Nor is it true, as many people believe, that more jobs are created in the United States only because real wages have stagnated or fallen since the 1970s. A growing proportion of wages is now paid in non-money forms, such as health insurance, stocks, 401(k) contributions, day care, and so forth, to avoid taxation. When these benefits are included, American wages have risen right along with productivity. Among poor Americans, the proportion of consumption devoted to food, clothing, and housing has fallen since the 1970s from 52 to 37 percent, which clearly shows that they have money to spare for much more than the bare necessities of life.

The type of protectionism most fashionable among intellectuals today holds that we should not permit trade with countries that have poor working conditions or environmental protections. Since President Clinton proposed this kind of boycott at the World Trade Organization talks in Seattle in 1999, trade liberalization has largely deadlocked. Developing countries refuse to negotiate under such threats.

Whatever affluent demonstrators in economically powerful countries may believe, low wages and poor environmental conditions in developing countries are not the result of lack of enlightenment or stinginess. Generally the problem is that employers cannot afford to pay higher wages and provide better working conditions, because worker productivity is much lower in undeveloped countries. Wages can be raised as labor becomes more valuable, and that can be achieved only through better infrastructure, more education, new machinery, better organization, and increased investment. If we force these countries to raise wages before productivity has been improved, firms and consumers will be asked to pay more for their manpower than it is currently worth—denying those citizens any chance to work next to the more productive, better-paid workers of the Western world. Unemployment among the world's poor would swiftly rise.

Jesus Reyes-Heroles, Mexican ambassador to the United States, has explained, "In a poor country like ours the alternative to low-paying jobs isn't high-paying jobs—it's no jobs at all." In effect, labor and environmental provisions tell the developing countries: You are too poor to trade with us, and we are not going to trade with you until you have grown rich. The problem is that only through trade can they grow richer and thereby, step by step, improve their living standards and their social conditions. This is a catch-22.

Suppose this idea had been in vogue at the end of the nineteenth century. Britain and France would have noted that Swedish wages were only a fraction of theirs, that Sweden had a 12- or 13-hour working day and a six-day week, and that Swedes were chronically undernourished. Child labor was widespread in spinning mills, glassworks, and match and tobacco factories. One factory worker out of 20 was under 14 years old. Britain and France, accordingly, would have refused to trade with Sweden and closed their frontiers to Swedish cereals, timber, and iron ore.

Would Swedes have benefited? Hardly. Such a decision would have robbed them of earnings and blocked their industrial development. They would have been left with intolerable living conditions, children would have stayed in the factories, and perhaps to this day they would be eating tree-bark bread when the harvest failed. But that didn't happen. Trade with Sweden was allowed to grow uninterrupted, industrialization got under way, and the Swedish economy was revolutionized. This didn't just help Sweden, it also gave Britain and France a new peer for prosperous exchange.

If today, as a condition for trading with the developing countries, we require their mining industries to be as safe as the West's are now, we make demands that we ourselves did not meet when our own mining industries were developing. It was only after raising our incomes that we were able to develop the technology and afford the safety equipment we take for granted today. If we require developing countries to adopt those practices and equipment before they can afford them, their industries will be knocked out. That will not help the world's poor. And it will not help us gain new prosperous, stable, clean trading partners.

Advocates of protectionism often complain of "sweatshops" allegedly run by multinational corporations in the Third World. Let's look at the evidence: Economists have compared the conditions of people employed in American-owned facilities in developing countries with those of people employed elsewhere in the same country. In the poorest developing countries, the average employee of an American-affiliated company makes eight times the average national wage! In middle income countries, American employers pay three times the national average. Even compared with corresponding modern jobs in the same country, the multinationals pay about 30 percent higher wages. Marxists maintain that multinationals exploit poor workers. Are much higher wages "exploitation"?

The same marked difference can be seen in working conditions. The International Labor Organization has shown that multinationals, especially in the footwear and garment industries, are leading the trend toward better working conditions in the Third World. When multinational corporations accustom workers to better-lit, safer, and cleaner factories, they raise the general standard. Native firms then also have to offer better conditions, otherwise no one will work for them. Zhou Litai, one of China's foremost labor attorneys, has pointed out that Western consumers are the principal driving force behind the improvements of working conditions in China, and worries that "if Nike and Reebok go, this pressure evaporates."

One of the few Western participants in the globalization debate to have actually visited Nike's Asian subcontractors to find out about conditions there is Linda Lim of the University of Michigan. She found that in Vietnam, where the annual minimum wage was $134, Nike workers got $670. In Indonesia, where the minimum wage was $241, Nike's suppliers paid $720. If Nike were to withdraw on account of boycotts and tariffs imposed from the West, these employees would be put out of work and would move to more dangerous and less lucrative jobs in native industry or agriculture.

There are of course rogues among entrepreneurs, just as there are in politics, and all parts of life. But bad behavior by a few is no reason for banning large corporations from investing overseas. That would make no more sense than disbanding the police because we find instances of police brutality.

It is commonly supposed that in order to cope with rising competition from the Third World and increasingly efficient machinery, we in the United States and Western Europe must work harder and put in longer and longer hours. Actually, the time we spend working has diminished, as rising prosperity resulting from growth has enabled us to earn the same pay by doing less work—if we want to. Compared with our parents' generation, most of today's workers go to work later, come home earlier, have longer lunch and coffee breaks, longer vacations, and more public holidays. In the U.S., working hours today are only about half of what they were a hundred years ago, and have diminished by about 10 percent just since 1973—a reduction equaling 23 days per year. On average, American workers have acquired five extra years of waking leisure time since 1973.

We start working progressively later in life, and retire earlier. Calculated over his lifetime, a Western worker in 1870 had only two hours off for each hour worked. By 1950 that figure had doubled to four hours off. Today it has doubled again to about eight hours off for each hour worked. Economic development, closely linked to an expansion of trade that has enabled us to specialize, makes it possible for us to reduce our working hours sharply even as we raise our material living standard.

It is natural for us in the affluent West to complain about stress. This is caused mostly by the fantastic growth of options. Pre-industrial citizens, spending all their lives in one place and perhaps meeting a hundred people in a lifetime, were less likely to feel that they did not have time for everything they wanted to do. People spent a lot of their non-working time sleeping. Today, we can travel the world, read newspapers, see films from every corner of the globe, and meet a hundred people every day. We used to go to the mailbox and wait for the postman. Now e-mail sits in our inbox, waiting for us. We have a huge entertainment industry that offers an almost infinite number of ways to pass the time. No wonder the result is a certain frustration over not finding hours enough to do everything.

But compared with the problems that earlier people had, and that most people in developing countries still have today, this kind of worry should be recognized for what it is—a luxury. Stress and burnout at work can be real. But they're not caused by globalization.

Corporations have not acquired more power through free trade. Indeed, they used to be far more powerful—and still are in dictatorships and controlled economies. Large corporations have chances to corrupt or manipulate when power is distributed by public officials who can be hobnobbed over luncheons to give protection through monopolies, tariffs, or subsides. Free trade, on the other hand, exposes corporations to competition. Above all, it lets consumers ruthlessly pick and choose across national borders, rejecting companies that don't measure up.

People living in isolation are dependent on local enterprises and are forced to buy what they offer at the price they demand. Free trade and competition are the best guarantees of alternatives penetrating the market if the dominant firm does not satisfy. The only reason a lump of sugar costs two to three times as much in the E.U. as on the world's open markets is because European governments block trade with sugar tariffs. In this way, bureaucrats create far more monopolies than does capitalism.

Capitalists, however, aren't necessarily loyal adherents of capitalism: Often, they are happy to collaborate with governments to protect their privileges. A market economy and free trade are the best ways to force them to offer a better return for our resources. Free trade does not confer coercive power on anyone. The freedom of a business in a free market economy is like a waiter's freedom to offer the menu to a restaurant patron. And it entitles other waiters—foreign ones, even!—to come running up with rival menus. The loser in this process, if anyone, is the waiter who once had a monopoly.

Nothing forces people to accept new products. If they gain market share, it is because people want them. Even the biggest companies survive at the whim of customers. Mega-corporation Coca-Cola has to adapt the recipe for its drinks to different regions in deference to local variations in tastes. McDonald's sells mutton burgers in India, teriyaki burgers in Japan, and salmon burgers in Norway. TV mogul Rupert Murdoch failed to create a pan-Asian channel and instead has been forced to build different channels to suit local audiences.

Companies in free competition can grow large and increase their sales only by being better than others. Companies that fail to do so quickly go bust or get taken over by someone who can make better use of their capital, buildings, machinery, and employees. Capitalism is very tough—but mainly on firms offering outdated, poor-quality, or expensive goods and services. Fear of established companies growing so large as to become unaccountable has absolutely no foundation in reality. In the U.S., the most capitalist large country in the world, the market share of the 25 biggest corporations has steadily dwindled over recent decades.

Freer markets make it easier for small firms with fresh ideas to compete with big corporations. Between 1980 and 1993, the 500 biggest American corporations saw their share of the country's total employment diminish from 16 to 11 percent. During the same period, the average personnel strength of American firms fell from 17 to 15 people, and the proportion of the population working in companies with more than 250 employees fell from 37 to 29 percent.

Of the 500 biggest enterprises in the United States in 1980, one third had disappeared by 1990. Another 40 percent had evaporated five years later. Whether they failed to grow enough to stay on the list, died, merged, or broke up, the key lesson is that big corporations have much less power over consumers than we sometimes imagine. Even the most potent corporation must constantly re-earn its stripes, or tumble fast.

Half the firms operating in the world today have fewer than 250 employees. The biggest brand logos are always flashed before our eyes, but we forget that they are constantly being joined by new ones. How many people recall that just a few years ago Nokia was a small Finnish firm making tires and boots?

Many people fear a "McDonaldization" or "Disneyfication" of the world, a creeping global homogeneity that leaves everyone wearing the same clothes, eating the same food, and seeing the same movies. But this portrayal does not accurately describe globalization. Anyone going out in the capitals of Europe today will have no trouble finding hamburgers and Coca-Cola, but he will just as easily find kebabs, sushi, Tex-Mex tacos, Peking duck, Thai lemongrass soup, and cappuccino. We know that Americans listen to Britney Spears and watch Adam Sandler films, but it's worth remembering that the United States is also a country with 1,700 symphony orchestras, 7.5 million annual trips to the opera, and 500 million museum visits a year. Globalization doesn't just give the world shlocky reality TV and overplayed music videos, but also classic films and documentaries on many new channels, news on the Internet, and masterpieces of music and literature in stores and on the Web.

The world is indeed moving toward a common objective, but that objective is not the predominance of a particular culture. Rather it is pluralism, the freedom to choose from a host of different paths and destinations. The market for experimental electronic music or film versions of novels by Dostoevsky may be small in any given place, so musicians and filmmakers producing such material could never produce anything without access to the much larger audience provided by globalization.

This internationalization is, ironically, what makes people believe that differences are vanishing. When you travel abroad, things look much the same as in your own country: The people there also have goods and chain stores from different parts of the globe. This phenomenon is not due to uniformity and the elimination of differences, but by the growth of pluralism everywhere.

Such opportunity can have negative effects in certain situations. When traveling to another country, we want to see something unique. Arriving in Rome and finding Hollywood films, Chinese food, Japanese Pokemon games, and Swedish Volvos, we miss the local color. And national specialties like pizza, pasta, and espresso are already familiar to us because we have them at home, too.

Bu this is another one of those luxury problems. I know a man from Prague who was visited by Czech friends who had settled abroad. The expatriates deplored the arrival of McDonald's in Prague because it threatened the city's distinctive charm. This response made the man indignant. How could they regard his home city as a museum, a place for them to visit now and then in order to avoid fast food restaurants? He wanted a real city, including the convenient and inexpensive food that these exile Czechs themselves had access to. A real, living city cannot be a "Prague summer paradise" for tourists. Other countries and their populations do not exist in order to give us picturesque holiday experiences. They, like us, are entitled to choose what suits their own tastes.

There's nothing new about cultures changing, colliding with each other, and cross-pollinating. Even the traditions we think of as most "authentic" have often resulted from cultural imports. One of the most sacred Christmas traditions in my home country of Sweden celebrates an Italian saint by adorning the hair of blond girls with lighted candles. Change and renewal are an inherent part of civilization. If we try to freeze certain cultural patterns in time they cease to be culture and instead become museum relics and folklore. There is nothing wrong with museums—but we can't live in one.

In the age of globalization, the ideas of freedom and individualism have attained tremendous force. There are few concepts as inspiring as that of self-determination. When people in other countries glimpse a chance to set their own course, it becomes almost irresistible. If there is any elimination of differences throughout the world, it has been the convergence of societies on the practice of allowing people to choose the sort of existence they please.

Global commerce does undermine old economic interests, challenge cultures, and erode some traditional power centers. Advocates of globalization have to show that greater gains and opportunities counterbalance such problems. The anti-globalists are right that we could reject globalized trade if we adamantly insisted. Capital can be locked up, commercial flows blocked, and borders barricaded.

This happened across our planet at least once before. Decades of expanding economic liberty and globalization during the nineteenth century were replaced with nationalistic saber-rattling, centralization, and closed borders at the beginning of the twentieth century. The outbreak of World War I marked a new era.

Globalization resumed after World War II, but countries like Burma and North Korea show that it is still possible to cut oneself off from international trade, so long as one is prepared to pay heavily for doing so with political oppression and economic deprivation. It is not "necessary" that we follow the trend toward expanding global trade. It is highly desirable.

Globalization makes an excellent scapegoat. It contains all the anonymous forces that have served this purpose throughout history: foreign countries, other races and ethnic groups, the uncaring market. Globalization does not speak up for itself when politicians blame it for overturning economies, increasing poverty, enriching a tiny minority, polluting the environment, or cutting jobs. And globalization doesn't usually get any credit when good things happen—an economy running at high speed, poverty diminishing, clearing skies. So if the trend toward greater global interchange and liberty is to continue, an ideological defense is needed.

In 25 years there are likely to be 2 billion more of us on this planet, and 97 percent of that population increase will occur in the developing world. There are no automatic processes deciding what sort of world they will experience. Most of that will depend on what people like you and me believe, think, and fight for.

Lasse Berg and Stig Karlsson record Chinese villagers' descriptions of the changes they experienced since the 1960s: "The last time you were here, people's thoughts and minds were closed, bound up," stated farmer Yang Zhengming. But as residents acquired power over their own livelihoods they began to think for themselves. Yang explains that "a farmer could then own himself. He did not need to submit. He decided himself what he was going to do, how and when. The proceeds of his work were his own. It was freedom that came to us. We were allowed to own things for ourselves."

Coercion and poverty still cover large areas of our globe. But thanks to globalizing economic freedom, people know that living in a state of oppression is not natural or necessary. People who have acquired a taste of economic liberty and expanded horizons will not consent to be shut in again by walls or fences. They will work to create a better existence for themselves. The aim of our politics should be to give them that freedom.

NO

Walden Bello

Justice, Equity and Peace Are the Thrust of Our Movement

The World Trade Organization (WTO) is the supreme institution of corporate-driven globalization, and the collapse of its fifth ministerial in Cancun on September 14 this year has dramatically underlined the deepening crisis of legitimacy of the globalist agenda.

Less than 10 years ago, our movement was marginalized. The founding of the WTO in 1995 seemed to signal that globalization was the wave of the future, and that those who opposed it were destined to suffer the same fate as the Luddites that fought against the introduction of machines during the industrial revolution [Luddites were British laborers who rioted (1811–1816), destroying textile machines, which they believed cause unemployment and low wages.] Globalization was going to bring prosperity in its wake, and how could one oppose the promise of the greatest good for the greatest number that the transnational corporations, guided by the invisible hand of the market, were going to shower the world?

But the movement stood firm in the face of the scorn of the establishment during the 1990s, when the boom in the world's mightiest capitalist engine—the U.S. economy—appeared to be destined to go on and on. It was steadfast in its prediction that, driven by the logic of corporate profitability, the liberalization and deregulation of trade and finance would bring about crises, widen inequalities within and across countries, and increase global poverty.

The Asian financial crisis in 1997 [in which foreign investment fled many Asian countries, seriously undermining their currencies and economies] provided sudden, savage proof of the destabilizing impact of eliminating controls from the flow of global capital. Indeed, what could be more savage than the fact that the crisis would bring 1 million people in Thailand and 22 million people in Indonesia below the poverty line in the space of a few weeks in the fateful summer of 1997?

The Asian financial crisis was one of those momentous events that removed the scales from people's eyes and enabled them to see the cold, brutal realities. And one of those realities was the fact that the free market policies that the International Monetary Fund and the World Bank imposed on some 100 developing and transitional economies between 1980 and 2000 had

Right Livelihood Award Ceremonies, December 8, 2003.

induced, in all but a handful of them, not a virtuous circle of growth, prosperity and equality but a vicious cycle of economic stagnation, poverty and inequality. The year 2001 brought us not only September 11, 2001 was also the year of reckoning for free-market fundamentalism—the year that the Argentine economy, the poster boy of neoliberal economics, crashed, and the U.S. stock market collapsed owing to the contradictions of finance-driven, deregulated global capitalism, wiping out $4.6 trillion in investor wealth—half of the U.S.' gross domestic product—and inaugurating a period of stagnation and rising unemployment. [Neoliberal economics is the belief that capitalism/free enterprise along with international economic integration provide the best hope of increasing the prosperity of all countries.]

As global capitalism moved from crisis to crisis, people organized in the streets, in work places, in the political arena to counter its destructive logic. In December 1999, massive street resistance by over 50,000 demonstrators combined with revolt of the developing governments inside the Seattle Convention Centre to bring down the third ministerial [summit meeting] of the WTO. Global protests also eroded the legitimacy of the IMF and the World Bank, the two other pillars of global economic governance, albeit in less dramatic fashion. Anti-neoliberal regimes came to power in Venezuela, Argentina, Brazil and Ecuador. The fifth ministerial meeting in Cancun [Mexico in September 2003], an event associated in many people's minds with the altruistic suicide of the Korean farmer Lee Kyung-Hae at the barricades, became Seattle II. And, just three weeks ago, in Miami, the same alliance of civil society and developing country governments forced Washington to retreat from the neoliberal programme of radical liberalization of trade, finance and investment that it had threatened to impose in the western hemisphere via the Free Trade Area of the Americas (FTAA).

Justice and equity has been one thrust of our movement. The other has been peace. For we never believed the pro-globalization argument that accelerated globalization would bring about the reign of "perpetual peace." Indeed, we warned that as globalization proceeded, its economically and socially-destabilizing effects would multiply conflicts and insecurities. Driven by corporate logic, globalization, we warned, would herald an era of aggressive imperialism that would seek to batter down opposition, seize control of natural resources, and secure markets.

It gave us no pleasure that we were proved right. Instead, the movement swung into action, becoming a global force for justice and peace that mobilized tens of millions of people throughout the world on February 15 of this year against the planned invasion of Iraq. We did not succeed in stopping the American and British invasion, but we have surely contributed to delegitimizing the Occupation and made it increasingly difficult for invaders that brazenly violated international law and many rules of the Geneva Convention to remain in Iraq.

The *New York Times*, on the occasion of the February 15 march, said that there are only two superpowers left in the world today, the United States and global civil society. Let me add that I have no doubt that the forces of justice and peace will prevail over the contemporary incarnation of empire, blood, terror and greed that is the U.S.A.

Our movement is on the ascendant. But our agenda is massive, our tasks formidable. To name just a few: We have to drive the U.S. out of Iraq and Afghanistan. We must stop Israel from destroying the Palestinian people. We must impose the rule of law on outlaw rogue states like the U.S., Britain and Israel.

But above all, we must change the rules of the global economy, for it is the logic of global capitalism that is the source of the disruption of society and of the environment. The challenge is that even as we deconstruct the old, we dare to imagine and win over people to our visions and programmes for the new.

Contrary to the claims of the ideologues of the establishment, the principles that would serve as the pillars of a new global order are present. The primordial principle is that instead of the economy, the market, driving society, the market must be—to use the image of the great Hungarian social democrat Karl Polanyi—"reembedded" in society and governed by the overarching values of community, solidarity, justice and equity. At the international level, the global economy must be deglobalized or rid of the distorting, disfiguring logic of corporate profitability and truly internationalized, meaning that participation in the international economy must serve to strengthen and develop rather than disintegrate and destroy local and national economies.

The perspective and principles are there; the challenge is how each society can articulate these principles and programmes in unique ways that respond to their values, their rhythms, their personality as societies. Call it post-modern, but central to our movement is the conviction that, in contrast to the belief common to both neoliberalism and bureaucratic socialism, there is no one shoe that will fit all. It is no longer a question of an alternative but of alternatives.

But there is an urgency to the task of articulating credible and viable alternatives to the global community, for the dying spasms of old orders have always presented not just great opportunity but great risk. At the beginning of the 20th century, the revolutionary thinker Rosa Luxemburg made her famous comment about the possibility that the future might belong to "barbarism." Barbarism in the form of fascism nearly triumphed in the 1930s and 1940s. Today, corporate-driven globalization is creating so much of the same instability, resentment and crisis that are the breeding grounds of fascist, fanatical and authoritarian populist movements. Globalization not only has lost its promise but it is embittering many. The forces representing human solidarity and community have no choice but to step in quickly to convince the disenchanted masses that, indeed, as the banner of the World Social Forum in Porto Alegre [Brazil,] proclaims, "Another world is possible." For the alternative is, as in the 1930s, to see the vacuum filled by terrorists, demagogues of the religious and secular Right, and the purveyors of irrationality and nihilism.

The future, dear friends, is in the balance.

POSTSCRIPT

Is Capitalism the Best Model for the Global Economy?

This debate like the others in this volume features a yes and a no and represents the reality that tough choices often have to be made. Nevertheless, it would be wrong to view this debate as an "either-or" issue for two reasons. First, if the answer to global prosperity is not capitalism, than what is the alternative? As noted, there are two major alternative economic theories—economic nationalism and economic structuralism—and each has many variations and nuances. It is also the case that this debate is not a simple dichotomy because, at least in practice, none of these theories are pure. For instance, pure communism has never existed in a complex society any more than pure capitalism.

As such, it is best to begin with your basic philosophical inclinations after reading these debates, then to do more exploration of the subject with an open mind to test your intellectual leanings. As with some of the debates, you can hone your ideas and learn a great deal about the views behind the views expressed in the readings by finding out more about the individuals whose views you have read. Two recommended sources on Walden Bellow are the Web site of the organization, Focus on the Global South, that he heads at http://www.focusweb.org/ and an interview regarding his background and work published in the *New Left Review* (July/August 2002) and available on the Web at http://www.newleftreview.com/. In much the same way, more about Johan Norberg can be gained by visiting the Web site for Timbro, the think tank where he works at http://www.timbro.com/ and by accessing his personal Web page at http://www.johannorberg.net/. One of the things you will find is that the ideas of both authors have evolved. Norberg in his teens started his own political party, the Anarchist Front, but later abandoned its anti-industrial views. Bello was once a member of the Philippine Communist Party, but later found its prescriptions and practices wanting and left it. You can also read Norberg's book, *In Defense of Global Capitalism* (Cato Institute, 2003) and Walden Bello's *Deglobalization: Ideas for a New World Economy* (Zed Books, 2005).

There are a number of other works that will also help you understand the nature of capitalism and its role in global development. A good beginning point is Hugo Radice, *Political Economy of Global Capitalism* (Routledge, 2005). Then to explore the development of global capitalism, read Berch Berberoglu, *Globalization and Change: The Origins, Development, and Transformation of Global Capitalism* (Lexington Books, 2004). Commentary on how to reform global capitalism can be found in John H. Dunning, *Making Globalization Good: The Moral Challenges of Global Capitalism* (Oxford University Press, 2005).

ISSUE 13

Should the Rich Countries Forgive All the Debt Owed by the Poor Countries?

YES: Romilly Greenhill, from "The Unbreakable Link—Debt Relief and the Millennium Development Goals," a Report from Jubilee Research at the New Economics Foundation (February 2002)

NO: William Easterly, from "Debt Relief," *Foreign Policy* (November/ December 2001)

ISSUE SUMMARY

YES: Romilly Greenhill, an economist with Jubilee Research at the New Economics Foundation, contends that if the world community is going to achieve its goal of eliminating world poverty by 2015, as stated in the UN's Millennium Declaration, then there is an urgent need to forgive the massive debt owed by the heavily indebted poor countries.

NO: William Easterly, a senior adviser in the Development Research Group at the World Bank, maintains that while debt relief is a popular cause and seems good at first glance, the reality is that debt relief is a bad deal for the world's poor.

L̲ike all countries, the world's poor states, or less developed countries (LDCs), are determined to improve their economic circumstances. However, most LDCs find it difficult to raise capital internally because their poor economic bases leave little for them to tax and leave almost nothing for savings and investment by the countries' citizens. Moreover, a great deal of the goods and services that these countries need to develop has to be imported, and most businesses and countries will not accept payment in the generally unstable currencies of the LDCs.

Therefore, to move forward the LDCs not only need significant amounts of development capital, but those funds have to be American dollars, Japanese yen, European euros, or one of the other "hard currencies" acceptable in world markets. A key concern for LDCs is where to obtain hard currency to use for development. There are four main sources: loans, investment, trade, and aid.

This issue takes up some of the controversy raised by problems with the first of these four sources: loans.

Loans flow to LDCs from international organizations (such as the World Bank); from other, wealthier countries; and from private investors (such as banks). There is nothing unusual about such transactions. Indeed, throughout the 1800s American businesses borrowed huge sums to finance the building of the nation's railroads and steel mills and for other means of economic development. And when it is running a deficit, the U.S. government still borrows tens of billions of dollars annually from foreign investors to finance the national debt.

The background to the current debt problem can be traced to the 1970s, when a number of factors converged to persuade the LDCs to borrow heavily to finance their development needs. The reasons are complex, but suffice it to say that LDCs were in dire need of funds and the lenders in the economically developed countries (EDCs) had surplus capital, which they urged the LDCs to borrow. By 1982 LDC international debt had sharply risen to $849 billion, and by 2002 it stood at $2.2 trillion.

For a variety of reasons, some of which were the fault of the LDCs (poor planning, corruption, political instability) and some of which were not the fault of the LDCs (a lagging world economy and other barriers to their exports), many of the LDCs have found themselves severely burdened by their debt service (principal and interest payments). There are two levels to this problem. One level has to do with the drain that the debt service puts on struggling LDCs. Export earnings are one source of funds to pay off debt, but in 2002 the LDCs owed more than 135 percent of their export revenue. In dollar figures, this mean that the LDCs paid out $325 billion (14 percent of their annual export earnings) that year to meet their principal and interest obligations. Supporters of debt reduction or total forgiveness argue that those funds should be reinvested in the LDCs rather than sent abroad.

The second level occurs when countries cannot meet their debt service. In essence, they are bankrupt. Such collapses harm the borrowers (whose credit is ruined and whose economies sometimes collapse), the lenders (who lose their money), and, arguably, the global community, which must continue to deal with all the ills that are rooted in the grinding poverty of a good percentage of the world's people and countries.

Some analysts, including Romilly Greenhill, the author of the first of the following selections, argue that in both the short and long runs it would be best to wipe out the debt of the heavily indebted poor countries. There are various ways to do this. One limited debt-relief program that was instituted in the early 1990s was the Brady Plan (after Secretary of the Treasury Nicholas Brady). Under it, banks forgave over $100 billion of what the LDCs owed (and took deductions on their corporate taxes), lowered interest rates on continuing loans, and made new loans at low rates. In return, international organizations (the World Bank, the International Monetary Fund) and the governments of the capital exporting countries guaranteed the loans and increased their own lending to the LDCs.

Not everyone agrees with such across-the-board debt-relief plans. William Easterly is one such opponent, and in the second selection, he maintains that forgiving debts will only encourage the poor practices that led to the debt crisis in the first place.

YES

Romilly Greenhill

The Unbreakable Link—Debt Relief and the Millennium Development Goals

At the start of the new millennium, the world's leaders met in the United Nations General Assembly to set out a new global vision for humanity. In their Millennium Declaration, the statesmen and women recognised their 'collective responsibility to uphold the principles of human dignity, equality and equity at the global level.' They pledged to 'spare no effort to free our fellow men, women and children from the abject and dehumanising conditions of extreme poverty.'

From these fine words, a set of goals was born: to eliminate world poverty by the year 2015; to achieve universal primary education; to promote gender equality and empower women; to reduce child mortality; improve maternal health; to combat HIV/AIDs and other diseases; and to ensure environmental sustainability. . . .

Since then, the Millennium Development Goals [MDGs]—as they were subsequently named—have been adopted by all major donor agencies as guiding principles for their strategies for poverty eradication. . . .

Moreover, since the adoption of the MDGs in the year 2000, events have conspired to reinforce the urgent need for poverty reduction in the world. According to Gordon Brown, the aftermath of September 11th has shown that 'the international community must take strong action to tackle injustice and poverty. . . . [and to] achieve our 2015 Millennium Development Goals.'

But meeting the 2015 targets requires resources. Ernest Zedillo, in his report of the High Level Panel for Financing for Development, has assessed that total additional resources of $50bn per year will be needed to meet these targets worldwide, over and above the current level of spending in key areas. This estimate is based on detailed costings in some of the key goal areas by UN bodies such as UNICEF, the World Health Organisation, and others such as the World Bank.

The UN Millennium Declaration was not the only remarkable event of the year 2000. Equally notable—though perhaps more poignant—was the winding down of the Jubilee 2000 campaign, described by Kofi Annan as 'the voice of the world's conscience and indefatigable fighters for justice.' The Jubilee 2000 coalition had campaigned for the cancellation of the un-payable debts of the poorest countries by the end of 2000, under a fair and transparent process. Their petition—the largest ever—had been signed by 24 million people worldwide.

The central message of the Jubilee 2000 campaign was that human rights should not be subordinated to money rights. Poor countries prepared to commit resources to meeting the basic needs and economic rights of their populations should not be prevented from doing so because of the need to pay back debts to rich creditor countries and institutions.

The Jubilee 2000 campaign had won a commitment to a $110bn write off of un-payable debts. This was to be achieved partly through an extension of the World Bank's Heavily Indebted Poor Countries (HIPC) initiative, and partly through additional bilateral commitments from creditors such as the UK.

But it is now clear that the HIPC initiative is not delivering enough either to produce the promised 'robust exit' from unsustainable debts or to meet internationally agreed poverty reduction goals. . . . [B]y the end of 2001—a full year after the millennium deadline called by the Jubilee 2000 coalition—only four countries had passed through all the hoops of the HIPC initiative. Out of the 42 countries included in the process, almost half of these had not even reached 'decision point', after which they receive some interim relief on their debt service payments. Moreover, even when relief is provided, research by Jubilee Plus has shown that debt burdens remain unsustainable. . . .

We show in this report that if poor country governments are to have suffi-cient resources to meet the MDGs, as well as to meet other essential expenditure needs and pro-poor investments, *the 42 HIPC countries as a whole cannot afford to make any debt service payments.* In fact, we find that *even if all the debts of these 42 countries are cancelled, the HIPCs will need an additional $30bn in aid each year if there is any hope of meeting goal 1 while for the other goals, a total of $16.5bn will be needed.*

These figures are based on actual debt service payments for 1999—before most of the HIPCs had received any substantial debt service relief from the HIPC initiative. But as the latest figures from the World Bank have shown, even when all 42 countries have fully passed through the HIPC initiative, the savings will only amount to a paltry $3.5bn per year. It is clear that much deeper, and faster, debt relief must be provided. . . .

Debt Service Payments Take Resources From the MDGs

Calculating the resources needed to meet the MDGs in each country is no easy task. Data on the number of poor people in each country, the current level of indicators such as HIV and malarial prevalence, or even the number of children in school, is often not available, or not reliable. Moreover, working out the exact amount that will need to be spent across different countries to meet common objectives requires making heroic assumptions about costs in each country. Some of the goals—such as 'reversing the loss of environmental resources' are inherently very difficult to evaluate. . . .

Goal 1: Eradicate extreme poverty and hunger

- Halve, between 1990 and 2015, the proportion of people whose income is less than one dollar a day
- Halve, between 1990 and 2015, the proportion of people who suffer from hunger.

Eradicating mass poverty is often seen as the most fundamental of the MDGs. In the simplistic world of the donor community, extreme poverty is defined as living on less than one dollar a day. . . .

Of all the MDGs, this goal is also the most difficult to relate to debt service payments. It is clear that debt repayments are taking resources that could be spent to reduce poverty, but quantifying the exact linkages is much more difficult. . . .

Our overall point is simply that by using standard and widely accepted economic models, we can show that in a world of finite development resources, debt repayments will be traded off with limited poverty reduction expenditures.

Goal 2: Achieving universal primary education

- Ensure that, by 2015, children everywhere, boys and girls alike, will be able to complete a full course of primary schooling

Access to primary education is a basic human right. Education benefits individuals, their families, and also society as a whole, by enabling greater participation in democratic processes. Education serves to empower individuals, helps them to take advantage of economic opportunities, and improves their health and that of their family.

Yet, in 2000, *one in three* children across the developing world did not complete the 5 years of basic education which UNICEF believes is the minimum required to achieve basic literacy. We are clearly a long way from achieving the Millennium Development Goal of achieving Universal Primary Education by 2015.

UNICEF has calculated the amount that countries will need to spend in order to meet the MDGs. They found that almost all of the HIPCs will need to increase spending on education. . . .

[T]he HIPC countries will only need to spend $6.5bn each year in order to ensure that every child gets an education sufficient to ensure basic literacy. While large relative to the incomes of HIPCs, on a global scale this figure is miniscule—representing, for example, less than half of one percent of the projected U.S. defence budget of $1,600bn over the next five years. And only $1.2bn of this is additional to what governments are currently spending—although [some] countries . . . will need much larger increases in spending than some of the other HIPCs.

Goal 4: Reducing child mortality

- Reduce, by two-thirds, between 1990 and 2015, the under-five mortality rate

Goal 5: Improving maternal health

- Reduce by three quarters, between 1990 and 2015, the maternal mortality ratio

Goal 6: Combating HIV/AIDS, malaria and other diseases

- Have halted by 2015, and begun to reverse, the spread of HIV/AIDS
- Have halted by 2015, and begun to reverse, the incidence of malaria and other related diseases.

A tragedy is unfolding in Africa. Within the last 24 hours, 5,500 Africans were killed by HIV/AIDS. One in five of all adults in Africa are infected by the virus, while 17 million Africans have died from AIDS since the start of the epidemic. AIDS has so far left 13 million children orphaned, a figure which will grow to 40 million by 2010 if no action is taken.

Moreover, AIDS is not the only killer. Other diseases such as malaria, TB, childhood infectious diseases, maternal and prenatal conditions and micronutrient deficiencies abound. Average life expectancy in Africa has *fallen* since 1980, from 48 to 47—and in individual countries, the fall is much more extreme. Life expectancy in Zambia is now only 38 years, down from 50 years in 1980, while Sierra Leone has a life expectancy of only 37 years. And even these figures mask the catastrophic impact on children. In Africa, 161 children out of every 1,000 children will die before their fifth birthday; in Niger, this figure is as high as one in four.

Yet, the Global Commission on Macroeconomics and Health has estimated that *eight million* lives could be spared each year if a simple set of health interventions needed to meet the MDGs were put in place.

. . . [T]he 39 HIPCs will between them need to spend $20bn each year on health if the MDGS are to be met—almost three times their 1999 levels of debt service. This figure may sound large, but it is only slightly more than the $17bn spent each year in Europe and the US on pet food. As with education, larger countries will need bigger increases in health spending. . . .

The need for more debt relief is evident. The . . . vast improvements in the lives of millions of people in poor countries are achievable, with an increase in expenditure totalling only 0.1% of GDP of the rich donor and creditor countries. Yet, despite this overwhelming imperative, the poorest countries are still paying debt service of $8bn per year.

Target 10: Halve, by 2015, the proportion of people without sustainable access to safe drinking water.   Like education and health care, access to safe water is a basic right. Safe water is vital for proper health and hygiene, including the prevention of water borne diseases. Distances travelled to fetch water result in a huge loss of time for poor people, particularly women and children. Yet, *one billion* people currently lack safe drinking water and almost *three billion—half the world's population*—lack adequate sanitation. *Two million* children die each year from water-related diseases. As the Vision 21 Framework for Action states, this situation is 'humiliating, morally wrong and oppressive.' . . .

In order to meet the MDG of halving the proportion of people without sustainable access to safe drinking water, . . . the HIPCs would have to spend only $2.4bn per year on water and sanitation—less than Europe spends on alcohol over ten days.

Target 11: By 2020, to have achieved a significant improvement in the lives of at least 100 million slum dwellers. . . . Hundreds of millions of the urban poor in developing countries currently live in unsafe and unhygienic environments where they face multiple threats to their health and security. . . . [T]he HIPCs will need to spend 1% of GDP annually on improving slum conditions. In total, this comes to *$1.7bn* for all the 39 HIPCs considered.

Other Goals and Targets

Goal 3: Promoting gender equality and empowering women

- Eliminate gender disparity in primary and secondary education preferably by 2005 and to all levels of education no later than 2015

Target 9: Integrate the principles of sustainable development into country policies and programmes and reverse the loss of environmental resources. Providing basic health, education and water to the populations of poor countries is clearly vital and should be given preference over debt service payments. But at the same time, other dimensions of development—such as promoting gender equality and protecting environmental resources, are also needed if development is to be sustainable in the long run.

Unfortunately, however, these goals are inherently difficult to cost, and are therefore difficult to compare with debt service payments.

Total Required to Meet MDGs

. . . [T]he total funds required each year to meet MDGs 2 to 7 are not exorbitant. . . . [A] mere $30.6bn per year is required. This figure may be small in global terms. But . . . it represents 18% of GDP for the 42 HIPCs as a whole, and a staggering 355% of their debt service. . . .

Linking Debt Servicing to the Millennium Development Goals

Even without servicing their external debts, it is clear that the 39 HIPCs face a formidable challenge if they are to raise the level of resources required to meet the MDGs.

While it is true that governments can raise their own revenues by taxing their domestic populations, in most of the HIPC countries the extreme poverty experienced means that governments find it very difficult to raise the kind of resources needed.

Debt servicing worsens this position by diverting preciously needed resources, which could be used for saving lives, and educating children, towards rich country creditors.

Moreover, governments cannot be spending all their revenues on social expenditures. Crucial expenditures such as maintaining law and order, public administration, essential infrastructures such as roads, policing and defence are also needed. . . .

Given the current ability of the HIPC governments to raise revenues, and given the essential expenditures needed to meet the MDGs and for other essential expenditures, we now ask: how much can the HIPCs afford to pay to their rich country creditors in debt service payments? . . .

Our analysis shows that, as a whole, *the HIPCs have no spare resources available that could be used for debt servicing*. In fact, even with 100% debt cancellation, *the HIPCs will require an additional $16.5bn if goals 2 to 7 are to be met*, and this is without the additional $30bn needed for goal 1. . . .

Conclusion: The Need for a Sabbath Economics

Our conclusion is clear. If the Millennium Development Goals are to be met, all of the HIPCs will need full cancellation of all of their debts. This is not an act of charity, but a moral imperative. While eight million die each year for want of the funds spent by the rich countries on their pets; when millions of children stay out of school for want of half a percent of the U.S. defence budget; and when the amount spent on alcohol in a week and a half in Europe would be adequate to provide sanitation to half the world's population, something is very wrong.

Maybe it is time, once again, to call on biblical principles. The 'Jubilee' principle—which provides ways of reversing the relentless flow of resources from the poor to the rich, and narrowing the gap between—formed the foundation of one of the most successful global campaigns ever. But there are others. The central tenets of 'Sabbath Economics' are that the world is abundant and provides enough for everyone—but that human communities must restrain their appetites and live within limits. For Sabbath Economics, disparities in wealth and power are not natural, but come about through sin, and must be mitigated within the community through redistribution.

We do not have to believe in God—or indeed any religion—to accept these principles. It is enough for us to recognise that more than a billion people do not need to live in poverty while their debts continue to be repaid. The current HIPC initiative does not and cannot do enough to bring down the unsustainable debt burden of the world's poorest countries. If the Millennium Development Goals are to be met, there is no alternative but to provide a new framework for debt relief—one which respects the human rights of the poor.

NO

William Easterly

Debt Relief

"Jubilee 2000 Sparked the Debt Relief Movement"

No Sorry, Bono, but debt relief is not new. As long ago as 1967, the U.N. Conference on Trade and Development argued that debt service payments in many poor nations had reached "critical situations." A decade later, official bilateral creditors wrote off $6 billion in debt to 45 poor countries. In 1984, a World Bank report on Africa suggested that financial support packages for countries in the region should include "multi-year debt relief and longer grace periods." Since 1987, successive G-7 [Group of Seven—Canada, France, Germany, Great Britain, Italy, Japan, and the United States] summits have offered increasingly lenient terms, such as postponement of repayment deadlines, on debts owed by poor countries. . . . In the late 1980s and 1990s, the World Bank and International Monetary Fund (IMF) began offering special loan programs to African nations, essentially allowing governments to pay back high-interest loans with low-interest loans—just as real a form of debt relief as partial forgiveness of the loans. The World Bank and IMF's more recent and well-publicized Highly Indebted Poor Countries (HIPC) debt relief program therefore represents but a deepening of earlier efforts to reduce the debt burdens of the world's poorest nations. Remarkably, the HIPC nations kept borrowing enough new funds in the 1980s and 1990s to more than offset the past debt relief: From 1989 to 1997, debt forgiveness for the 41 nations now designated as HIPCs reached $33 billion, while new borrowing for the same countries totaled $41 billion.

So by the time the Jubilee 2000 movement began spreading its debt relief gospel in the late 1990s, a wide constituency for alleviating poor nations' debt already existed. However, Jubilee 2000 and other pro–debt relief groups succeeded in raising the visibility and popularity of the issue to unprecedented heights. High-profile endorsements range from Irish rock star Bono to Pope John Paul II and the Dalai Lama to Harvard economist Jeffrey Sachs; even retiring U.S. Sen. Jesse Helms has climbed onto the debt relief bandwagon. In that respect, Jubilee 2000 (rechristened "Drop the Debt" before the organization's campaign officially ended on July 31, 2001) should be commended for putting the world's poor on the agenda—at a time when most people in rich

From *Foreign Policy*, November/December 2001, pp. 20–26. Copyright © 2001 by Foreign Policy. Reproduced with permission of Foreign Policy as conveyed via Copyright Clearance Center.

nations simply don't care—even if the organization's proselytizing efforts inevitably oversimplify the problems of foreign debt.

"Third World Debts Are Illegitimate"

Unhelpful idea Supporters of debt relief programs have often argued that new democratic governments in poor nations should not be forced to honor the debts that were incurred and mismanaged long ago by their corrupt and dictatorial predecessors. Certainly, some justice would be served if a legitimate and reformist new government refused to repay creditors foolish enough to have lent to a rotten old autocracy. But, in reality, there are few clear-cut political breaks with a corrupt past. The political factors that make governments corrupt tend to persist over time. How "clean" must the new government be to represent a complete departure from the misdeeds of an earlier regime? Consider President Yoweri Museveni of Uganda, about the strongest possible example of a change from the past—in his case, the notorious past of Ugandan strongman Idi Amin. Yet even Museveni's government continues to spend money on questionable military adventures in the Democratic Republic of the Congo. Would Museveni qualify for debt relief under the "good new government" principle? And suppose a long-time corrupt politician remains in power, such as Kenyan President Daniel Arap Moi. True justice would instead call for such leaders to pay back some of their loot to development agencies, who could then lend the money to a government with cleaner hands—a highly unlikely scenario.

Making debt forgiveness contingent on the supposed "illegitimacy" of the original borrower simply creates perverse incentives by directing scarce aid resources to countries that have best proved their capacity to mismanage such funds. For example, Ivory Coast built not just one but two new national capitals in the hometowns of the country's previous rulers as it was piling up debt. Then it had a military coup and a tainted election. Is that the environment in which aid will be well used? Meanwhile, poor nations that did not mismanage their aid loans so badly—such as India and Bangladesh—now do not qualify for debt relief, even though their governments would likely put fresh aid resources to much better use.

Finally, the legitimacy rationale raises serious reputation concerns in the world's financial markets. Few private lenders will wish to provide fresh financing to a country if they know that a successor government has the right to repudiate the earlier debt as illegitimate. For the legitimacy argument to be at all convincing, the countries in question must show a huge and permanent change from the corruption of past regimes. Indeed, strict application of such a standard introduces the dread specter of "conditionality," i.e., the imposition of burdensome policy requirements on developing nations in exchange for assistance from international financial institutions. Only rather than focusing solely on economic policy conditions, the international lending agencies granting debt relief would now be compelled to make increasingly subjective judgments regarding a country's politics, governance structures, and adherence to the rule of law.

"Crushing Debts Worsen Third World Poverty"

Wrong in more ways than one   Yes, the total long-term debt of the 41 HIPC nations grew from $47 billion in 1980 to $159 billion in 1990 to $169 billion in 1999, but in reality the foreign debt of poor countries has always been partly fictional. Whenever debt service became too onerous, the poor nations simply received new loans to repay old ones. Recent studies have found that new World Bank adjustment loans to poor countries in the 1980s and 1990s increased in lock step with mounting debt service. Likewise, another study found that official lenders tend to match increases in the payment obligations of highly indebted African countries with an increase in new loans. Indeed, over the past two decades, new lending to African countries more than covered debt service payments on old loans.

Second, debt relief advocates should remember that poor people don't owe foreign debt—their governments do. Poor nations suffer poverty not because of high debt burdens but because spendthrift governments constantly seek to redistribute the existing economic pie to privileged political elites rather than try to make the pie grow larger through sound economic policies. The debt-burdened government of Kenya managed to find enough money to reward President Moi's home region with the Eldoret International Airport in 1996, a facility that almost nobody uses.

Left to themselves, bad governments are likely to engage in new borrowing to replace the forgiven loans, so the debt burden wouldn't fall in the end anyway. And even if irresponsible governments do not run up new debts, they could always finance their redistributive ways by running down government assets (like oil and minerals), leaving future generations condemned to the same overall debt burden. Ultimately, debt relief will only help reduce debt burdens if government policies make a true shift away from redistributive politics and toward a focus on economic development.

"Debt Relief Allows Poor Nations to Spend More on Health and Education"

No   In 1999, Jubilee 2000 enthused that with debt relief "the year 2000 could signal the beginning of dramatic improvements in healthcare, education, employment and development for countries crippled by debt." Unfortunately, such statements fail to recognize some harsh realities about government spending.

First, the iron law of public finance states that money is fungible: Debt relief goes into the same government account that rains money on good and bad uses alike. Debt relief enables governments to spend more on weapons, for example. Debt relief clients such as Angola, Ethiopia, and Rwanda all have heavy military spending (although some are promising to make cuts). To assess whether debt relief increases health and education spending, one must ask what such spending would have been in the absence of debt relief—a difficult question. However, if governments didn't spend the original loans on helping the poor, it's a stretch to expect them to devote new fiscal resources toward helping the poor.

Second, such claims assume that the central government knows where its money is going. A recent IMF and World Bank study found that only two out of 25 debt relief recipients will have satisfactory capacity to track where government spending goes within a year. At the national level, an additional study found that only 13 percent of central government grants for non-salary education spending in Uganda (another recipient of debt relief) actually made it to the local schools that were the intended beneficiaries.

Finally the very idea that the proceeds of debt relief should be spent on health and education contains a logical flaw. If debt relief proceeds are spent on social programs rather than used to pay down the debt, then the debt burden will remain just as crushing as it was before. A government can't use the same money twice—first to pay down foreign debt and second to expand health and education services for the poor. This magic could only work if health and education spending boosted economic growth and thus generated future tax revenues to service the debt. Unfortunately, there is little evidence that higher health and education spending is associated with faster economic growth.

"Debt Relief Will Empower Poor Countries to Make Their Own Choices"

Not really   Pro–debt relief advocacy groups face a paradox: On one hand, they want debt relief to reach the poor; on the other, they don't want rich nations telling poor countries what to do. "For debt relief to work, let the conditions be set by civil society in our countries, not by big world institutions using it as a political tool," argued Kennedy Tumutegyereize of the Uganda Debt Network. Unfortunately, debt relief advocates can't have it both ways. Civil society remains weak in most highly indebted poor countries, so it would be hard to ensure that debt relief will truly benefit the poor unless there are conditions on the debt relief package.

Attempting to square this circle, the World Bank and IMF have made a lot of noise about consulting civil society while at the same time dictating incredibly detailed conditions on debt relief. The result is unlikely to please anyone. Debt relief under the World Bank and IMF's current HIPC initiative, for example, requires that countries prepare Poverty Reduction Strategy Papers. The World Bank's online handbook advising countries on how to prepare such documents runs well over 1,000 pages and covers such varied topics as macroeconomics, gender, the environment, water management, mining, and information technology. It would be hard for even the most skilled policymakers in the advanced economies to follow such complex (no matter how salutary) advice, much less a government in a poor country suffering from scarcity of qualified managers. In reality, this morass of requirements emerged as the multilateral financial institutions sought to hit on all the politically correct themes while at the same time trying hard to make the money reach the poor. If the conditions don't work—and of course they won't—the World Bank and IMF can simply fault the countries for not following their advice.

"Debt Relief Hurts Big Banks"

Wrong   During the 1970s and early 1980s, large commercial banks and official creditors based in rich nations provided substantial loans at market interest rates to countries such as Ivory Coast and Kenya. However, they pulled out of these markets in the second half of the 1980s and throughout the 1990s. In fact, from 1988 to 1997, such lenders received more in payments on old loans than they disbursed in new lending to high-debt poor countries. The multilateral development banks and bilateral lenders took their place, offering low-interest credit to poor nations. It's easy to understand why the commercial and official creditors pulled out. Not only did domestic economic mismanagement make high-debt poor countries less attractive candidates for potential loans, but with debt relief proposals in the air as early as 1979, few creditors wished to risk new lending under the threat that multilateral agencies would later decree loan forgiveness.

The IMF and World Bank announced the HIPC initiative of partial and conditional forgiveness of multilateral loans for 41 poor countries in September 1996. By the time the debt relief actually reached the HIPCs in the late 1990s, the commercial banks and high-interest official creditors were long gone and what was being forgiven were mainly "concessional" loans—i.e., loans with subsidized interest rates and long repayment periods. So really, debt relief takes money away from the international lending community that makes concessional loans to the poorest nations, potentially hurting other equally poor but not highly indebted nations if foreign aid resources are finite (as, of course, they are). Indeed, a large share of the world's poor live in India and China. Neither nation, however, is eligible for debt relief.

"Debt Relief Boosts Foreign Investment in Poor Nations"

A leap of faith   It is true that forgiving old debt makes the borrowers more able to service new debt, which in theory could make them attractive to lenders. Nevertheless, the commercial and official lenders who offer financing at market interest rates will not want to come back to most HIPCs any time soon. These lenders understand all too well the principle of moral hazard: Debt relief encourages borrowers to take on an excessive amount of new loans expecting that they too will be forgiven. Commercial banks obviously don't want to get caught with forgiven loans. And even the most charitable official lenders don't want to sign their own death warrants by getting stuck with forgiven debt. Both commercial and official lenders may want to redirect their resources to safer countries where debt relief is not on the table. Indeed, in 1991, the 47 least developed countries took in 5 percent of the total foreign direct investment (FDI) that flowed to the developing world; by 2000 their portion had dropped to only 2.5 percent. (Over the same period, the portion of global FDI captured by all developing nations dropped as well, from 22.3 to 15.9 percent.) Even capital flows to now lightly indebted "safe" countries might suffer from the perception that their debts also may be forgiven at some point.

Ultimately, only the arms of multilateral development banks that provide soft loans—with little or no interest and very long repayment periods—are going to keep lending to HIPCs, and only then under very stringent conditions.

"Debt Relief Will Promote Economic Reform"

Don't hold your breath   During the last two decades, the multilateral financial institutions granted "structural adjustment" loans to developing nations, with the understanding that governments in poor countries would cut their fiscal deficits and enact reforms—including privatization of state-owned enterprises and trade liberalization—that would promote economic growth. The World Bank and IMF made 1,055 separate adjustment loans to 119 poor countries from 1980 to 1999. Had such lending succeeded, poor countries would have experienced more rapid growth, which in turn would have permitted them to service their foreign debts more easily. Thirty-six poor countries received 10 or more adjustment loans in the 1980s and 1990s, and their average percentage growth of per capita income during those two decades was a grand total of zero. Moreover, such loans failed to produce meaningful reforms, and developing countries now cite this failure as justification for debt relief. Yet why should anyone expect that conditions on debt forgiveness would be any more effective in changing government policies and behavior than conditions on the original loans?

Partial and conditional debt forgiveness is a *fait accompli.* Expanding it to full and unconditional debt forgiveness—as some groups now advocate— would simply transfer more resources from poor countries that have used aid effectively to those that have wasted it in the past. The challenge for civil society, the World Bank, IMF, and other agencies is to ensure that conditional debt forgiveness really does lead to government reforms that enhance the prospects of poor countries.

How can we promote economic reform in the poorest nations without repeating past failures? The lesson of structural adjustment programs is that reforms imposed from the outside don't change behavior. Indeed, they only succeed in creating an easy scapegoat: Insincere governments can simply blame their woes on the World Bank and IMF's "harsh" adjustment programs while not doing anything to fundamentally change economic incentives and ignite economic growth. It would be better for the international financial institutions to simply offer advice to governments that ask for it and wait for individual countries to come forward with homegrown reform programs, financing only the most promising ones and disengaging from the rest. This approach has worked in promoting economic reform in countries such as China, India, and Uganda. Rushing through debt forgiveness and imposing complex reforms from the outside is as doomed to failure as earlier rounds of debt relief and adjustment loans.

POSTSCRIPT

Should the Rich Countries Forgive All the Debt Owed by the Poor Countries?

It is easy to say, "Well, don't borrow if you cannot pay it back" or "If you borrowed it, you have to pay it back even if you suffer." Those who would disagree with that approach might say it contains something of Victor Hugo's *Les Miserables* (1862). The novel's main character, Jean Valjean, is convicted of theft and spends 19 years doing hard labor in the quarries. The injustice is that all he stole was some bread, and he only did so to feed the children of his impoverished sister. Advocates of debt relief would argue that, like Jean Valjean, the LDCs cannot be faulted too harshly for borrowing to try to meet rudimentary development needs.

The debt relief advocates would also argue that if the LDCs were allowed to invest what they pay in debt service in their economies, then they might be able to develop more quickly and become better contributing members of the global community. Here again, the "charity" of forgiving the debt echoes *Les Miserables*. Returning to that story, the charity of a priest helps Jean Valjean, and he builds on that to become the effective and revered mayor of a small French town. The point is that whether Greenhill or Easterly is correct, the story of heavily indebted poor countries is not one of deadbeat nations simply looking to get out of their debts. A worthwhile study on the debt issue is Tony Addison, Henrik Hansen, and Finn Tarp, eds., *Debt Relief for Poor Countries* (Palgrave Macmillian, 2004).

What to do about the debt also has to be seen in the context of the poverty that a great deal of the world experiences and the immense gap between the 15 percent of so of the world's population who live in the wealthy EDCs and those who live in the LDCs. For example, the two dozen wealthiest countries in 2000 had an average per capita gross national product (GNP) of $27,680; the LDCs' per capita GNP was $1,230. That figure somewhat obscures the fact that there are over 50 countries where the annual per capita GNP is $750 or less. Another stark figure is that over 1 billion people, or more than 20 percent of the combined populations of the LDCs, live on the functional equivalent of a dollar a day. An analysis of the causes and implications of the gap between the LDCs and EDCs is Mitchell A. Seligson and John T. Passé, *Development and Underdevelopment: The Political Economy of Global Inequality* (Lynne Rienner, 2003).

What to do about the gulf between the rich and the poor and how to relieve the suffering of the billions who live below or near the poverty line are complex issues, and there is an extensive and important literature on development. A recent study is Akira Kohsak, *New Development Strategies: Beyond the Washington Consensus* (Palgrave Macmillan, 2004).

Disarmament Diplomacy

This site, maintained by the Acronym Institute for Disarmament Diplomacy, provides up-to-date news and analysis of disarmament activity, with a particular focus on weapons of mass destruction.

```
http://www.acronym.org.uk
```

The Center for Security Policy

The Web site of this Washington, D.C.–centered "think tank" provides a wide range of links to sites dealing with national and international security issues.

```
http://www.centerforsecuritypolicy.org
```

National Defense University

This leading center for joint professional military education is under the direction of the Chairman of the U.S. Joint Chiefs of Staff. Its Website is valuable for general military thinking and for material on terrorism.

```
http://www.ndu.edu/
```

Office of the Coordinator for Counterterrorism

This worthwhile site explores the range of terrorist threats and activities, albeit from the U.S. point of view, and is maintained by the U.S. State Department's Counterterrorism Office.

```
http://www.state.gov/s/ct/
```

Centre for the Study of Terrorism and Political Violence

The primary aims of the Centre for the Study of Terrorism and Political Violence are to investigate the roots of political violence; to develop a body of theory spanning the various and disparate elements of terrorism; and to recommend policy and organizational initiatives that governments and private sectors might adopt to better predict, detect, and respond to terrorism and terrorist threats.

```
http://www.st-and.ac.uk/academic/intrel/
              research/cstpv/
```

PART 4

Issues About Violence

*W*hatever we may wish, war, terrorism, and other forms of physical coercion are still important elements of international politics. Countries calculate both how to use the instruments of force and how to implement national security. There can be little doubt, however, that significant changes are under way in this realm as part of the changing world system. Strong pressures exist to expand the mission and strengthen the security capabilities of international organizations and to gauge the threat of terrorism. This section examines how countries in the international system are addressing these issues.

- Is Preemptive War an Unacceptable Doctrine?

- Is the War on Terrorism Succeeding?

ISSUE 14

Is Preemptive War an Unacceptable Doctrine?

YES: High Level Panel on Threats, Challenges, and Change, from "A More Secure World: Our Shared Responsibility," A Report to the Secretary General of the United Nations (December 2, 2004)

NO: Steven L. Kenny, "The National Security Strategy Under the United Nations and International Law," Strategy Research Project, U.S. Army War College (March 19, 2004)

ISSUE SUMMARNY

YES: The High Level Panel on Threats, Challenges, and Change, which was appointed by United Nations Secretary-General Kofi Annan in response to the global debate on the nature of threats and the use of force to counter them, concludes that in a world full of perceived potential threats, the risk to the global order posed by preemptive war is too great for its legality to be accepted.

NO: Colonel Steven L. Kenny argues in a research report he wrote at the U.S. Army War College, Carlisle Barracks, Pennsylvania, that substantial support from the acceptability of preemptive war results from such factors as the failure of the UN to enforce its charter, customary international law, and the growing threat of terrorists and weapons of mass destruction.

In September 2002 President George W. Bush issued "The National Security Strategy of the United States of America," an annual report to Congress required by the National Security Act of 1947. Seldom has one ignited the sort of fiery debate that the 2002 report sparked.

Undoubtedly the most controversial part of the president's policy, as noted in the introduction to Issue 2 in this volume, was his assertion that the United States had the right to strike enemies before they attacked Americans. Bush argued that the growing threat of terrorism and weapons of mass destruction had extended the meaning of self-defense to include preemptive action. As Bush put it in his report:

> . . . Our enemies have openly declared that they are seeking Weapons of mass destruction. . . . As a matter of common sense and self-defense, America will act

against such emerging threats before they are fully formed. . . . History will judge harshly those who saw the coming danger but failed to act. In the new world we have entered, the only path to peace and security is the path of action.

This claimed right to strike first and do so unilaterally was at the heart of what was soon dubbed the Bush Doctrine. As a statement of principle, the Bush Doctrine set off an intense debate, but it reached a crescendo when the United States applied the doctrine as part of its rationale for war against Iraq. When Bush addressed the nation on March 19, 2003 to announce the beginning of military operations against Iraq, he explained:

> The people of the United States . . . will not live at the mercy of an outlaw regime that threatens the peace with weapons of mass murder. We will meet that threat now, with our [military forces] so that we do not have to meet it later with armies of fire fighters and police and doctors on the streets of our cities.

As you will see in the following articles by the UN-based High Level Panel on Threats, Challenges, and Changes and by Steven L. Kenny, the doctrine of preemption is debated at both a practical level and legal level. At the pragmatic level it is important to consider both near-term and long-term consequences. Certainly, attacking an enemy country or terrorist organization before it attacks you has its obvious attractions. But what if your sense of threat is wrong, and people on both sides are killed and wounded? Additionally, the "golden rule" has to be considered, and asserting the right to preemptive war means that others also have the right to strike first if they feel threatened.

Legally, there are two main points to consider. One is the traditional "just war" theory that holds in part that *jus ad bellum* (just cause of war) exists only when war is (1) a last resort, (2) declared by legitimate authority, (3) waged in self-defense or to establish/restore justice, and (4) fought to bring about peace. Some might dismiss these standards a mere philosophy, but the United States applied them as part of the charges against Japanese and German leaders after World War II, tried those individuals in international tribunals, and imprisoned and even executed them after they were convicted. Moreover, just war theory is part of what underpins the definitions of illegal warfare that is subject to the newly established International Criminal Court (see Issue 18). The legal aspect of preemptive war also involves the obligations of all countries, including the United States, that signed the UN Charter. Under Article 2 of that treaty, signatories pledge to "refrain in their international relations from the threat or use of force against the territorial integrity or political independence of any state." And while Article 51 stipulates, "Nothing in the present Charter shall impair the inherent right of individual or collective self-defense if an armed attack occurs," Article 39 specifies that in all other cases, the "Security Council shall determine the existence of any threat to the peace, breach of the peace, or act of aggression and . . . decide what measures shall be taken . . . to maintain or restore international peace and security."

Collective Security and the Use of Force

What happens if peaceful prevention fails? If none of the preventive measures so far described stop the descent into war and chaos? If distant threats do become imminent? Or if imminent threats become actual? Or if a non-imminent threat nonetheless becomes very real and measures short of the use of military force seem powerless to stop it?

We address here the circumstances in which effective collective security may require the backing of military force, starting with the rules of international law that must govern any decision to go to war, if anarchy is not to prevail. It is necessary to distinguish between situations in which a State claims to act in self-defense; situations in which a State is posing a threat to others outside its borders; and situations in which the threat is primarily internal and the issue is the responsibility to protect a State's own people. In all cases, we believe that the Charter of the United Nations, properly understood and applied, is equal to the task: Article 51 needs neither extension nor restriction of its long-understood scope, and Chapter VII fully empowers the Security Council to deal with every kind of threat that States may confront. The task is not to find alternatives to the Security Council as a source of authority but to make it work better than it has.

That force *can* legally be used does not always mean that, as a matter of good conscience and good sense, it *should* be used. We identify a set of guidelines— five criteria of legitimacy—which we believe the Security Council (and anyone else involved in these decisions) should always address in considering whether to authorize or apply military force. The adoption of these guidelines (serious-ness of threat, proper purpose, last resort, proportional means and balance of consequences) will not produce agreed conclusions with push-button predict-ability, but should significantly improve the chances of reaching international consensus on what have been in recent years deeply divisive issues.

We also address here the other major issues that arise during and after violent conflict, including the needed capacities for peace enforcement, peace-keeping and peacebuilding, and the protection of civilians. A central recurring theme is the necessity for all members of the international community, devel-oped and developing States alike, to be much more forthcoming in providing and supporting deployable military resources. Empty gestures are all too easy to

Secretary General of the United Nations, December 2, 2004.

make: an effective, efficient and equitable collective security system demands real commitment.

Using Force: Rules and Guidelines

The framers of the Charter of the United Nations recognized that force may be necessary for the "prevention and removal of threats to the peace, and for the suppression of acts of aggression or other breaches of the peace." Military force, legally and properly applied, is a vital component of any workable system of **collective** security, whether defined in the traditional narrow sense or more broadly as we would prefer. But few contemporary policy issues cause more difficulty, or involve higher stakes, than the principles concerning its use and application to individual cases.

The maintenance of world peace and security depends importantly on there being a common global understanding, and acceptance, of when the application of force is both legal and legitimate. One of these elements being satisfied without the other will always weaken the international legal order— and thereby put both State and human security at greater risk.

A. The Question of Legality

The Charter of the United Nations, in Article 2, expressly prohibits Member States from using or threatening force against each other, allowing only two exceptions: self-defense under Article 51, and military measures authorized by the Security Council under Chapter VII [Articles 39–51] (and by extension for regional organizations under Chapter VIII [Articles 52–54]) in response to "any threat to the peace, breach of the peace or act of aggression."

For the first 44 years of the United Nations, Member States often violated these rules and used military force literally hundreds of times, with a paralyzed Security Council passing very few Chapter VII resolutions and Article 51 only rarely providing credible cover. Since the end of the cold war, however, the yearning for an international system governed by the rule of law has grown. There is little evident international acceptance of the idea of security being best preserved by a balance of power, or by any single—even benignly motivated— superpower.

But in seeking to apply the express language of the Charter, three particularly difficult questions arise in practice: first, when a State claims the right to strike preventively, in self-defense, in response to a threat which is not imminent; secondly, when a State appears to be posing an external threat, actual or potential, to other States or people outside its borders, but there is disagreement in the Security Council as to what to do about it; and thirdly, where the threat is primarily internal, to a State's own people.

Article 51 of the Charter of the United Nations and self-defense The language of this article is restrictive: "Nothing in the present Charter shall impair the inherent right of individual or collective self-defense if an armed attack occurs against a member of the United Nations, until the Security Council has taken measures to maintain international peace and security." However, a threatened State, according to long established international law, can take military action as long as the threatened attack is *imminent,* no other means would deflect it and the action is proportionate. The problem arises where the threat in question is not imminent but still claimed to be real: for example the acquisition, with allegedly hostile intent, of nuclear weapons-making capability.

Can a State, without going to the Security Council, claim in these circumstances the right to act, in anticipatory self-defense, not just preemptively (against an imminent or proximate threat) but preventively (against a non-imminent or non-proximate one)? Those who say "yes" argue that the potential harm from some threats (e.g., terrorists armed with a nuclear weapon) is so great that one simply cannot risk waiting until they become imminent, and that less harm may be done (e.g., avoiding a nuclear exchange or radioactive fallout from a reactor destruction) by acting earlier.

The short answer is that if there are good arguments for preventive military action, with good evidence to support them, they should be put to the Security Council, which can authorize such action if it chooses to. If it does not so choose, there will be, by definition, time to pursue other strategies, including persuasion, negotiation, deterrence and containment—and to visit again the military option.

For those impatient with such a response, the answer must be that, in a world full of perceived potential threats, the risk to the global order and the norm of non-intervention on which it continues to be based is simply too great for the legality of unilateral preventive action, as distinct from collectively endorsed action, to be accepted. Allowing one to so act is to allow all. We do not favour the rewriting or reinterpretation of Article 51.

Chapter VII of the Charter of the United Nations and external threats
In the case of a State posing a threat to other States, people outside its borders or to international order more generally, the language of Chapter VII is inherently broad enough, and has been interpreted broadly enough, to allow the Security Council to approve any coercive act ion at all, including military action, against a State when it deems this "necessary to maintain or restore international peace and security." That is the case whether the threat is occurring now, in the imminent future or more distant future; whether it involves the State's own actions or those of non-State actors it harbors or supports; or whether it takes the form of an act or omission, an actual or potential act of violence or simply a challenge to the Council's authority.

We emphasize that the concerns we expressed about the legality of the preventive use of military force in the case of self-defense under Article 51 are not applicable in the case of collective action authorized under Chapter VII. In the world of the twenty-first century, the international community does have to be concerned about nightmare scenarios combining terrorists, weapons of mass destruction and irresponsible States, and much more besides, which may conceivably justify the use of force, not just reactively but preventively and before a latent threat becomes imminent. The question is not whether such action can be taken: it can, by the Security Council as the international community's collective security voice, at any time it deems that there is a threat to international peace and security. The Council may well need to be prepared to be much more proactive on these issues, taking more decisive action earlier, than it has been in the past.

Questions of legality apart, there will be issues of prudence, or legitimacy, about whether such preventive action *should* be taken: crucial among them is whether there is credible evidence of the reality of the threat in question (taking into account both capability and specific intent) and whether the military response is the only reasonable one in the circumstances. We address these issues further below.

It may be that some States will always feel that they have the obligation to their own citizens, and the capacity, to do whatever they feel they need to do, unburdened by the constraints of collective Security Council process. But however understandable that approach may have been in the cold war years, when the United Nations was manifestly not operating as an effective collective security system, the world has now changed and expectations about legal compliance are very much higher. One of the reasons why States may want to bypass the Security Council is a lack of confidence in the quality and objectivity of its decision-making. The Council's decisions have often been less than consistent, less than persuasive and less than fully responsive to very real State and human security needs. But the solution is not to reduce the Council to impotence and irrelevance: it is to work from within to reform it, including in the ways we propose in the present report. The Security Council is fully empowered under Chapter VII of the Charter of the United Nations to address the full range of security threats with which States are concerned. The task is not to find alternatives to the Security Council as a source of authority but to make the Council work better than it has.

Chapter VII of the Charter of the United Nations, internal threats and the responsibility to protect The Charter of the United Nations is not as clear as it could be when it comes to saving lives within countries in situations of mass atrocity. It "reaffirm(s) faith in fundamental human rights" but does not do much to protect them, and Article 2 prohibits intervention "in matters which are essentially within the jurisdiction of any State." There has been, as a result, a long-standing argument in the international community between those who insist on a "right to intervene" in man-made catastrophes and those who argue that the Security Council, for all its powers under Chapter VII to "maintain or restore international security," is prohibited from authorizing any coercive action against sovereign States for whatever happens within their borders.

Under the Convention on the Prevention and Punishment of the Crime of Genocide (Genocide Convention), States have agreed that genocide, whether committed in time of peace or in time of war, is a crime under international law which they undertake to prevent and punish. Since then it has been understood that genocide anywhere is a threat to the security of all and should never be tolerated. The principle of non-intervention in internal affairs cannot be used to protect genocidal acts or other atrocities, such as large-scale violations of international humanitarian law or large-scale ethnic cleansing, which can properly be considered a threat to international security and as such provoke action by the Security Council.

The successive humanitarian disasters in Somalia, Bosnia and Herzegovina, Rwanda, Kosovo and now Darfur, Sudan, have concentrated attention not on the immunities of sovereign Governments but their responsibilities, both to their own people and to the wider international community. There is a growing recognition that the issue is not the "right to intervene" of any State, but the "responsibility to protect" of *every* State when it comes to people suffering from avoidable catastrophe—mass murder and rape, ethnic cleansing by forcible expulsion and terror, and deliberate starvation and exposure to disease. And there is a growing acceptance that while sovereign Governments have the primary responsibility to protect their own citizens from such catastrophes, when they are unable or unwilling to do so that responsibility should be taken up by the wider international community—with it spanning a continuum involving prevention, response to violence, if necessary, and rebuilding shattered societies. The primary focus should be on assisting the cessation of violence through mediation and other tools and the protection of people through such measures as the dispatch of humanitarian, human rights and police missions. Force, if it needs to be used, should be deployed as a last resort.

The Security Council so far has been neither very consistent nor very effective in dealing with these cases, very often acting too late, too hesitantly or not at all. But step by step, the Council and the wider international community have come to accept that, under Chapter VII and in pursuit of the emerging norm of a collective international responsibility to protect, it can always authorize military action to redress catastrophic internal wrongs if it is prepared to declare that the situation is a "threat to international peace and security," not especially difficult when breaches of international law are involved. We endorse the emerging norm that there is a collective international responsibility to protect, exercisable by the Security Council authorizing military intervention as a last resort, in the event of genocide and other large-scale killing, ethnic cleansing or serious violations of international humanitarian law which sovereign Governments have proved powerless or unwilling to prevent.

B. The Question of Legitimacy

The effectiveness of the global collective security system, as with any other legal order, depends ultimately not only on the legality of decisions but also on the common perception of their legitimacy—their being made on solid evidentiary grounds, and for the right reasons, morally as well as legally. If the Security Council is to win the respect it must have as the primary body in the collective security system, it is critical that its most important and influential decisions, those with large-scale life-and-death impact, be better made, better substantiated and better communicated. In particular, in deciding whether or not to authorize the use of force, the Council should adopt and systematically address a set of agreed guidelines, going directly not to whether force *can* legally be used but whether, as a matter of good conscience and good sense, it *should* be. The guidelines we propose will not produce agreed conclusions with push-button predictability. The point of adopting them is not to guarantee that the objectively best outcome will always prevail. It is rather to maximize the possibility of achieving Security Council consensus around when it is appropriate or not to use coercive action, including armed force; to maximize international support for whatever the Security Council decides; and to minimize the possibility of individual Member States bypassing the Security Council.

In considering whether to authorize or endorse the use of military force, the Security Council should always address—whatever other considerations it may take into account—at least the following five basic criteria of legitimacy:

a. *Seriousness of threat.* Is the threatened harm to State or human security of a kind, and sufficiently clear and serious, to justify prima facie the use of military force? In the case of internal threats, does it involve genocide and other large-scale killing, ethnic cleansing or serious violations of international humanitarian law, actual or imminently apprehended?
b. *Proper purpose.* Is it clear that the primary purpose of the proposed military action is to halt or avert the threat in question, whatever other purposes or motives may be involved?
c. *Last resort.* Has every non-military option for meeting the threat in question been explored, with reasonable grounds for believing that other measures will not succeed?
d. *Proportional means.* Are the scale, duration and intensity of the proposed military action the minimum necessary to meet the threat in question?
e. *Balance of consequences.* Is there a reasonable chance of the military action being successful in meeting the threat in question, with the consequences of action not likely to be worse than the consequences of inaction?

The above guidelines for authorizing the use of force should be embodied in declaratory resolutions of the Security Council and General Assembly. We also believe it would be valuable if individual Member States, whether or not they are members of the Security Council, subscribed to them.

Peace Enforcement and Peacekeeping Capability

When the Security Council makes a determination that force must be authorized, questions remain about the capacities at its disposal to implement that decision. In recent years, decisions to authorize military force for the purpose of enforcing the peace have primarily fallen to multinational forces. Blue helmet peacekeepers—in United Nations uniform and under direct United Nations command—have more frequently been deployed when forces are authorized with the consent of the parties to conflict, to help implement a peace agreement or monitor ceasefire lines after combat.

Discussion of the necessary capacities has been confused by the tendency to refer to peacekeeping missions as "Chapter VI operations" and peace enforcement missions as "Chapter VII operations"—meaning consent-based or coercion-based, respectively. This shorthand is also often used to distinguish missions that do not involve the use of deadly force for purposes other than self-defense, and those that do.

Both characterizations are to some extent misleading. There *is* a distinction between operations in which the robust use of force is integral to the mission from the outset (e.g., responses to cross-border invasions or an explosion of violence, in which the recent practice has been to mandate multinational forces) and operations in which there is a reasonable expectation that force may not be needed at all (e.g., traditional peacekeeping missions monitoring and verifying a ceasefire or those assisting in implementing peace agreements, where blue helmets are still the norm).

But both kinds of operation need the authorization of the Security Council (Article 51 self-defense cases apart), and in peacekeeping cases as much as in peace-enforcement cases it is now the usual practice for a Chapter VII mandate to be given (even if that is not always welcomed by troop contributors). This is on the basis that even the most benign environment can turn sour—when spoilers emerge to undermine a peace agreement and put civilians at risk—and that it is desirable for there to be complete certainty about the mission's capacity to respond with force, if necessary. On the other hand, the difference between Chapter VI and VII mandates can be exaggerated: there is little doubt that peacekeeping missions operating under Chapter VI (and thus operating without enforcement powers) have the right to use force in self-defense—and this right is widely understood to extend to "defense of the mission."

The real challenge, in any deployment of forces of any configuration with any role, is to ensure that they have (a) an appropriate, clear and well understood mandate, applicable to all the changing circumstances that might reasonably be envisaged, and (b) all the necessary resources to implement that mandate fully.

The demand for personnel for both full-scale peace-enforcement missions and peacekeeping missions remains higher than the ready supply. At the end of 2004, there are more than 60,000 peacekeepers deployed in 16 missions around the world. If international efforts stay on track to end several long-standing wars in Africa, the numbers of peacekeepers needed will soon substantially increase. In the absence of a commensurate increase in available personnel, United Nations peacekeeping risks repeating some of its worst failures of the 1990s. At present, the total global supply of personnel is constrained both by the fact that the armed forces of many countries remain configured for cold war duties, with less than 10 per cent of those in uniform available for active deployment at any given time, and by the fact that few nations have sufficient transport and logistic capabilities to move and supply those who are available. For peacekeeping, and in extreme cases peace enforcement, to continue to be an effective and accepted instrument of collective security, the availability of peacekeepers must grow. The developed States have particular responsibilities here, and should do more to transform their existing force capacities into suitable contingents for peace operations.

Prompt and effective response to today's challenges requires a dependable capacity for the rapid deployment of personnel and equipment for peacekeeping and law enforcement. States that have either global or regional air- or sea-lift capacities should make these available to the United Nations, either free of charge or on the basis of a negotiated fee-based structure for the reimbursement of the additional costs associated with United Nations use of these capacities.

Member States should strongly support the efforts of the Department of Peacekeeping Operations of the United Nations Secretariat, building on the important work of the Panel on United Nations Peace Operations (see A/55/305–S/2000/809), to improve its use of strategic deployment stockpiles, standby arrangements, trust funds and other mechanisms to meet the tighter deadlines necessary for effective deployment.

However, it is unlikely that the demand for rapid action will be met through United Nations mechanisms alone. We welcome the European Union decision to establish standby high readiness, self-sufficient battalions that can reinforce United Nations missions. Others with advanced military capacities should be encouraged to develop similar capacities at up to brigade level and to place them at the disposal of the United Nations.

Steven L. Kenny

The National Security Strategy Under the United Nations and International Law

We make war so that we may live in peace. (Aristotle)

In response to an international order of growing terrorism, trans-national crime, "rogue" and "failed" states potentially armed with WMD [weapons of mass destruction] and willing to use them, the [U.S. document] *National Security Strategy* has invoked an escalation of the right of self-defense as it prosecutes the Global War on Terrorism. Termed preemption, it is in fact a policy of preventive self-defense.

The *National Security Strategy* policy of preventive self-defense has been generally condemned throughout the international arena and also within the U.S. However, this condemnation is not universal. This study will show that a significant amount of validity can be conferred on the National Security Strategy due to: (1) the failure of the UN enforce its charter, essentially abandoning the purposes of the UN (2) the continued use and threat of use of preventive self-defense by many states and previous U.S. administrations (3) state practice (4) customary international law (5) the slowly changing body of international law that is responding to and inferring more significance due to the rise of transnational terrorists and WMD proliferation over state sovereignty.

The National Security Strategy of the United States of America

In the *Overview of The National Security Strategy of the United States of America,* September 2002, President George W. Bush put forth a number of idealistic aspirations [regarding spreading democracy and securing peace throughout the world]. . . . While quite laudable, . . . these generated a muted level of interest and discussion. The overwhelming attention of both the United States and

Strategy Research Project, U.S. Army War College, March 19, 2004.

the international community focused almost singularly on another significant tenet espoused throughout the document. [This included:]

- Identifying and destroying the threat before it reaches our borders. . . . we will not hesitate to act alone if necessary, to exercise our right of self-defense by acting preemptively against such terrorists.
- The United States has long maintained the option of preemptive actions to counter a sufficient threat to our national security. The greater the threat, the greater is the risk of inaction—and the more compelling the case for taking anticipatory action to defend ourselves, even if uncertainty remains as to the time and place of the enemy's attack. To forestall or prevent such hostile acts by our adversaries, the Untied States will, if necessary, act preemptively.
- The United States will not use force in all cases to preempt emerging threats, nor should nations use preemption as a pretext for aggression. Yet in an age where the enemies of civilization openly and actively seek the world's most destructive technologies the United States cannot remain idle while dangers gather.

Anticipatory? Preemption? Preventive?

The words anticipatory, preemptive, and preventive when associated with the self-defense of a nation generated extensive debate before the United Nations was even a dream. However, there is no need for an exhaustive review and discussion of this history to discern an opinion or conclusion on what these terms have come to mean today within the international community and the United Nations. Current publications . . . provide definitions quite acceptable to the vast majority of international legal scholars and members of the United Nations.

From the Department of Defense, Dictionary of Military and Associated Terms:

> "Preemptive Attack—An attack initiated on the basis of incontrovertible evidence that an enemy attack is imminent."
>
> "Preventive War—A war initiated in the belief that military conflict, while not imminent, is inevitable, and that to delay would involve greater risk."

From the United States Army, Judge Advocate General's School, Operational Law Handbook, 2002:

> "Anticipatory self-defense finds its roots in the 1842 Caroline case and a pronouncement by then Secretary of State Daniel Webster that a state need not suffer an actual armed attack before taking defensive action, but may engage in anticipatory self-defense if the circumstances leading to the use of force are "instantaneous, overwhelming, and leaving no choice of means and no moment for deliberation."

From these definitions, one can discern an obvious hierarchy based on the level of imminence the threat presents. Anticipatory self-defense associated with an "instantaneous" or truly, imminent threat. Preemptive attack associated with "incontrovertible evidence that an enemy attack is imminent." Preventive war associated with an "inevitable" future threat, but not linked in any way with the concept of an imminent threat.

One can form an association between anticipatory self-defense and preemptive attack based on their respective references to a requirement for some level of an imminent threat. Based on this requirement of imminence, the distinction between anticipatory self-defense and preemptive attack has become blurred and these terms are often used interchangeably. However, the lack of any reference to an imminent threat in the definition of preventive war would clearly distinguish it from anticipatory self-defense and preemptive attack.

Interestingly, a review of the use of the words anticipatory and preemptive in the *National Security Strategy* reveals an obvious disconnect with the Department of Defense and United States Army, Judge Advocate General definitions. In most cases "preventive" can be substituted for anticipatory and preemption within the National Security Strategy and the document is transformed to agree with these definitions.

For the purposes of this paper, it will be stipulated that when the *National Security Strategy of the United States* uses the words anticipatory and preemption in the context of the nation's self-defense, it is in fact referring to concepts that are more commonly accepted as preventive self-defense.

While the legality of initiating the use of force in self-defense remains an area of much debate within the United Nations and international law, one can clearly delineate a significant difference in this arena when comparing the use of anticipatory/preemptive to preventive. In fact, it is quite evident that most of the world (including much of the United States) would support the argument that the use of preventive in the context of self-defense is not a matter of self-defense at all. The vast majority of legal debate, argument, and opinion declares that the concept of preventive self-defense is illegal under international law and the Charter of the United Nations.

One might easily dismiss the validity of the National Security Strategy based on the above conclusion. However, international law and the United Nations have been and remain a dynamic entity. Taking a stance in this arena is an open invitation for a debate. Perhaps there is a future for the National Security Strategy.

Use of Force in Self-Defense Under the Provisions of the Charter of the United Nations

A nation and its right of self-defense is a controversial and active part of the international legal debate, even more than 50 years since most of the world's nations became signatories of the charter of the United Nations. Why? Because the world has suffered many conflicts in the past 50 plus years and self-defense is claimed as a factor in most of them. Self-defense of a nation remains the most common legal justification under international law and the United Nations for the use of coercive force between states.

Under the charter of the United Nations, the generally accepted sections applicable to the use of force in self-defense are:

Chapter One, Article 2 (Principles), Paragraph 4

"All members shall refrain in their international relations from the threat or use of force against the territorial integrity or political independence of any state, or in any other manner inconsistent with the Purposes of the United Nations."

Chapter 7 (actions with respect to threats to the peace, breaches of peace, and acts of aggression), Article Fifty-One

"Nothing in the Charter shall impair the inherent right of individual or collective self-defense if an armed attack occurs against a Member of the United Nations, until the Security Council has taken measures necessary to maintain international peace and security. Measures taken by Members in the exercise of this right of self-defense shall be immediately reported to the Security Council and shall not in any way affect the authority and responsibility of the Security Council under the present Charter to take at any time such action as it deems necessary in order to maintain or restore international peace and security."

These two articles appear fairly straightforward. They could be boiled down to no use of force except in self-defense after an attack and then only until the Security Council takes necessary measures to "restore international peace and security." There exists a substantial amount of legal opinion in the arena of international law that supports this simple, somewhat literal, interpretation of these articles. Any use of force outside of this interpretation would be considered a violation of international law and the charter of the United Nations. Is it really this simple?

At least in practice, no. Columbia University international security policy expert Richard K. Betts wrote,

I am aware of no case in which international law has blocked a decision to wage war—that is, a case in which a government decided that strategic necessity required war yet refrained because international law was deemed to forbid it. "He further notes that once the decision is made by a state to go to war," they find a lawyer to tell the world that international law allows it. . . .

United Nations Charter and the Use of Force—Framer's Intent

Professor Timothy Kearly, University of Wyoming School of Law, conducted an interesting analysis of the 1967 book, *Foreign Relations of the United States,* which contained the minutes of the United States delegation to United Nations Conference on International Organization (UNCIO), held in San Francisco,

April 25–June 25, 1945. The UN Charter's final form was constructed at this conference. The minutes covered the U.S. delegation's internal meetings and meetings with the other "Great Powers" who would eventually become the five permanent members of the United Nations Security Council. His article, "Regulation of Preventive and Preemptive Force in the United Nations Charter: A Search for Original Intent," conducted an extensive review of these minutes, focusing on the discussions concerning the development of the provisions on the use of force in self-defense under the UN Charter. His purpose was to determine if one could ascertain the intent of the framers of the UN Charter from these minutes. The intent of the framers is important because it is often invoked in the international arena to support a position in the contentious arguments over the use of force in self-defense. . . .

The analysis reveals that there were significant differences among the Great Powers. They all brought their own concerns that didn't always coincide with the other's concerns. This is reflected in Kearly's conclusion that, "there were substantial, unresolved disagreements . . . about the circumstances under which states should be able to use force without Security Council approval . . . combined with time pressures . . . [this] resulted in Charter use of force provisions that are imprecise, somewhat inconsistent, and open to interpretation." He goes on to note "the drafters were not concerned with their lack of precision because they assumed the permanent members of the Security Council would negotiate judgments concerning uses of force case by case in good faith.

This conclusion confirms that considerable ambiguity is built into the charter. The ambiguity was acceptable to the framers because they wanted an agreement; without the ambiguity, it is likely there would be no agreement due to the differing concerns and motives of the Great Powers. The fact that the framers introduced ambiguity reveals exactly what the framers intended. . . .

This indicates that the framers intended the law applying organ, i.e. the Security Council to make decisions case by case in good faith as the need arose in ambiguous situations. Conversely, it does not appear the framers intended international jurists to interpret the charter and establish new legal principles. . . .

Preventive Self-Defense

Critics of the *National Security Strategy* typically renounce the U.S. as embarking on unilateral, hegemonic mission to overturn the guarantee of international peace secured in 1945 by the UN Charter. They cry that the concept of preventive self-defense was eliminated for good on that day. Noam Chomsky, a prominent MIT professor and political dissident recently wrote, "Preventive war is, very simply, the 'supreme crime' condemned at Nuremburg." A rather harsh condemnation considering that the concept of preventive self-defense is part of the UN Charter. Critics of the National Security Strategy such as Chomsky typically ignore or summarily dismiss the fact that in 1945, the UN Charter actually contained a provision authorizing the very crime they denounce so vehemently, preventive self-defense.

Chapter 17, Article 107

> Nothing in the present Charter shall invalidate or preclude action, in relation to any state which during the Second World War has been an enemy of any signatory to the present Charter, taken or authorized as a result of that war by the Governments having responsibility for such action.

Article 107 was created due to the desires of the Soviet Union, Great Britain and France. They had suffered greatly in the two World Wars at the hands of aggressive powers that did not respond to peaceful efforts to resolve matters. Kearly in his previously mentioned review concluded, "With respect to preventive force . . . the charter's drafters intended to prohibit such assertive action except as specifically authorized in the form of . . . Article 107 actions against former enemy states." They wanted "an explicit authorization under the charter to use force preventively against their most likely foes, but did not want other states to have that authorization."

There is some ambiguity in Article 107 and a restrictionist [one who argues the UN charter strictly limits when countries can use force] might propose to limit Article 107 to actions "taken or authorized as a result of that war." This was addressed in 1951 by Hans Kelsen in a legal analysis of fundamental problems with the UN Charter. He concluded that it would be impossible to limit the use of Article 107 precisely due to the very ambiguity of this phrase. An additional conclusion from Kearly's review, the framers "assumed the permanent members of the Security Council would negotiate judgments concerning uses of force case by case in good faith." The framers understood the concerns of the European allies and they supported this article based on good faith. Although Article 107 does not apply today, it has to be recognized that preventive self-defense is not a new concept to the UN and its members.

Preventive self-defense as expressed in the National Security Strategy could be fairly evaluated on a case-by-case basis within the Security Council to determine if it is in concert with the purposes of the UN. Just as it was when Article 107 was created. For the UN and international community to laconically declare that any use of preventive self-defense is a violation of the UN Charter simply ensures they won't be consulted and opens the door to the National Security Strategy option of unilateral action.

Interestingly, it would be a mistake to believe the U.S. is the sole keeper of the preventive self-defense flame. In September 2003, the French Ministry of Defense stated, "Outside our borders, within the framework of prevention and projection-action, we must be able to identify and prevent threats as soon as possible. Within this framework, possible preemptive action is not out of the question, where an explicit and confirmed threat has been recognized." Although the French Ministry calls for "preemptive action," the use of "explicit and confirmed threat" only indicates only a clear, unambiguous threat. There is no requirement expressed for the threat to be of an imminent nature. Thus, this statement appears to be indicating that France is prepared to take actions of a preventive nature in self-defense. More similarity to the National Security Strategy is found in the French 2003–2008 Military Program Bill of Law, "especially as transnational terrorist networks develop and organize outside our territory, in areas not governed by states, and even at times with the help of enemy states."

State Practice

In 1970 Thomas Franck, a [scholar] . . . in the field of international law . . . , quipped "the high-minded resolve of Article 2 (4) mocks us from its grave." He pointed out that in the first 25 years of the UN, there were "one hundred separate outbreaks of hostility between states. In only one occasion was the UN "able to mount a collective enforcement action. The eternal failure of the UN to enforce its mandate results in an endemic use of force self-defense. There is no way to establish aggressor and aggrieved in the international system resulting in both parties claiming to have used force only in self-defense. Thirty years later, Franck's observation remains the status quo. In 1999 alone, there were 44 countries involved in conflict. Of the 44 countries, 20 suffered fatalities of at least 1000 and many suffered even higher numbers. . . .

Professor of Law and the Fletcher School of Law and Diplomacy, Michael Glennon, updated the record regarding hostility between states in 2002. "Between 1945 and 1999, two-thirds of the members of the United Nations—126 States out of 189—fought 291 interstate conflicts in which over 22 million people were killed."

Glennon related congressional testimony by former Secretary of Defense, William Perry, who stated "we will attack the launch sites of any nation that threatens to attack the U.S. with nuclear or biological weapons." The reservation of the right of first use of nuclear weapons has always been the stated policy of the members of the Security Council. This would clearly violate Article 51 as an act of preventive self-defense. The very threat itself is a violation of Article 2 (4).

Glennon concluded that "international 'rules' concerning use of force are no longer regarded as obligatory by states," declaring that Article 2 (4) and Article 51 are invalidated by state practice.

Franck recognized that the UN Charter did not have the mechanisms required for the modern world. The framers of the UN Charter were building on their experience, "large military formations preceded by mobilization and massing of troops." This allowed time for preparation and negotiation and was the type of aggression the framers intended to address by Article 2 (4) and Article 51. Franck noted "Modern warfare, however, has inconveniently by-passed these Queensberry-like practices. One too small and the other too large to be encompassed effectively . . . first, wars of agitation, infiltration and subversion carried on by proxy through national liberation movements; and second, nuclear wars involving the instantaneous use, in a first strike, of weapons of near-paralyzing destructiveness."

The *National Security Strategy* is making the same argument today. However, the environment is even more dangerous as the real fear is the "One too small" (terrorists) could come to posses and use the "weapons of near-paralyzing destructiveness." Article 2 (4) and Article 51 simply aren't designed to address this threat and the Security Council has refused to consider it.

International Law

Customary international law is not necessarily what is written down, but what states actually practice. Accepting that preventive self-defense is illegal under the UN Charter, use of preventive self-defense as proposed under the *National Security Strategy* would be considered illegal. However, if preventive self-defense reflects customary international law, it could be considered lawful.

Georgetown University professor, author and international law expert Anthony Clark Arend states, "International law is created through consent of states expressed through treaties and customs." If conflicts arise concerning treaties (such as the UN Charter) and customs, the "conflict is resolved by determining the rules to which states consent at the present time." Arend lists 19 incidents from 1948 to 1999 where force has been used against "the political independence and territorial integrity of states" without the authorization of the Security Council and where no reasonable claim of self-defense could be made. Arend notes incidents such as the Soviet invasion of Czechoslovakia in 1948, the Argentine invasion of the Falkland Islands in 1982, the U. S. invasion of Grenada in 1983, the Iraqi attack on Kuwait in 1990, the NATO/U.S. actions against Yugoslavia in the Kosovo situation in 1999 and states that there have been "numerous acts of intervention in domestic conflict, covert actions, and other uses of force" throughout this period. According to Arend, "Given this historical record of violations, it seems very difficult to conclude that the charter framework is truly controlling of state practice, and if it is not controlling, it cannot be considered to reflect existing international law."

Arend concludes that a customary prohibition on the use of force solely for annexation of territory such as Kuwait in 1990 remains under current customary law. However, current customary international law otherwise bears no resemblance to any prohibition contained in Article 2 (4) and that "For all practical purposes, the UN Charter framework is dead." He adds that since the Article 2 (4) is not reflected in state practice, "the Bush doctrine of preemption does not violate international law."

The Future of International Law

Michael N. Schmitt, Professor of International Law at the George C. Marshall European Center for International Affairs, conducted an extensive study on the response of international law to conflict over time. Law is not static; it is dynamic, responding and adjusting to the community on whose behalf it operates. It does not respond on a case-by-case basis, but moves in a general direction which can be predictive of its future. Professor Schmitt offers a compelling analysis that indicates the international community may already be moving in a direction that will accommodate the *National Security Strategy* under international law. . . .

[He] proposes a legal basis for the violation of the territorial integrity in the pursuit of terrorists, . . . [suggesting] a modification to the concept of imminency. First, he holds that imminency must accommodate the principle of self-defense. He submits that imminency should be defined as "the last viable window of opportunity, the point at which any further delay would render a viable defense ineffectual. . . . Any other interpretation would gut the right of self-defense."

On the subject of preemption, he states that the condemnation of such a policy is based on the fact that terrorist attacks are mischaracterized as isolated incidents. Considering al-Qa'ida for example, which has been involved in a terror campaign since 1993. "Once a terrorist campaign is launched, the issue of preemption becomes moot because an operation already underway cannot, by definition, be preempted." Nor would a response be considered preventive in nature.

Schmitt's conclusion, "There is little doubt that events of the last five years are signaling a sea of change in jus ad bellum [just cause of war]. Slowly but surely this body of law is becoming more permissive in response to the demise of nuclear armed bipolar competition and the rise of both transnational terrorists and WMD proliferation."

Conclusion

> It is revolting to have no better reason for a rule of law than that it was laid down in the time of Henry IV. It is still more revolting if the grounds upon which it was laid down have vanished long since, and the rule simply persists from blind imitation.
>
> Oliver Wendell Holmes

Since the formation of the UN, there have been nearly 300 interstate conflicts resulting in the deaths of 22 million people. The hope of the joint declaration of President [Franklin D.] Roosevelt and Prime Minister [Winston S.] Churchill "that all nations of the world must come to the abandonment of the use of force" has never materialized.

Success of the UN depended on a Security Council that that makes decisions on a case-by-case basis in good faith and enforces them. The UN never matured into the enforcement organization the framers intended and that was required for its success. It is essentially a political organization and as Secretary of State Colin Powell warned, it is close to becoming "a feckless debating society."

In response to an international order of growing terrorism, trans-national crime, "rogue" and "failed" states potentially armed with WMD and will to use them, the National Security Strategy has invoked an escalation of the right of self-defense. Termed preemption, it is in fact a policy of preventive self-defense.

The National Security Strategy policy of preventive self-defense has been condemned throughout the international arena and also within the U.S. However, this condemnation is not universal. It has been shown that a significant amount of validity can be conferred on the National Security Strategy due to: (1) the failure of the UN enforce its charter, essentially abandoning the purposes of the UN (2) the continued use and threat of use of preventive self-defense by many states and previous U.S. administrations (3) state practice (4) customary international law (5) the slowly changing body of international law that is responding to and inferring more significance on the rise of transnational terrorists and WMD proliferation over state sovereignty.

There is no doubt that this is a path fraught with peril. The Global War on Terrorism will go on for decades. Any use of preventive self-defense must retain the principles of jus ad bellum and jus in bello [just conduct of war]. It should be a tool of last resort utilized only after careful consideration combined with efforts exercising all elements of national power. However, it is a tool that will be required as the U.S engages and defeats its enemies in the Global war on Terrorism.

POSTSCRIPT

Is Preemptive War
an Unacceptable Doctrine?

The report by the High Level Panel on Threats, Challenges, and Change to UN Secretary General Kofi Annan rejecting preemptive war reinforced his existing view. In September 2003, Annan had told the UN General Assembly that the argument behind preemptive war:

> . . . represents a fundamental challenge to the principles on which, however imperfectly, world peace and stability have rested for the last fifty-eight years. My concern is that, if it were to be adopted, it could set precedents that resulted in a proliferation of the unilateral and lawless use of force, with or without justification.

Perhaps the secretary general is correct, but there is a relatively small percentage of people around the world who would agree with him categorically. A survey conducted in 20 countries in 2003 found that only 32 percent of the respondents thought that preemptive was "never justified." But when the 22 percent who answered, "rarely justified" are added to this total, it is clear that most people are either opposed to or very wary of the principle enunciated in the Bush Doctrine. More at ease with it are the 14% who replied "often justified" and the 27% who said "sometimes justified," with the remaining 5% unsure. Among Americans, the replies were 22 percent often justified, 44 percent sometimes justified, 17 percent rarely justified, 13 percent never justified, and 3 percent unsure. More on preemptive war within the changing threat environment of the early twenty-first century can be found in Michael Walzer, *Arguing About War* (Yale University Press, 2004), Joel Rosenthal, "New Rules for War," *Naval War College Review* (Fall 2004), and the several articles included in "Symposium: War and Self-Defense," *Ethics & International Affairs* (Winter 2004).

At least part of the problem is that many people believe that the United Nations too seldom acts with decisiveness to meet threats, part of the evaluation of the world body addressed in Issue 17. Whatever the cause of the UN's inability or unwillingness to act, some argue that it creates a security gap that justifies national action even without UN authorization. A 2004 poll in nine countries found that in only three was there a majority of people willing to say their country should wait for UN approval to act militarily to deal with an international threat. In the other six, more respondents felt that their country could act because getting UN approval was too difficult. Among Americans, 48 percent rejected waiting for UN authorization to act, 41 percent favoring waiting, and 10 percent unsure.

As suggested in the report by the High Level Panel on Threats, Challenges, and Change, one answer to the new spectrum of threats may be to substantially improve both the decision-making process in the UN and the capabilities of the forces that it can utilize. The panels entire 95-page report is available at `http://www.un.org/secureworld/`. Secretary General Annan clearly recognized the need for change in his September 2003 address, commenting, "It is not enough to denounce [preemptive force] unless we also face up squarely to the concerns that make some states feel uniquely vulnerable, since it is those concerns that drive them to take unilateral action. We must show that those concerns can, and will, be addressed effectively through collective action." Keep this matter in mind when pondering the points in Issue 18.

ISSUE 15

Is the War on Terrorism Succeeding?

YES: Douglas J. Feith, from "On the Global War on Terrorism," Address to the Council on Foreign Relations (November 13, 2003)

NO: John Gershman, from "A Secure America in a Secure World," Report of the Foreign Policy in Focus Task Force on Terrorism (September 1, 2004)

ISSUE SUMMARY

YES: Douglas J. Feith, U.S. undersecretary of defense for policy, tells his audience that in the global war on terrorism, the United States is succeeding in defeating the terrorist threat to the American way of life and argues that the terrorists are on the run, that the world is safer and better for what has been accomplished, and that Americans have much of which to be proud.

NO: John Gershman, who is co-director of Foreign Policy in Focus for the Interhemispheric Resource Center and teaches at the Robert F. Wagner School for Public Service at New York University, contends that the "war on terrorism" being waged by the administration of President George W. Bush reflects a major failure of leadership and makes Americans more vulnerable rather than more secure.

Although the use of terrorism extends far back into history, recent decades have seen a rise in the practice for numerous reasons. Some of these have more to do with modern technology than politics. One such factor is that more than in the past, governments are armed with aircraft and other high-tech weapons that are unavailable to opposition forces, making it nearly suicidal for dissidents to use conventional tactics. Second, terrorist targets are now more readily available than in the past because people are more concentrated in large buildings, in airliners, and other locations. Third, modern television and satellite communications makes it easy for terrorists to gain an audience. This is important because terrorism is not usually directed at its victims as such. Rather, it is intended to frighten others. Fourth, technology has led to the creation of increasingly lethal weapons that terrorists can use to kill and injure large numbers of people. These include biological, chemical, nuclear, and radiological weapons.

Other causes cited for the rise of terrorism are more political and more controversial. Most generally, there is the grinding poverty that billions of people, especially in less developed countries, experience. Reflecting this view, President Alejandro Toledo of Peru advised a UN conference, "To speak of development is to speak also of a strong and determined fight against terrorism." The rapid pace of change in the world, which is causing cultural dislocation and feeling of alienation in many people (see Issue 2), is also cited as a cause of the type of personal alienation and anger felt by many terrorist groups. For example, a survey found that 78 percent of the Muslims in the Middle East and 46% of Muslims globally felt that their religion was under attack. Many analysts also believe that terrorism by Muslim groups is sparked by the overwhelming view among Muslims that Israel is oppressing the Palestinians and that the United States favors Israel. Also cited is the presence of U.S. forces in the Middle East, especially those in Saudi Arabia near the holiest sites of Islam in Mecca and Medina. Prominent in the view of the Bush administration is the argument that terrorists want to impose their religious or political ideas on others and therefore hate democracy.

Understanding the roots of terrorism is critical. In the short-term, it is of course wise to guard against the symptoms, which are the terrorists acts perpetrated on innocent victims. Everyone agrees this should be done. The split is on the long-term. One view favors measures meant to kill or capture terrorists and so destroy their organizations in the belief that they can be decimated to the point that they no longer present a threat. The other view is that the most effective long-term counterterrorism approach is to eliminate or ameliorate as many of the causes as possible. As you will see in the debates that follow, disagreement about the causes of terrorism is important because differing views lead to differing prescriptions for countering it.

Another conundrum is how to know whether or not one is winning? The easiest way is counting attacks and casualties. But even here there are problems. The mostly widely cited data source, the U.S. State Department, is controversial. In addition to technical problems, the line between military action and terrorism is not precise. Some people argue that the State Department tends to count the action of enemies more readily than allies. Additionally, actions conducted by uniformed military forces arguably sometimes fall into the category of terrorism, and these are also usually not reflected in the State Department's data. With this in mind, it is still worthwhile to note that according to the State Department, terrorist attacks are declining, although still a relatively regular event. The record number of international attacks (666) occurred in 1987. There were 208 attacks in 2003 causing 625 deaths (including 35 Americans) and over 3,600 other casualties.

Not long ago, Americans were mostly unconcerned about terrorism. The September 11, 2001, terrorist attacks shattered this sense of security. In response President Bush declared war on terror, and that campaign is still prominent in Americans' sense of priorities. If we are winning the war on terrorism, as Douglas Feith argues we are in the first reading, then the voters' choice presidential choice in 2004 was wise. If John Gershman is correct in the second reading, then major policy changes are needed.

Douglas J. Feith  **YES**

On the Global War on Terrorism

My talk is on the global war on terrorism and I'd like to start with a personal story. On September 11, 2001, I was in Moscow with my colleague [Assistant Secretary of Defense for International Security Policy] J.D. Crouch, discussing the Anti-Ballistic Missile Treaty, an ancient text. As we were leaving the defense ministry in the late afternoon, the world entered a new era, for that was when the first plane hit the World Trade Center.

We asked the U.S.-European command [EUCOM] for the means to get back to Washington despite the general shutdown of U.S. air traffic, and EUCOM provided us a with a KC-135 tanker, which met us in Germany. And we collected there a handful of stray Defense Department officials who were also stranded by the suspension of the commercial air traffic. . . . All of us were frustrated to be away at such a moment and grateful to be getting back to the Pentagon fast, which was of course still smoldering.

In the KC-135, we conferred and wrote papers about how to comprehend the September 11th attack as a matter of national security policy. President Bush's statements even then showed that he thought of the attack, in essence, as an act of war rather than a law enforcement matter.

Now, this point may seem unremarkable, but think back to the 1993 World Trade Center bombing and to the attacks on Khobar Towers in 1996, on the U.S. East African embassies in '98, on the U.S.S. Cole in Yemen in 2000. When such attacks occurred over the last decades, U.S. officials avoided the term "war." The primary response was to dispatch the FBI to identify individuals for prosecution. Recognizing the September 11th attack as war was a departure from the established practice. It was President [George W.] Bush's seminal insight the wisdom of which I would say is attested by the fact that it looks so obvious in retrospect.

We in the KC-135 chewed over such questions as what it means to be at war not with a conventional enemy, but with a network of terrorist organizations and their state sponsors. We talked about how to formulate our war aims, how to define victory, what should be our strategy.

And as we were mulling all of this, the airplane's crew invited us to the cockpit to look down on the southern tip of Manhattan, and we saw smoke rising from the ruins of the twin towers. Aside from sadness and anger, the smoke engendered an enduring sense of duty to prevent the next big attack.

Council on Foreign Relations, November 13, 2003.

When we landed in Washington on September 12th, we were primed to join the work the President had already gotten underway to develop a strategy for the war.

That work has held up well since September 2001.

The President and his advisors considered the nature of the threat. If terrorists exploited the open nature of our society to attack us repeatedly, the American people might feel compelled to change that nature, to close it, to defend ourselves. Many defensive measures come at a high price. That is, interference with our freedom of movement, intrusions on our privacy, inspections, and an undesirable, however necessary, rebalancing of civil liberties against the interests of public safety. In other words, at stake in the war in terrorism are not just the lives and limbs of potential victims, but our country's freedom.

It isn't possible to prevent all terrorist attacks. There are simply too many targets in the United States   too many tall buildings. It's possible, however, to fight terrorism in a way that preserves our freedom and culture. So the conclusion was that our war aim should be to eliminate terrorism as a threat to our way of life as a free and open society.

Because the United States can't count on preserving our way of life by means of a defensive strategy, there was and is no practical alternative to a strategy of offense. We have to reach out and hit the terrorists where they reside, plan and train, and not wait to try to defeat their plans while they are executing them on U.S. soil. To deal with the threat from the terrorists we have to change the way we live or change the way they live.

Accordingly, the President's strategy in the war on terrorism has three parts. One is disrupting and destroying terrorists and their infrastructure. This involves direct military action, but also intelligence, law enforcement and financial regulatory activity. The list of senior members of al-Qaida and affiliated groups who've been killed or captured since 9-11 is impressive, and includes such figures as Khalid Shaykh Mohammad, Abu Zubaydah, Hambali, Mohammad Atef.

These and other successes against the terrorists demonstrate that international cooperation is alive, well and effective. We've worked jointly with the Philippines, Pakistan, Saudi Arabia, Turkey, Spain, France, Jordan, Morocco and Egypt among others. From our interrogations of detainees, we know that the absence of large-scale attacks on the United States since 9-11 has not been for want of bad intentions and efforts on the terrorists' part. We have been disrupting their plans and operations. Our strategy of offense, which is to say forcing the terrorists to play defense, is sound.

The second part of our strategy targets the recruitment and indoctrination of terrorists. The objective is to create a global intelligence and moral environment hostile to terrorism. We refer to this part as "the battle of ideas." As the President's national strategy for combating terrorism puts it, "We want terrorism viewed in the same light as slavery, piracy or genocide   behavior that no respectable government can condone or support, and all must oppose."

This requires a sustained effort to de-legitimate terrorism and to promote the success of those forces, especially within the Muslim world, that are working to build and preserve modern, moderate and democratic political and educational institutions.

And the third part of the strategy of course is securing the homeland. The Bush Administration has created the Department of Homeland Security, while the Defense Department has organized a new Northern Command in which for the first time a combatant commander has the entire continental United States within his area of responsibility. By the way, it's a matter of some pride that the U.S. Northern Command managed to achieve full operational capability—quite appropriately on September 11th, 2003—in less than a year. And we are in the process also for the first time of fielding defenses against ballistic missiles of all ranges. Our strategy envisions international cooperation. The war is global. We have forged formidable, adaptable partnerships—a rolling set, because some coalition partners are comfortable helping in some areas but not in others.

After 9-11, nearly 100 nations joined us in one or more aspects of the war on terrorism, in military operations against al-Qaida and the Taliban in Afghanistan, in maritime interdiction operations, in financial crackdowns against terrorist funding, and in law enforcement actions, as well as intelligence sharing and diplomatic efforts. In Operation Enduring Freedom Afghanistan, there are 71 members of the coalition, including contributors to the International Security Assistance Force; 37 have contributed military assets. In Iraq, 32 countries are now contributing forces.

As President Bush noted early on, the war's greatest strategic danger remains the possibility that terrorists will obtain chemical, biological or nuclear weapons. The list of states that sponsor terrorism correlates obviously and ominously with the list of those that have programs to produce such weapons of mass destruction.

The nexus of terrorist groups, state sponsors of terrorism and WMD [weapons of mass destruction] is the security nightmare of the 21st century. It remains our focus. We are treating this threat as a compelling danger in the near term. We are not waiting for it to become imminent, for we cannot expect to receive unambiguous warning of, for example, a terrorist group's acquisition of biological weapons agents. We know the list of terrorist-sponsoring states with WMD programs—Iran, Syria, Libya and North Korea. Iraq used to be in that category but no longer is.

Iraq, under Saddam Hussein, was a sadistic tyranny that developed and used weapons of mass destruction, launched aggressive attacks and wars against Iran, Kuwait, Israel and Saudi Arabia, and supported terrorists by providing them with safe harbor, funds, training and other help. It had defied a long list of legally binding U.N. Security Council resolutions. It undid the U.N. inspection regime of the 1990s. It eviscerated the economic sanctions regime and it shot virtually daily at the U.S. and British aircraft patrolling Iraq's northern and southern no-fly zones. In sum, containment of Saddam Hussein's Iraq was a hollow hope. The best information available from intelligence sources said that, one, Saddam Hussein had chemical and biological weapons and was pursuing nuclear weapons; and, two, if Saddam Hussein obtained fissile material from outside Iraq as opposed to producing it indigenously, he could have had a nuclear weapons within a year.

Those assessments, and most of the underlying information, were not recent products of the intelligence community. They were consistent with the intelligence that predated the administration of George W. Bush, and they were consistent with the intelligence from cooperative foreign services and with the United Nations' estimates of weapons unaccounted for.

It was reasonable—indeed necessary—for the U.S. government to rely on the best information it had available. And while we haven't yet found, and may not find, stockpiles of chemical or biological weapons in Iraq, David Kay reports that the Iraq survey group has obtained corroborative evidence of Saddam's nuclear, chemical and biological programs, covert laboratories, advanced missile programs, and Iraq's program active right up to the start of the war to conceal WMD-related developments from the U.N. inspectors.

The Iraqi dictator posed a serious threat. Given the nature of that threat, seen in light of our experience with the 9-11 surprise attack, and the crumbling one after another of the pillars of containment, it would have been risky in the extreme to have allowed him to remain in power for the indefinite future. Intelligence is never perfect, but that's not grounds for inaction in the face of the kind of information the President had about Saddam Hussein's Iraq.

Saddam's demise has freed Iraqis of a tyrant, deprived terrorists of a financier and supporter, eliminated a threat to regional stability, taken Iraq off the list of rogue states with WMD programs, and created a new opportunity for free political institutions to arise in the Arab world. All of this serves our cause in the global war on terrorism.

In Iraq and Afghanistan, democratization has begun. Success will strengthen the forces of moderation in the Muslim world. It could create a new era in the Middle East. Already since Iraq's liberation talk of reform and democracy is more common and more intense in the Arab world. It would be desirable if the Middle East reached a political turning point similar to the points in history when Asian democracy and Latin American democracy blossomed and spread rapidly. . . .

Extremism of the type that fuels terrorism is a political phenomenon. It's driven by ideology, and ideologies we know can be defeated. Like Soviet communism and Nazism, radical Islamism can be discredited by failure. When the Soviet system collapsed it helped demonstrate that our nation's positive message—individual liberty, the rule of law, tolerance and peace—has global appeal. Soviet communism was discredited, practically and morally, by its ultimately undeniable failures to deliver goodness or happiness. Radical Islamism, an ideological stew of historical resentments, political hatreds, religious intolerance and violence, can be expected to have a similar end. Like communism, it promises a Utopia that it can't deliver. . . .

From its inception in the days following 9-11, the President and his team have implemented their strategy for the war on terrorism with steadiness, prudence and good results. The plans for our combat and post-combat operations in Afghanistan and Iraq get challenged from time to time, as is inevitable and good in a democracy. Though these plans have by and large worked well, we review and revise them continually, as Jerry Bremer's recent visit to Washington highlights. Those plans were and are the product of much cooperation across the U.S. government and with key allies. They helped us avert many ills. For example, Iraq has not found itself with masses of internally displaced persons and international refugees, starvation, a collapse of the currency, destruction of the oil fields, the firing of Scud missiles against Israel or Saudi Arabia, or widespread inter-communal violence. There's value in pausing and reflecting on the anticipated catastrophes that we were spared through a combination of foresight, military skill, and the kind of luck that tends to favor forces that plan and work hard and wisely.

The United States and its coalition partners are on sound courses in Afghanistan and Iraq, though much remains to be done in both places. As long as we are making progress in rebuilding the infrastructure, in allowing normal life to return, and most important in helping the Afghans and Iraqis develop political institutions for the future, we are on the path to success, despite the attacks of the terrorists and former regime supporters.

Staying the course won't be easy or cheap. We're reminded of this every time we hear of another attack on U.S. or coalition forces. The President asked Congress to make available the necessary resources, and Congress has done so. To crown our military victories with strategic victories, we will have to succeed in both the civil and the military aspects of our efforts in Afghanistan and Iraq.

In the global war on terrorism we're succeeding in our goal. We are defeating terrorism as a threat to our way of life. Our coalitions are on the offensive. The terrorists are on the run. And the United States has preserved our freedom. The world is safer and better for what we have accomplished. Americans have much to be proud of.

NO

John Gershman

A Secure America in a Secure World

The [George W.] Bush administration's "war on terrorism" reflects a major failure of leadership and makes Americans more vulnerable rather than more secure. The administration has chosen a path to combat terrorism that has weakened multilateral institutions and squandered international goodwill. Not only has Bush failed to support effective reconstruction in Afghanistan, but his war and occupation in Iraq have made the United States more vulnerable and have opened a new front and a recruiting tool for terrorists while diverting resources from essential homeland security efforts. In short, Washington's approach to homeland security fails to address key vulnerabilities, undermines civil liberties, and misallocates resources.

The administration has taken some successful steps to counter terrorism, such as improved airline and border security, a partial crackdown on terrorist financing, improved international cooperation in sharing intelligence, the arrest of several high-level al-Qaida figures, and the disruption of a number of planned attacks. But these successes are overwhelmed by policy choices that have made U.S. citizens more rather than less vulnerable. The Bush White House has undermined the very values it claims to be defending at home and abroad—democracy and human rights; both Washington's credibility and its efforts to combat terrorism are hampered when it aids repressive regimes. Furthermore, the administration has weakened the international legal framework essential to creating a global effort to counter terrorism, and it has failed to address the political contexts—failed states and repressive regimes—that enable and facilitate terrorism.

Six factors explain the failure of the Bush administration's approach:

A. **Overemphasis on Military Responses:** The Bush administration has used everyone's legitimate concerns about terrorism to justify a massive increase in military spending that has little or nothing to do with combating terrorism. According to the Center for Defense Information, only about one-third of the increase in the FY2003 Pentagon budget over pre-Sept. 11 budgets funds programs and activities closely related to homeland security or counterterrorism operations.

Foreign Policy in Focus Task Force on Terrorism, September 1, 2004.

In addition, by enshrining preventive war in the national security strategy both as a general policy doctrine and for countering terrorism in particular, the administration has further reduced everyone's security.

B. **Failure in Intelligence Sharing:** The White House has failed to develop better mechanisms to share critical information both among intelligence agencies and between federal and local agencies. The recently created Terrorist Threat Intelligence Center is unaccountable to Congress and fails to place the coordination of intelligence gathering in the hands of those who must act on the findings.

C. **Undermining Democracy and Civil Liberties:** The Bush administration has undermined democracy at home through increased government secrecy. On the civil liberties front, the USA PATRIOT Act [Uniting and Strengthening America by Providing Appropriate Tools Required to Intercept and Obstruct Terrorism Act of 2001] imposes guilt by association on immigrants, expands the government's authority to conduct criminal searches and wiretaps, and undermines fundamental freedoms guaranteed by the Bill of Rights—none of which have proved necessary or effective in tracking down terrorists.

D. **Undermining Homeland Security:** Bush's approach to homeland security has two key flaws. First, his administration has been far too laissez-faire in its approach to ensuring the security of the 85 percent of the nation's critical infrastructure owned or controlled by the private sector. Second, it has failed to meet the basic needs of emergency responders, has underfunded key national agencies like the Coast Guard and the Bureau of Customs and Border Protection, and has created new unfunded mandates for local governments, forcing them to transfer scarce funds from social services and public safety to homeland security tasks.

E. **Weakening International Institutions:** The Bush administration has been hostile to a whole set of multilateral institutions that are central to enhancing international law and security, from the International Criminal Court to nearly all multilateral arms control and disarmament efforts, including the Biological and Chemical Weapons Conventions, the ABM Treaty, and the Comprehensive Test Ban Treaty.

F. **Failure to Attack Root Causes:** The Bush White House has failed to address the root causes of international terrorism and the social and political contexts in which such terrorism thrives, including repressive regimes, failed states, and the way in which poverty and inequality can create conditions of support for terrorist acts. Addressing the basic causes and conditions that facilitate terrorism in no way implies appeasement. Rather, it reflects both a pragmatic commitment to diffuse terrorism's political roots and a normative commitment to respect the values the United States preaches. Yet, heedless to the time bomb of widening global wealth disparity, the Bush administration has taken advantage of the crisis surrounding the Sept. 11 terrorist attacks to justify its pursuit of an expanded trade and investment liberalization agenda. This agenda fails to address the central challenges of reducing poverty and inequality and of promoting sustainable growth in developing countries.

A New Framework

A different approach would not fight a "war on terrorism." Rather, it would treat terrorism as an ongoing threat that needs to be tackled through a strong, co-ordinated strategy focused on strengthening civilian public sectors and enhancing the international cooperation necessary to prevent and respond to terrorist attacks. Although the military has a clear role to play, it is a supporting actor in the fight against terrorism and Washington must restructure the military in ways that enhance its capacities to respond to the threat posed by international terrorism. The safety challenge of terrorism exposes the weakness of Washington's conventional ideas of national security and the folly of traditional responses—typically military—to threats against U.S. citizens.

America needs a new agenda for combating terrorism, one that secures citizens against attacks and that situates the use of force within an international legal and policy framework. This agenda must bring international terrorists to justice, debilitate their capacity to wage terrorism, and undermine the political credibility of terrorist networks by addressing related political grievances and injustices. Below, we outline a four-part framework for a new agenda to counter terrorism.

A. Strengthen Homeland Security

To do this, the emphasis needs to be on preventing terrorist attacks and mitigating the effects of terrorist violence. Specific initiatives should:

- Improve Intelligence Gathering and Oversight: The coordination of intelligence gathering related to domestic security should be based within the Department of Homeland Security, since this is the agency responsible for acting on the information. The CIA—current home of the Terrorist Threat Intelligence Center—has proven unable to coordinate well with other intelligence agencies. The key issue facing the improvement of domestic counterterrorism intelligence capabilities does not involve a choice of organizational form (i.e., boosting the FBI's capabilities or creating a new domestic intelligence body) but rather an effort to reinstate civil liberties and reinforce judicial and congressional oversight of intelligence operations.
- Strengthen Border Security: Adequately fund key border security programs and agencies such as the Container Security Initiative, the Coast Guard, and the Bureau of Customs and Border Protection.
- Protect Critical Infrastructure: It is essential that government step up security for critical infrastructure, especially regarding:
  - Nuclear Power Plants: Spent reactor fuel pools at U.S. commercial nuclear power plants represent potentially the most consequential vulnerability to terrorist attacks. The most important step that can be taken to significantly reduce this vulnerability is to learn from several European nations that have placed all spent fuel older than five years into thick-walled, dry storage modes.
  - The Chemical Industry: The Department of Homeland Security needs to establish and enforce minimum requirements for the improvement

of security and the reduction of potential hazards at chemical plants and other industrial facilities that store large quantities of hazardous materials.

- Food and Agriculture Safety: There is a need for a comprehensive national plan both to defend against the intentional introduction of biological agents in an act of terror and to create a network of laboratories to coordinate the detection of bioterror agents in the event of an attack.
- Information Technology: There are numerous serious proposals to better secure information technology in virtually all of the nation's critical infrastructure, from the air-traffic-control system to aircraft themselves, from the electric-power grid to financial and banking systems, and from the Internet to communication systems.
- Support Emergency Responders: In addition to improving emergency preparedness plans, the administration needs to provide training, equipment, and increased support to all levels of government to strengthen emergency response capabilities by fire, police, and rescue departments as well as public health systems, all of which will be front-line emergency responders in case of a terrorist attack.
- Prevent Terrorists from Obtaining Weapons: To prevent terrorists from obtaining conventional or other weapons of mass destruction, specific initiatives should:
  - Strengthen International Conventions: There is a need to fortify the conventions for the control, nonproliferation, and elimination of weapons of mass destruction and their delivery systems.
  - End the National Missile Defense Program (also known as "Star Wars"): The Sept. 11 attacks highlight how imminent security threats are posed not from missiles but from other types of delivery systems. Combined with concerns about the destabilizing effects of the missile defense system and the false promise of security it offers, the United States should end efforts to build a National Missile Defense system and redirect those monies toward arms control and disarmament efforts.
  - Control Weapons in Russia: There is a need for increased funding for the Defense Threat Reduction Agency and other efforts to monitor and control weapons material in Russia and the former Soviet Union.

B. Strengthen International and National Legal Systems to Hold Terrorists Accountable

An effective response to terrorism requires bolstering the national and international legal infrastructure necessary to identify and prosecute the individuals and organizations that facilitate, finance, perpetrate, and profit from terrorism.

Specific initiatives should:

- Expand international police cooperation;
- Adopt the Princeton Principles on Universal Jurisdiction for prosecutions of crimes against humanity;
- Strengthen the institutions of international law by supporting the creation of a specialized tribunal for judging international terrorists; and,

- Provide technical assistance to countries to implement all the recommendations of the Financial Action Task Force with respect to money laundering and terrorist financing.

In those instances where military force is necessary to combat nonstate actors like al-Qaida, working through international institutions is justified on both normative and pragmatic grounds. The use of force should require specific authorization from the United Nations Security Council that includes specific goals and a time line, and military operations would preferably be under U.N. control. In any event, the exercise of such force should adhere to international humanitarian law and the principles of the "just war" tradition.

C. Defend and Promote Democracy at Home and Abroad

Antiterrorist efforts should not sacrifice the very values that Americans are trying to defend. Washington must listen closely to the mounting concerns of civil libertarians and constitutional rights groups who caution that the new counter-terrorism campaign may lead to a garrison state that undermines all that America stands for while doing little to protect citizens against unconventional threats. The USA PATRIOT Act is perhaps the greatest threat to civil liberties in the country today, and we applaud the numerous states, cities, towns, and counties that have passed resolutions demanding that local law enforcement not implement the provisions of those regulations that infringe on basic rights.

In forging international coalitions against terrorism, the administration should strengthen restrictions on the provision of military aid, weapons, and training to regimes that systematically violate human rights. Proactively, the White House and Congress should more rigorously condition such programs on adherence to internationally recognized human rights standards. In addition, the United States should support efforts to strengthen international legal and human rights norms, conventions, and organizations and should evaluate its own foreign policies in light of those norms.

D. Attack Root Causes

Combating terrorism requires looking beyond any one terrorist event—horrific as it may be—to address the broader socioeconomic, political, and military contexts from which international terrorism emerges. Because terrorism is a particular kind of violent act aimed at achieving a political objective, a preventive strategy must address its political roots.

U.S. policy must recognize a distinction between international terrorism in general and the specific threat posed by al-Qaida and other extremist Islamist movements, so as not to be perceived as waging a war on Islam. The 9/11 Commission Report, for example, is careful to make such a distinction. This requires that U.S. policymakers learn to distinguish between illegitimate demands and legitimate demands pursued through illegitimate means. The anti-democratic and jihadist character of al-Qaida's ideology suggests that even if the United States were to pursue the kinds of alternative policies outlined here, Americans would still be the target of attacks by committed members of al-Qaida and similar groups. Addressing root causes is one way of insuring that terrorist group efforts to mobilize support meet as inhospitable a social, economic, and political climate as possible.

The success of these policies will only be fully realized when there are no more breeding grounds for terrorist politics. These political contexts include: repressive political regimes, which spawn terrorism; failed and failing states, which can provide terrorists with arenas for operations; poverty and inequality, which can enhance support for terrorist acts and provide a source of recruits, even though poverty itself does not cause terrorism; and efforts by the United States to institutionalize its positions of global dominance, including through alliances with repressive regimes.

Specific initiatives should:

- Strengthen and Democratize International Bodies for Effective Global Governance: By proclaiming global dominance as its overarching strategic objective, the United States has made itself a target. Bush's pursuit of the preventive war doctrine as the foundation of such dominance—embodied in the invasion and occupation of Iraq—can be used to justify the argument that the current "war on terrorism" is in fact a war on Islam. And Washington's current foreign policy has further reinforced the beliefs of those who argue that the United States is an imperial power intent on holding itself above the law.

  In addition to strengthening the U.N. and other multilateral institutions, the United States must reconfigure its approach to security. We suggest a dual focus: on the cooperative arrangements necessary to insure our protection in an era of international terrorist networks with global reach, and on deterrence against possible threats from state antagonists. Such efforts require a vibrant network of global, regional, and bilateral alliances whereby the security of the world strengthens the security of America.
- End Support for Repressive Regimes: The United States must, in both word and deed, make a clean break with its history of support for repressive regimes throughout the world. Such a move would entail curbing military aid, expanding human rights and democracy, and reducing the dependence of the United States and its allies on oil imports from repressive regimes. Additional steps would include: (1) withholding military aid and opposing weapons sales to countries that systematically violate basic human rights, and (2) increasing support for human rights and democracy in North Africa, the Middle East, Central Asia, Southeast Asia, Colombia, and elsewhere through bilateral and multilateral initiatives.
- Deal with Failed States: The Afghanistan situation, and the broader reality that weak and failing states can provide enabling conditions for the operations of terrorist networks, has highlighted the need for increasing the U.N.'s capacity to engage in peace enforcement, peacekeeping, and other "nation-building" activities.

Reorient U.S. Policy in the Middle East and Central Asia: A broader U.S. policy along the lines of respecting basic human rights and democratic freedoms in the Middle East and elsewhere could still contribute to easing—though not eradicating—the conditions associated with terrorism. Such efforts would involve eliminating weapons of mass destruction and addressing the political grievances behind continuing unrest in the region. This includes opposing the bigotry embodied in both al-Qaida's and other extremist groups' opposition to Israel's existence. The United States should continue its strategic and moral commitment to Israeli sovereignty, but there is a distinction between Israel's right to exist and support for the occupation in the West Bank and Gaza. Washington's tacit approval of the occupation plays a major role in fueling anti-American extremism, sentiments that al-Qaida has opportunistically used to its own advantage.

Specific initiatives should:

- End U.S. financial and military backing for the Israeli occupation of the West Bank and Gaza;
- Advocate Palestinian self-determination and a negotiated settlement as outlined in U.N. Security Council resolutions;
- Promote efforts to create a zone free from weapons of mass destruction in the Middle East;
- Strengthen the multilateral forces involved in Afghanistan to provide the security necessary for reconstruction and development; and
- Set an immediate timetable for the withdrawal of U.S. troops from Iraq and channel support primarily through the United Nations to promote reconstruction and development.

Address Poverty and Inequality: An expansion of broad-based development can, under certain conditions, weaken local support for terrorist activities and discourage terrorist recruits. Since approval of some organizations engaged in terrorist acts is due in part to the social services and financial incentives that those organizations provide, an expansion of economic opportunities can decrease direct participation in those organizations or dampen enthusiasm for their activities.

Development policies that weaken states' capacities to insure access to, or provision of, basic services can create conditions in which terrorist groups can more easily mobilize support. At the global level, the Bush administration should end its promotion of trade and investment agreements that reinforce the discredited policies of the Washington Consensus. Instead, the United States should reorient discussions at bilateral, regional, and global economic organizations and meetings toward creating a multilateral framework more conducive to the development of poor countries. Washington should also reduce the debt owed to it by developing countries, champion debt reduction efforts at the international financial institutions, and seek an end to structural adjustment lending by the World Bank and the International Monetary Fund.

Promote Clean Energy: The United States should pursue an energy policy at home and abroad that emphasizes conservation, energy efficiency, and renewables and that makes itself and its allies less reliant on imported oil supplies.

Changing Course

No single component of this framework is an adequate response to terrorism. Only by joining all four strategies—pursuing prevention and preparedness, strengthening the international framework for multilateral action, defending and promoting civil rights, and addressing root causes—will the U.S. government be able to truthfully tell the American people that it is doing all that it can to prevent future terrorist attacks. Our proposed security strategy would be more effective at making the U.S. a safer place for all its citizens. It would also have the added advantages of improving the nation's quality of life by improving public safety, health care, and air quality.

The 9/11 Commission has accomplished a great deal by placing this debate at the forefront of policy debates. But its recommendations focus somewhat narrowly on intelligence operations and congressional oversight without addressing the broader foreign policy, military, and homeland security issues that are equally important to constructing an effective response to terrorism. Its contribution, while important, remains inadequate to forging the comprehensive strategy necessary to effectively combat terrorism.

The challenge is to construct a national security policy that demonstrates America's new commitment to protecting U.S. citizens by incorporating effective counterterror measures into the national security strategy. At the same time, American citizens must demand and U.S. foreign policy must assert a renewed commitment to constructing an international framework of peace, justice, and security that locks terrorists out in the cold—with no home, no supporters, no money, and no rallying cry. With that response, the events of September 11, 2001, will indeed have changed America and the world.

POSTSCRIPT

Is the War on Terrorism Succeeding?

As the preceding readings indicate, the debate on whether or not the war of terrorism is succeeding can be divided into two basic parts. One is measurement. At least from an American perspective, it is possible to argue that as of this writing in January 2005, success can be demonstrated by the fact that there have been no successful attacks on Americans within the United States sine 9/11. Moreover, if the bomb attacks on U.S. forces in Iraq using car bombs and other convert methods are not counted, there have been no major terrorist attacks on U.S. forces akin in severity to the boat bomb attack in October 2000 that wreaked havoc on the USS *Cole* and killed 17 sailors while the ship was in Port Aden, Yemen, taking on fuel.

From a less optimistic perspective, it is myopic to claim progress if the most notorious terrorist leader, Osama bin Laden remains at large and active, creating tapes aired on Muslim and other news broadcasts in December 2004. One called for the overthrow the Saudi monarchy, the other declared a holy war against U.S. forces in Iraq and on cooperating Iraqis and called for a boycott of the planned elections. Moreover, terrorist attacks in Saudi Arabia and elsewhere were widely attributed to either al-Qaeda directly or to bin Laden's influence. Many other aspects of the effort to thwart terrorism are even less certain. Little or no public information exists to gauge the degree to which the flow of finances to and among terrorists has been hindered. Knowing whether any progress is being made in the struggle for "the hearts and minds" of Muslims and others is also very difficult. Among those who believe that the war on terrorism is going well is Ediwn J. Feulner, "Winning the War," Heritage Foundation Commentary (June 18, 2004) at `http://www.heritage.org/Press/Commentary/`. Taking the opposite view is Charles V. Peña, Winning the Un-War: Strategy for the War on Terrorism," *Journal of Military and Strategic Studies* (Fall 2004).

The second and related aspect of the debate has to do with causes and tactics. The U.S. approach to the war on terrorism has tended to stress military and economic strategies designed to disrupt and destroy terrorist organization and their ability to operate. Illustrating this orientation, is the U.S. State Department's 2004 report on the previous year, *Patterns of Global Terrorism 2003*, which points out that "four enduring policy principles guide our counterterrorism strategy." These are, "First make no concession to terrorists and strike no deals. . . . Second, bring terrorists to justice for their crimes. . . . Third, isolate and apply pressure on states that sponsor terrorism to force them to change their behavior. . . . Fourth, bolster the counterterrorist capabilities of those countries that work with the Untied States and require assistance." The annual report is on the State Department's Web site, with the report cited here at `http://`

www.state.gov/s/ct/rls/pgtrpt/2003/. For a multimedia presentation of U.S. policy, go to http://www.whitehouse.gov/infocus/achievement/chap1-nrn.html.

What the four U.S. principles do not mention and what receives scant attention in balance of the report address the causes of terrorism. John Gershman and other critics charge that this almost surely condemns the war on terrorism to failure. In the end, almost everyone agrees that it will require a major effort before terrorism becomes a concern rather than a major threat. The question is how to allocate resources between, on the one hand, attacking the terrorists and building defense against them, and, on the other hand, targeting the political, social, and economic roots of terrorism. One analysis of the various approaches to countering terrorism is Adam Garfinkle, ed., *A Practical Guide to Winning the War on Terrorism* (Hoover Institution, 2004).

ISSUE 16

Is Government-Ordered Assassination Sometimes Acceptable?

YES: Bruce Berkowitz, from "Is Assassination an Option?" *Hoover Digest* (Winter 2002)

NO: Margot Patterson, from "Assassination as a Weapon," *National Catholic Reporter* (September 6, 2002)

ISSUE SUMMARY

YES: Bruce Berkowitz, a research fellow at the Hoover Institution at Stanford University, argues that while government-directed political assassinations are hard to accomplish and are not a reliably effective political tool, there are instances where targeting and killing an individual is both prudent and legitimate.

NO: Margot Patterson, a senior writer for *National Catholic Reporter*, contends that assassinations are morally troubling, often counterproductive, and have a range of other drawbacks.

Political assassination by internal plotters has existed seemingly as long as politics have. Ancient kings and other rulers were poisoned, stabbed to death, or otherwise killed with some regularity. Assassinations of heads of state and government have continued into modern times. In the United States, Presidents Abraham Lincoln, James Garfield, William McKinley, and John F. Kennedy were all felled by an assassin's bullet.

Much less common throughout history has been the assassination or attempted assassination of the leader of a country by a national government. Certainly this has occurred, especially in times of particular stress. In the United States, for example, extraordinary hostility and a sense of threat occasioned by the cold war promoted assassination. A report issued in 1975 by the U.S. Senate's Select Committee on Intelligence chaired by Senator Frank Church (D-Idaho)—and thus known as the Church Committee—disclosed that the Central Intelligence Agency had allegedly attempted to assassinate at least five foreign leaders during the previous 15 years: President Salvador Allende of Chile, Cuban president Fidel Castro, Congolese prime minister Patrice Lumumba, Iraqi dictator Abdul Karim Qassem, and the Dominican Republic's Rafael Trujillo. The Soviet Union and other countries also reportedly engaged in attempted assassinations.

The Church Committee report was issued during a period of American post–Vietnam War reaction against the excesses of the cold war. Reflecting this changed mood, President Gerald R. Ford issued Executive Order 11905 in 1976. The directive ordered that "no employee of the United States Government shall engage in, or conspire to engage in, political assassination." Since then, every president through President Bill Clinton reaffirmed the U.S. stance against government-directed assassinations, and two presidents (Jimmy Carter and Ronald Reagan) issued supplementary executive orders banning the practice.

According to some analysts, this stance does not mean that the United States did not try to assassinate foreign leaders during this time. Perhaps the clearest example occurred when President Reagan authorized a military attack using warplanes and conventional bombs on the home of Libyan leader Muammar Qaddafi in 1986. He was not home that evening, but his infant daughter was killed.

One of the difficulties of governments' authorizing assassinations is where to draw the line. Consider, for example, the reasonably neutral definition of assassination offered by Bruce Berkowitz in the first of the following readings. He characterizes assassination as "deliberately killing a particular person to achieve a military or political objective, using the element of surprise to gain an advantage." That would cover a range of activities, from bribing the cook of a head of state during peacetime to specifically targeting leaders such as Qaddafi in conventional military attacks. The Reagan administration argued that the attack on Qaddafi's residence was part of a larger strike in retaliation for Libyan activities that had led to a terrorist bombing of a popular nightspot in Germany, killing many off-duty American soldiers. In this and similar cases, governments have also made the case that the "targeted killing" of a leader using conventional military means is acceptable.

Attitudes in the United States toward assassinations changed in the aftermath of the September 11, 2001, terrorist attacks. For example, before the attacks, the White House reprimanded Israel for having bombed an apartment building in Gaza in an attempt to kill a leader of the terrorist organization Hamas. Not long after the September 11 attacks, however, President George W. Bush signed a so-called intelligence finding authorizing the CIA to engage in "lethal covert operations" to kill Osama bin Laden and to destroy Al Qaeda. The following November, Bush also authorized the CIA to kill Qaed Salim Sinan al-Harethi, who was suspected of masterminding the attack on the USS *Cole* in Yemen. He was killed when a hellfire missile from a CIA-controlled Predator drone destroyed his car in Yemen.

The issue here is whether or not such tactics are wise and moral. In the first selection, Berkowitz endorses, albeit cautiously, governments' carrying out political assassinations. In the second selection, Margot Patterson argues that whether referred to as "assassinations" or "targeted killing" (or some other euphemism), specifically setting out with forethought to kill an individual is neither practical nor moral.

YES

Bruce Berkowitz

Is Assassination an Option?

Soon after the September 11 terrorist attacks on New York and Washington, U.S. officials announced that they had evidence linking Osama bin Laden to the attacks. As Americans began to recover from their initial shock, many of them asked, "Why don't we just get rid of the guy?"

The terrorist tragedy reopened one of the most controversial issues in national security policy: assassination. Few topics raise more passion. Yet, despite the intense emotions assassination raises, assassination rarely gets the kind of dispassionate analysis that we routinely devote to other national security issues. That is what I will do here. When it comes to assassination, four questions are key: What is it? Is it legal? Does it work? And when, if ever, is assassination acceptable?

What Is It?

One reason assassination—or, for that matter, banning assassination—provokes so much disagreement is that people often use the term without a precise definition and thus are really arguing about different things. One needs to be clear. Depending on the definition, one can be arguing about activities that are really quite different.

For example, is killing during wartime assassination? Does assassination refer to killing people of high rank, or can anyone be the target of assassination? Does it matter if a member of the armed forces, a civilian government official, or a hired hand does the killing? Depending on the definition, killing a military leader during a bombing raid might be "assassination" but killing a low-level civilian official with a sniper might not.

For what it is worth, the *Merriam-Webster Dictionary* defines *assassination* by referring to the verb *assassinate,* which is defined as "to injure or destroy unexpectedly and treacherously" or "murder by sudden or secret attack usually for impersonal reasons." In other words, assassination is murder—killing a person—using secrecy or surprise. Assassination stands in contrast to murder without surprise (e.g., a duel). Also, assassination is not murder for personal gain or vengeance; assassinations support the goals of a government, organization, group, or cause.

Although people associate assassinations with prominent people, strictly speaking, assassination knows no rank. Leaders are often the targets of state-sponsored assassination, but history shows that generals, common soldiers, big-time crime bosses, and low-level terrorists have all been targets, too. Also, it does not seem to matter how you kill the target. It does not matter if you use a bomb or a booby trap; as long as you target a particular person, it's assassination.

For our purposes, assume that assassination is "deliberately killing a particular person to achieve a military or political objective, using the element of surprise to gain an advantage." We can call such a killing "sanctioned assassination" when a government has someone carry out such an action—as opposed to, say, "simple assassination," killing by an individual acting on his own. Then the question is, should we allow the United States to sanction such activities? And, if we allow the government to sanction assassination, when and how should do it?

Is It Legal?

You might be surprised to learn that there are no international laws banning assassination. The closest thing to a prohibition is the 1973 Convention on the Prevention and Punishment of Crimes Against Internationally Protected Persons, Including Diplomatic Agents. This treaty (which the United States signed) bans attacks against heads of state while they conduct formal functions, heads of government while they travel abroad, and diplomats while they perform their duties.

The Protected Persons Convention was intended to ensure that governments could function and negotiate even during war. Without it, countries might start a war (or get drawn into one) and then find themselves unable to stop because there was no leader at home to make the decision to do so and because their representatives were getting picked off on their way to cease-fire negotiations.

But other than these narrow cases, the Protected Persons Convention says nothing about prohibiting assassination. Even then it applies only to officials representing bona fide governments and "international organizations of an intergovernmental character." So presumably the convention shields the representatives of the United Nations, the World Trade Organization, the International Red Cross, and, probably, the PLO [Palestine Liberation Organization]. It does not protect bosses of international crime syndicates or the heads of terrorist groups such as Al Qaeda.

Another treaty that some might construe as an assassination ban is the Hague Convention on the "laws and customs" of war. The Hague Convention states that "the right of belligerents to adopt means of injuring the enemy is not unlimited." (This was a bold statement in 1907, when the convention was signed.)

The Hague Convention tried to draw a sharp line between combatants and noncombatants; combatants were entitled to the convention's protections but were also obliged to obey its rules. For example, the Hague Convention tried to distinguish combatants by requiring them to wear a "fixed distinctive emblem recognizable at a distance." Wear the emblem while fighting, and you are entitled to be treated as a POW [prisoner of war] if captured; fail to follow the dress code, and you might be hanged as a mere bandit.

Alas, maintaining this definition of a "combatant" proved a losing battle throughout the twentieth century. Guerrilla warfare transformed civilians into soldiers. Strategic bombing transformed civilians into targets. Headquarters staff, defense ministers, and civilian commanders in chief today are all more likely to wear suits than uniforms. Teenage paramilitary soldiers in Liberia are lucky to have a pair of Levis to go along with their AK-47s, let alone fatigues or insignia. That is why, practically speaking, a "combatant" today is anyone who is part of a military chain of command.

Yet the Hague Convention may be more interesting not for what it prohibits but for what it permits. The closest the convention comes to banning assassination is when it prohibits signatories from killing or wounding "treacherously individuals belonging to the hostile nation or army." But when it refers to "treachery," it is referring to fighting under false pretenses (e.g., flying the enemy's flag or wearing his uniform to lure him to death). The Hague Convention specifically permits "ruses of war." Snipers, land mines, deception, camouflage, and other sneaky tactics are okay. In fact, one might even argue that, since the convention prohibits *indiscriminate* killing, state-sanctioned assassination—the most precise and deliberate killing of all—during war is exactly what the treaty calls for.

The third international agreement that is relevant to assassination is the Charter of the United Nations, which allows countries to use military force in the name of self-defense. If a country can justify a war as "defensive," it can kill any person in the enemy's military chain of command that it can shoot, bomb, burn, or otherwise eliminate. And it can use whatever "ruses of war" it needs to get the job done. As a result, the main legal constraints on sanctioned assassination other than domestic law, which makes murder a crime in almost all countries, are rules that nations impose on themselves.

The U.S. government adopted such a ban in 1976, when President Ford—responding to the scandal that resulted when the press revealed CIA involvement in several assassinations—issued Executive Order 11905. This order prohibited what it called "political assassination" and essentially reaffirmed an often-overlooked ban that Director of Central Intelligence Richard Helms had adopted for the CIA four years earlier. Jimmy Carter reaffirmed the ban in 1978 with his own Executive Order 12036. Ronald Reagan went even further in 1981; his Executive Order 12333 banned assassination in toto. This ban on assassination remains in effect today.

Even so, there has been a disconnect between our policy and practice. The United States has tried to kill foreign leaders on several occasions since 1976, usually as part of a larger military operation.

For example, in 1986, U.S. Air Force and Navy planes bombed Libya after a Libyan terrorist attack against a nightclub frequented by American soldiers in Berlin. One of the targets was Muammar Qaddafi's tent. During Desert Storm in 1991, we bombed Saddam Hussein's official residences and command bunkers. After the United States linked Osama bin Laden to terrorist bombings of U.S.

embassies in Kenya and Tanzania in 1998, we launched a cruise missile attack at one of his bases in Afghanistan.

In each case, U.S. officials insisted that our forces were merely aiming at "command and control" nodes or at a building linked to military operations or terrorist activities. In each case, however, the same officials admitted off the record that they would not have been upset if Qaddafi, Saddam, or bin Laden had been killed in the process.

More recently, according to press reports, presidents have also approved so-called lethal covert operations—operations in which there is a good chance that an unfriendly foreign official might be killed. For example, the press reported a CIA-backed covert operation to topple Saddam in 1996 that probably would have killed him in the process, given the record of Iraqi leadership successions (no one has left office alive). After the September 11 terrorist strikes on New York and Washington, former Clinton officials leaked word to reporters that the CIA had trained Pakistani commandos in 1999 to snatch bin Laden. Given the record of such operations, bin Laden would likely not have survived.

In short, the unintended result of banning assassinations has been to make U.S. leaders perform verbal acrobatics to explain how they have tried to kill someone in a military operation without really trying to kill him. One has to wonder about the wisdom of any policy that allows officials to do something but requires them to deny that they are doing it. We would be better off simply doing away with the prohibition, at least as it applies to U.S. military operations.

Does It Work?

The effectiveness of assassination has depended much on its objectives. Most (but not all) attempts to change the course of large-scale political and diplomatic trends have failed. Assassination has been more effective in achieving small, specific goals.

Indeed, past U.S. assassination attempts have had great difficulty in even achieving the minimal level of success: killing the intended target. According to the available information, *every* U.S. effort to kill a high-ranking official since World War II outside a full-scale war has failed. This record is so poor that it would be hard to find an instrument of national policy that has been less successful in achieving its objectives than assassination. . . .

According to the [Senator Frank] Church Committee investigations of the 1970s, the CIA supported assassins trying to kill Patrice Lumumba of the Congo in 1961 and repeatedly tried to assassinate Fidel Castro between 1961 and 1963. In addition, American officials were either privy to plots or encouraged coups that caused the death of a leader (Rafael Trujillo of the Dominican Republic in 1961, Ngo Dinh Diem of South Vietnam in 1963, General René Schneider of Chile in 1970, and, later, President Salvador Allende in 1973). And, as noted, in recent years the United States has tried to do away with Qaddafi, Saddam, and bin Laden.

What is notable about this record is that it is remarkably free of success. Castro, Qaddafi, Saddam, and (at least at this writing) bin Laden all survived. (As this is being written, U.S. forces are hunting bin Laden as part of the larger war against the Taliban in Afghanistan.) What is more, Qaddafi continued to support terrorism (e.g., the bombing of Pan Am flight 103). Saddam has managed to outlast the terms of two presidents who wanted to eliminate him (George Bush and Bill Clinton), while continuing to support terrorism—and developing weapons of mass destruction.

One might have predicted this dismal record just by considering why American leaders have resorted to the assassination option. More often than not, assassination is the option when nothing seems to work but officials think that they need to do *something*. When diplomacy is ineffective and war seems too costly, assassination becomes the fallback—but without anyone asking whether it will accomplish anything.

This seems to have been the thinking behind the reported U.S. covert operation to eliminate Saddam in the mid-1990s. Despite a series of provocations—an assassination attempt against former president George Bush, violence against Shi'ite Muslims and Kurds, and violations of U.N. inspection requirements—the Clinton administration was unwilling to wage a sustained, full-scale war against him. Diplomacy was also failing, as the United States was unable to hold together the coalition that won Desert Storm. Covert support to Saddam's opponents in the military was the alternative. It was an utter failure.

True, some other countries have been more successful in that they have killed their target. For example, after the terrorist attack on Israeli athletes in the 1972 Munich Olympics, Israeli special services tracked down and killed each of the Palestinian guerrillas who took part (they also killed an innocent Palestinian in a case of mistaken identity). In 1988 Israeli commandos killed Khalil Al-Wazir, a lieutenant of Yasser Arafat's, in a raid on PLO headquarters in Tunisia. More recently, Israel has killed specifically targeted Palestinian terrorist leaders—for example, Yechya Ayyash, who was killed with a booby-trapped cell phone.

Other countries have also attempted assassinations with some degree of tactical success. During the Cold War, the KGB [Soviet political police] was linked to several assassinations. Most recently, the Taliban regime in Afghanistan was suspected of being involved in the assassination of Ahmed Shah Massoud, the leader of the Northern Alliance opposition.

But even "successful" assassinations have often left the sponsor worse off, not better. The murder of Diem sucked the United States deeper into a misconceived policy. The assassination of Abraham Lincoln (carried out by a conspiracy some believe to have links to the Confederate secret service) resulted in Reconstruction. German retribution against Czech civilians after the 1942 assassination of Nazi prefect Reinhard Heydrich by British-sponsored resistance fighters was especially brutal. The 1948 assassination of Mohandas Gandhi by Hindu extremists led to violence that resulted in the partition of India.

In short, assassination has usually been unreliable in shaping large-scale political trends the way the perpetrators intended (though the assassination of Yitzhak Rabin by a Zionist extremist in 1995 may be the exception). When it accomplishes anything beyond simply killing the target, it is usually by depriving an enemy of the talents of some uniquely skilled individual. For example, in 1943 U.S. warplanes shot down an aircraft known to be carrying Admiral

Isoroku Yamamoto—the architect of Japan's early victories in the Pacific. His loss hurt the Japanese war effort. The same could be said of the loss of Massoud to the Northern Alliance.

The problem is, picking off a talented individual is almost always harder than it looks. One paradox of modern warfare is that, although it is not that hard to kill many people, it can be very difficult to kill a particular person. One has to know exactly where the target will be at a precise moment. This is almost always hard, especially in wartime.

Should We Do It, and If So, How?

This is the most complex issue, of course. The morality of sanctioned assassination depends mainly on whether and when one can justify murder. Most religions and agnostic philosophies agree that individuals have the right to kill in self-defense when faced with immediate mortal danger. This principle is codified in American law. And, as we have seen, even international law seems to allow killing—even killing specific individuals—when it can be justified as armed self-defense.

Although most Americans do not like the idea of deliberate killing, they do not completely reject it, either. Most would agree that their government should be allowed to kill (or, more precisely, allow people to kill in its behalf) in at least two situations.

One situation is when a police officer must eliminate an immediate threat to public safety—for example, shooting an armed robber or apprehending a suspect who has proven dangerous in the past and who resists arrest. The other situation is when soldiers go to war to defend the country from attack. In addition, many—but not all—Americans believe that the government should be allowed to kill in the case of capital crimes.

It is probably not a coincidence that the U.S. Constitution also envisions these three—and only these three—situations in which the federal government might take a life: policing, going to war, and imposing capital punishment. Logically, then, assassination must fit into one of these three tracks. Assassination can be considered a police act, in which case it must follow the rules for protecting accused criminals. Or it can be considered a military act, in which case it must follow the rules that control how the United States wages war. Or it can be considered capital punishment, in which case it must follow the rules of due process.

Given this, when would we want to allow government to kill a particular foreign national? Clearly we should not use assassination as a form of de facto capital punishment. Unless the intended target presents a clear and immediate threat, there is always time to bring a suspect to justice, where we could guarantee due process. Similarly, although police should be able to protect themselves and others while making an arrest, we would not want police to pursue their targets with the expectation that they would routinely kill them.

The only time we should consider assassination is when we need to eliminate a clear, immediate, lethal threat from abroad. In other words, assassination is a military option. We need to understand it as such because the United States will face more situations in which it must decide whether it is willing, in effect, to go to war to kill a particular individual and how it will target specific individuals during wartime. Two factors make this scenario likely.

First, technology often makes it hard for one *not* to target specific people. Weapons are so accurate today that, when one programs their guidance systems, you aim not just for a neighborhood, or a building in the neighborhood, but for a particular *room* in a particular building. In effect, even bombing and long-range missile attacks have become analogous to sniping. You cannot always be sure you will hit your target—just as snipers often miss and sometimes hit the wrong target—but you still must aim at specific people.

Second, the nature of the threats we face today will likely require us to target-specific individuals. Terrorist organizations today use modern communications to organize themselves as worldwide networks. These networks consist of small cells that can group and regroup as needed to prepare for a strike. This is how the bin Laden organization has operated. Seeing how successful these tactics have been, many armies will likely often adopt a similar approach. To defeat such networked organizations, our military forces will need to move quickly, find the critical cells in a network, and destroy them. This inevitably will mean identifying specific individuals and killing them—in other words, assassination.

But when we do so, we should be clear in our own minds that, when the United States tries to assassinate someone, we are going to war—with all the risks and costs that war brings. These include, for example, diplomatic consequences, the danger of escalation, the threat of retaliation against our own leaders, the threat of retaliation against American civilians, and so on.

Because assassination is an act of war, such activities should always be considered a military operation. American leaders need to resist the temptation to use intelligence organizations for this mission. Intelligence organizations are outside the military chain of command. Intelligence operatives are not expected to obey the rules of war and thus are not protected by those rules. At the same time, intelligence organizations are also not law enforcement organizations. In many situations, having intelligence organizations kill specific individuals looks too much like a death sentence without due process.

Indeed, there is reason to question whether intelligence organizations are even technically qualified for assassination. In every publicly known case in which the CIA has considered killing a foreign leader, the agency has outsourced the job. In most cases, it has recruited a foreign intelligence service or military officials with better access. In some of the attempts to kill Castro, the CIA recruited Mafia hit men. Even in the more recent reported cases of lethal covert actions, foreigners would have done the actual killing. It is hard to maintain control and quality when you subcontract assassination services—as the record shows.

The United States did not ask for the threats we currently face, and killing on behalf of the state will always be the most controversial, most distasteful policy issue of all. That is why we need to use blunt language and appreciate exactly what we are proposing. Sugarcoating the topic only hides the tough issues we need to decide as a country. But if we do need to target specific people for military attack, it is important that we get it right.

Margot Patterson  **NO**

Assassination as a Weapon

Osama bin Laden Dead or Alive. Those were the words President [George W.] Bush used to describe U.S. policy toward the man believed to have masterminded the Sept. 11 attacks on the World Trade Center and Pentagon. So far Osama bin Laden appears to have eluded U.S. forces and their allies, but the war on terror that the United States unleashed after Sept. 11 has triggered far-reaching changes in the United States, ushering in an era of growing police powers at home and greater bellicosity abroad.

Increasingly, that bellicosity is triggering alarm, as the United States comes to be perceived as fighting fire with fire. A recent front-page story in *USA Today*, headlined "Global warmth for U.S. after 9/11 turns to frost," describes how the United States is coming to be seen by its allies as a rogue state. Internally, too, the United States faces criticism that it is becoming what it deplores: a society of men, not laws, operating without either internal or external brakes.

"Bush and his administration are pushing the edge on all moral fronts right now. We are dangerously treading on civil liberties in this country today," said Robert Ashmore, a professor emeritus of philosophy at Marquette University.

Civil rights advocates have numerous concerns, not least of which are the approximately 1,000 people picked up after Sept. 11 and held for months without charges being filed against them. The American Civil Liberties Union, which has filed a Freedom of Information Act request for information about the detainees, says it doesn't know how many people are still being held. Plans for an unprovoked attack on Iraq, the refusal of the United States to join the International Criminal Court after earlier recognizing its establishment, the embrace of pre-emptive strikes as new U.S. military doctrine, and greater discussion of assassination as a legitimate tactic in the war against terrorism indicate that the moral boundaries of U.S. policy are far different from what they were [in 2001].

For Robert Johansen, a professor of government and international studies at the University of Notre Dame, many of the tactics the United States has adopted in its war against terrorism are counterproductive.

"When we talk about terrorism, the single most important thing the United States can do in combating terrorism is to clarify the difference between the United States and terrorism," Johansen said. "Those engaging in terrorism want to deny that there is any difference. We do not intend to kill innocent people, and the moment we begin to move closer to killing innocent people the differentiation between us and terrorists begins to diminish."

From Margot Patterson, "Assassination as a Weapon," *National Catholic Reporter* (September 6, 2002). Copyright © 2002 by *National Catholic Reporter*. Reprinted by permission.

For Johansen, therefore, assassination is unacceptable policy. "To kill people without some assessment of guilt is morally inappropriate, and that would mean some kind of trial," he said.

Fundamentally Different

Johansen said assassination is fundamentally different from the conduct of war. Murder is not legitimate killing; killing in war is considered legitimate. One problem with declaring a war on terrorism and treating the terrorists as if they were conducting a war is that it gives the other side some legitimacy, Johansen said. In his view, the attacks of Sept. 11 would be better considered a crime, not an act of war.

Johansen said the most fundamental principle in international relations is reciprocity. "We will not do anything ourselves that we're unwilling to have done against us. If we don't want U.S. leaders to be targets of assassins, I don't think the U.S. can legitimately use a policy of assassination against others. Inevitably this will come back to haunt us," he said.

Is the United States trying to assassinate Osama bin Laden? Some would say undoubtedly; others argue that attempts on the life of Osama bin Laden can be viewed as part of an effort to disable the enemy's command and control center and are covered by the rules of wartime engagement.

"It is not against the laws of war in attacking an enemy force to look especially for a particular individual or individuals, who if they resist at all you're entitled to kill them," said Col. Dan Smith, an analyst at the Center for Defense Information. What you can't do, said Smith, is capture an enemy and then tie him up and execute him.

The Confines of Warfare

"The pursuit of Osama bin Laden and Mullah Omar are supposedly undertaken within the confines of warfare, within the definition of warfare, so the fact that we are after them in the same way we were after the high command of the Nazis or Japanese imperial forces . . . keeps this from becoming a legal problem," said Smith.

If this strikes some as a legalistic and linguistic nicety, it also underscores the fact that to decision-makers the word "assassination" can be as flexible and open to interpretation as the word "is" famously was to President [Bill] Clinton. Officially, Washington is still bound by President Gerald Ford's 1976 executive order prohibiting U.S. government employees from engaging in political assassination, but some analysts believe that the United States is moving closer to adopting policies that it previously condemned. In October, *The Washington Post* reported that President Bush had signed an intelligence finding authorizing preemptive covert lethal action against Osama bin Laden and Al Qaeda.

"Washington's previous position was that assassinations were basically wrong morally, and politically counterproductive," said Fawaz Gerges, a Mideast scholar at Sarah Lawrence College. According to Gerges, Sept. 11 represented a watershed in Washington's thinking, not just toward the Mideast but toward other conflicts as well.

"Now the new thinking in Washington is that the United States under certain conditions and in certain situations should be able to empower the CIA to assassinate terrorists or certain people who represent a threat to the United States.

"We are slowly and steadily injecting life into the policy of targeted or limited assassination of certain individuals," Gerges said.

The wisdom of such a policy can and is being debated. Israel, for instance, the only country that openly uses assassination, is the model and sometimes the inspiration for this debate. As such, Israel's strategy of what it calls "active defense" against suspected terrorists is a test case of the pros and, many would say, the cons of such a policy.

Since the start of the second intifada [Palestinian civil uprising] in the occupied territories in the fall of 2000, Israel has slain close to 100 Palestinians it alleges were involved in terrorism. Gerges said there are two narratives about Israel's targeting of Palestinian activists. According to the Israeli narrative, Israel does not target Palestinian leaders or leaders who do not use violence as part of struggle. According to the Palestinian narrative, the policy of assassination has no limits and no checks and balances. Scores of innocent people have been killed as collateral damage, and Palestinians assert that Israel's policy of pre-emptive strikes is intended to keep the two sides from ever sitting down at the negotiating table.

Counterproductive at Best

Gerges' own take on these different narratives is that Israel's assassination policy has been counterproductive at best. "Violence has not only not resolved the conflict, it has produced opposite results. It has led to more bloodshed and destruction on both sides."

Mark Regev, a spokesman for the Israeli Embassy in Washington, said Israel conducts surgical targeted strikes against a known terrorist operative. "We use the latest military technologies specifically to hit only the person we want to hit and to avoid any collateral damage," said Regev, who rejected the word "assassinate" to describe Israel's policy of targeting suspected terrorists.

But Israel's execution in Gaza of Hamas leader Saleh Shehada in a July missile strike that also killed 14 civilians shows that Israel's targeted killings are not always so tidy.

Richard Falk, a professor of international law at Princeton University who was one of three people on a United Nations commission appointed by U.N. Commissioner for Human Rights Mary Robinson to investigate human rights in Israel, said the commission concluded that Israel's execution policy was as much about terrorizing the Palestinian population as defending Israel against terrorism.

"Some of the individuals targeted were very inappropriate from a security perspective. They were people active in the peace movement, people with strong contacts with Israeli peace activists. Israel produced no evidence that validated their selection as dangerous terrorists," Falk said.

Proponents of adding assassination to the U.S. national security arsenal say assassination can be a kinder, gentler way of achieving certain objectives than, say, war. Jeffrey Richelson, a senior fellow at the National Security Archive, and the author of *The U.S. Intelligence Community* and *A Century of Spies: Intelligence in the Twentieth Century,* said it's absurd to argue that assassination is immoral in all circumstances. "It may be a solution to a particular problem and in some circumstances it may be morally justified," he said.

"Profound Failure"

Writing in a recent issue of *The Journal of Intelligence and CounterIntelligence,* Richelson examines the case for assassination in an essay titled "When Kindness Fails: Assassination as a National Security Option." While noting that the United States' experience in efforts at assassinating foreign leaders is one of "profound failure," Richelson said the United States should not preclude using assassination in certain circumstances. Assassination should be employed only, but not always, to deal with terrorists or with heads of rogue states who are developing weapons of mass destruction, he said.

"Assassination should not be used as attempted in the past—as a foreign policy tool to eliminate troublesome foreign leaders such as Fidel Castro or Rafael Trujillo," he wrote. Richelson argued that if the United States assassinates foreign leaders and terrorists, it should acknowledge it as the Israeli government does now both as a deterrent to similar individuals and as a way of making clear under what circumstances the United States will resort to lethal means. "If the president can order such an operation, he should be able to defend it publicly," Richelson concluded.

Others question just how efficacious assassination would be as a policy. "Even as a strategic tool, assassination isn't useful," said Ashmore, who said that there is no evidence that Israel's resort to extrajudicial executions has made life any more secure for Israelis. He pointed out that Israel has the ability to go into Palestinian villages and put Palestinian suspects on trial but refuses to do this. "It prefers to act in the manner of a terrorist," Ashmore said.

Ashmore said that too much attention is focused on the terrorism of insurgents, those who are opposing government for one reason or other, and not enough on the terrorist practices of states. "Throughout history the vast majority of the victims of terrorism are the victims of state terrorism. Our own American history, unfortunately, provides many examples of state terrorism on the part of the United States," Ashmore said.

"What's tragic about so much of our support for terrorism is that it's blown up in our face," Ashmore said. "We make enemies among the people when we support despotic powers. It's very short-range strategic thinking to support someone like Pinochet or Mobotu or the Shah. I think the same thing is going to happen to us in Saudi Arabia and Egypt. We are supporting repressive regimes."

According to one former CIA analyst who prefers not to be identified, not only terrorism but also technology is driving some of the new debate about assassination.

"The weapons are so accurate today that it's hard to often avoid putting a weapon in a very precise place," he said. "My problem with assassination is that usually it comes up in the context of covert action. People are unwilling to accept the risk of war and diplomacy doesn't seem to be working, so they kid themselves that they have this bad magic. They also fool themselves that they aren't committing an act of war. If you decide you need to take military action, you shouldn't fool yourself that you're not taking military action."

Not Easy to Accomplish

Greg Treverton, a staffer on the Church Committee, the Senate committee headed by Sen. Frank Church that investigated CIA assassination plots prior to 1970, is the author of *Reshaping National Intelligence for an Age of Information*.

Like many others, Treverton said he sees signs that the United States is now moving toward easing the ban on political assassination, which resulted in large part from the discoveries made by the Church Committee. The ban applies to assassination efforts by U.S. government employees, but not the military.

"The idea at the time was that a willful targeting of a foreign leader was not a very good idea—beneath us morally, and not always easy to accomplish. I think we are reconsidering whether the lesser of evils might not be targeting leaders," Treverton said. "If the alternative to killing Saddam Hussein is fighting a major war in which Iraqis die like flies but Hussein stays in power, wouldn't it be morally superior to kill him directly if we could? If our goal is regime change, then there must be [a] better way than fighting a major war in which some Americans die and an awful lot of Iraqis die."

But Treverton said as with any assassination effort, it's unclear whether Saddam Hussein's successor would be better or worse.

POSTSCRIPT

Is Government-Ordered Assassination Sometimes Acceptable?

After the articles by Berkowitz and Patterson were written, the United States went to war against Iraq. Virtually the first U.S. action was to attack a bunker in which intelligence placed Iraq's president. According to press reports, President George W. Bush mulled over the legality of such an attack and decided that since the compound where Saddam Hussein was located was a command-and-control facility, it was a legitimate target. At another point, an American commentator noted that Hussein was actively engaged in directing Iraqi forces and was therefore an acceptable—indeed desirable—target. The logic was that a so-called decapitation attack held the prospect of disrupting the ability and will of the Iraqi military to fight. The legality of such acts falls in significant part under the theory of just war and international law, a topic that is examined in Kristen Eichensehr, "On the Offensive: Assassination Policy Under International Law," *Harvard International Review* (Fall 2003).

Even if one assumes that a tacit state of war existed between the United States and Iraq at the time and, therefore, that the attack on Hussein was a military action, not an assassination attempt, there are still some troubling questions. One is whether it would have been equally legitimate for the Iraqis to try to kill President Bush. The logic of a decapitation attack could also be applied to an Iraqi attempt on the life of the U.S. commander in chief. The controversy over government-sanctioned assassination also raises the troubling issue of means. Would it make a difference if Hussein had been killed by explosives delivered by U.S. warplanes while President Bush had been poisoned to death by a cook paid by Iraqi intelligence? Would the former have been legitimate military action and the latter a terrorist assassination? Does it make any difference that the CIA carries out some of the U.S. "targeted" attacks? Is that taking the country back to the practices condemned by the Church Committee's report? That 1975 report may be found in its original form in some government depository libraries, and it has also been published by several presses, including under the title *Alleged Assassination Plots Involving Foreign Leaders* (Fredonia Books, 2001).

In many ways, these questions touch on the larger question of what defines terrorism. The tendency for the United States and other militarily powerful countries has been to define terrorism in a way that, to a degree, allows killings that can be carried out using conventional military means and condemns attacks carried out by weak states and organizations using a sniper's bullet or some other nonmilitary means. From the point of view of some people, such parameters are nothing more than self-serving distinctions that are irrelevant to the basic issue of whether or not it is moral to specifically target and kill an individual.

Yet another controversy relates to whether or not it can be moral for a government to commit an act that would be immoral for an individual. This is hardly a new debate. In 1793, when France was at war with Great Britain, Paris asked for U.S. assistance, as called for in the treaty that had brought France to America's aid during the American Revolution. The problem is that war with Great Britain was dangerous. Secretary of State Thomas Jefferson argued that the United States was bound by its word, just as an individual would be. He said, "Between society and society the same moral duties exist as between the individuals composing them." Secretary of the Treasury Alexander Hamilton disagreed, contending that "the rule of morality . . . is not precisely the same between nations as between individuals. . . . Millions [of people] are [affected by] . . . matters of government; while the consequences of the private actions of an individual ordinarily terminate with himself."

A good source for more on government-directed assassinations and their implications is the March 2003 issue of *University of Richmond Law Review* and John M. Collins," Assassination & Abduction: Viable Foreign Policy Tools?" *U.S. Naval Institute Proceedings* (April 2004).

The United Nations Department of Peacekeeping Operations

This UN site is the gateway to not only all the functions of the United Nations, but also to many associated agencies.

http://www.un.org/

International Monetary Fund

The International Monetary Fund (IMF) was established to promote international monetary cooperation, exchange stability, and orderly exchange arrangements; to foster economic growth and high levels of employment; and to provide temporary financial assistance to countries to help ease balance of payments adjustment. Learn more about the IMF at its home page.

http://www.imf.org

The International Law Association

The International Law Association, which is currently headquartered in London, was founded in Brussels in 1873. Its objectives, under its constitution, include the "study, elucidation and advancement of international law, public and private, the study of comparative law, the making of proposals for the solution of conflicts of law and for the unification of law, and the furthering of international understanding and goodwill."

http://www.ila-hq.org

United Nations Treaty Collection

The United Nations Treaty Collection is a collection of 30,000 treaties, addenda, and other items related to treaties and international agreements that have been filed with the UN Secretariat since 1946. The collection includes the texts of treaties in their original language(s) and English and French translations.

http://untreaty.un.org

Global Rights for Women

Women's Human Rights Resources (WHRR) is a free, accessible on-line library on international women's rights law. The purpose of the WHRR web site is to help researchers, students, teachers, and human rights advocates locate authoritative and diverse information on women's international human rights via the Internet. Site of the University of Toronto

http://www.law-lib.utoronto.ca/Diana/

PART 5

International Law and Organization Issues

*P*art of the process of globalization is the increase in scope and importance of both international law and international organizations. The issues in this section represent some of the controversies involved with the expansion of international law and organizations into the realm of military security. Issues here relate to increasing international organizations' responsibility for security, the effectiveness of international financial organizations, and the proposal to authorize international courts to judge those who are accused of war crimes.

- Is the United Nations Fundamentally Flawed?

- Should the United States Ratify the International Criminal Court Treaty?

- Is the Convention on the Elimination of All Forms of Discrimination Against Women Worthy of Support?

ISSUE 17

Is the United Nations Fundamentally Flawed?

YES: Brett D. Schaefer, "U.N. Requires Fundamental Reforms," *Heritage Lecture #842, Heritage Foundation* (June 16, 2004)

NO: Mary Robinson, from "Relevance of the United Nations," Address to the Plenary Session of the Conference on the Relevance of the United Nations (June 26–28, 2003)

ISSUE SUMMARY

YES: Brett D. Schaefer, the Jay Kingham Fellow in International Regulatory Affairs in the Center for International Trade and Economics at The Heritage Foundation, contends that the UN is not doing as well as it should in championing the principles set forth in its charter and that, therefore, fundamental UN reform is required.

NO: Mary Robinson, the United Nations high commissioner for human rights and a former president of Ireland, argues that despite all the United Nations' shortcomings and criticism, the UN is as relevant now as it was when created.

T he United Nations was established in 1945 as a reaction to the failure of its predecessor, the League of Nations to prevent World War II and the horrendous impact of that conflict. From this perspective, founding the UN in 1945 represented something akin to a sinner resolving to reform. In this case, a world that had just barely survived a horrendous experience pledged to organize itself to preserve the peace and improve humanity.

Six decades later, international violence continues, global justice and respect for international law remain goals rather than reality, and grievous economic and social ills still afflict the world. Further, people have come to recognize the environmental threats the world faces, but many of these continue to worsen rather than be arrested or reversed through international cooperation. Is the United Nations a failure?

Despite the many high-flown phrases in its Charter, the UN has been and remains principally a political organization in which the countries of the world maneuver to advance their political agendas. For example, the United States, which was the prime mover behind the creation of the UN, worked to make the Security Council the focus of UN collective security decision-making involving military action, economic sanctions, and other coercive measures. Then to ensure its central role, the United States (along with its wartime allies the British, Chinese, French, and Russians/Soviets) secured a permanent seat on the 15 country council and also established the ability of these permanent members to cast a veto and with their one vote halt UN action, block the selection of the UN secretary general, and otherwise bring many UN functions to a halt.

For an extended period, the United States was usually able to dominate the UN, although the Soviets had become enemies and often used their veto to block action. However, in time, the UN's membership changed, and with that shift, U.S. dominance eroded. There were only 50 UN members in 1945, and most of these countries were U.S. allies. In the 1960s and 1970s, dozens of former colonies in Africa, Asia, and elsewhere gained independence and joined the UN as sovereign countries. Now there are 191 members. Newer member countries often saw things differently than the United States, and by the 1980s the U.S. ability to almost always muster a majority in the UN General Assembly had waned. American influence in the Security Council also ebbed. China's seat in the UN and on the council was shifted from the government on Taiwan to the communist government in Beijing in 1972, and France became increasingly independent-minded and critical of U.S. policy.

The U.S. frustration with the UN reached a peak during the period preceding the U.S. invasion of Iraq in March 2003. The administration of President George W. Bush, supported primarily by the British under Prime Minister Tony Blair, pushed hard to get the Security Council to authorize them and other member countries to take military action against Saddam Hussein's regime. The determined opposition of France supported by elected member Germany on the Security Council doomed the effort given the French veto. But China and France might also have exercised their veto, and observers also believe that the majority of council members were opposed to the American-British initiative. "I think unless the United Nations shows some backbone and courage, it could render the Security Council irrelevant," an irritated President Bush warned.

There have also been other signs of official disenchantment in Washington with the UN. For instance Congress in 2002 mandated a reduction in the percentage of the basic UN budget contributed by the United States from 25 percent to 22 percent (the U.S. gross domestic product is 31 percent of the world gross domestic product). Moreover, Congress seldom even appropriates that much, leaving the Untied States as a country with by far the largest shortfall ($1.2 billion in late 2004) in its UN budget obligations.

What, if anything, should be done? In the first reading, Brett Schaefer calls for fundamental change in the United Nations and urges the United States not to hesitate or be apologetic about advancing its interests by pressing for a series of changes. Mary Robinson does not attempt to portray the UN as flawless, but she argues it is a vital organization and that its problems stem more from the unwillingness of its members, including the United States, to meet their obligations and participate cooperatively than from a fundamental flaw in the UN itself.

Brett D. Schaefer    **YES**

U.N. Requires
Fundamental Reforms

Concern over the United Nations' management and efficiency is nearly as old as the organization itself—with the U.S. initiating its first review of the organization only two years after its founding in 1945. I'm sure it will surprise no one that the review found problems with duplication, mushrooming mandates and programs, and poor coordination.

In the decades since, the U.S. has made repeated efforts to resolve these problems. One successful effort was the unilateral decision of the U.S. to reduce payments to the U.N. until it amended its budgeting rules to permit large contributors more say in budgeting decisions. This paved the way for consensus-based budgeting and gave the U.S. a theoretical veto over the U.N. budget.

More recently, the U.S. offered to pay its arrears to the U.N. if the organization adopted specific reforms. This deal, known as the Helms-Biden legislation [enacted in 1997], forced the U.N. to adopt—among other reforms—results-based budgeting. It also led the U.N. to reduce America's portion of the regular U.N. budget from 25 percent to 22 percent and the peacekeeping budget from 31 percent to 27 percent.

You may notice a trend in U.S. efforts to reform the U.N.: frequent use of America's financial leverage as the organization's largest contributor. The reason for this is that America really has few options to force reform on an unwilling organization. In the General Assembly (which approves the budget for the organization) each of the U.N.'s 191 members has only one vote—regardless of how much they contribute to the organization.

The one-vote structure inevitably creates inequities with small, poor nations gaining far more from the U.N. than they pay for. Obviously, most of these nations do not concur with America's priorities on reform. On the contrary, most nations see the U.N. as a source of patronage, jobs, financial resources, and a diplomatic force multiplier of sorts. These nations want a bigger U.N.—not a smaller, more efficient U.N. Given these conflicting priorities, it is hardly surprising that progress on reform has been slow and that progress has largely been achieved at the point of America's checkbook.

From Heritage Lecutre #842, June 16, 2004. Copyright © 2004 by Heritage Foundation. Reprinted by permission.

Yet U.S. criticism and reform efforts do have an impact. In 1997 and 2002, the U.N. announced its own reform agenda. . . . [T]he Secretary General has made some progress on these reform agendas. Despite this progress, the U.N. still suffers from a huge credibility problem in America, particularly among conservatives in Congress (who do not believe the organization is serious about reform).

The U.N. does itself no favors through its public relations blunders. A case in point is the recent flap over letters from Benon Sevan, former director of the U.N. Oil for Food program, to companies involved in that program. These letters instructed them to treat contracts and other information as confidential and turn them over only after receiving U.N. approval. Although this may be intended to ensure that Paul Volker has all the documents he needs to conduct the U.N. investigation, it feeds into the broad perception that the organization intends to obstruct any outside inquiry of the Oil-for-Food scandal—including investigations by Congress and the General Accounting Office.

The bottom line is that the U.N.'s credibility problem can only be overcome through greater transparency and accountability. Until these issues are addressed, the U.N. should continue to expect close scrutiny from the U.S. Congress and repeated attempts to use America's purse strings to impose reform. A case in point is Senator John Ensign's (R-NV) Oil for Food legislation that would cut funding for the U.N. unless it cooperates with the U.S. investigation.

Long-Term Vision

The reform efforts I've described thus far are inadequate if the U.N. is to fulfill its stated principles. They are the equivalent of fad dieting—irregular attempts to fix the obvious symptoms of failure. This is not to say that reform efforts focusing on the number of employees, budget growth, and improved efficiency are not important. They protect taxpayer funds and make the U.N. a more effective organization.

However, they are not the fundamental changes that are needed to resolve the underlying problems of the U.N.

A more fundamental approach to U.N. reform is required: one that answers key questions and defines an overall vision of what the end result of a reform process would be. Questions that need to be asked and answered include: What is the U.N. supposed to do? Is it doing it? Why not? What must be done to return the U.N. to first principles? What means are available for accomplishing this goal and what is the best option?

The first question can be answered by looking at the U.N. Charter, which clearly states the purposes of the organization. The U.N. was founded to:

- maintain international peace and security, including taking collective measures to remove threats to peace;
- promote equal rights and self-determination of peoples without distinction as to race, sex, language, or religion;
- help solve problems of an economic, social, cultural, or humanitarian character; and
- encourage "social progress and better standards of life in larger freedom."

I would argue that the U.N. is not doing as well as it should in championing the principles set forth in its Charter. Consider:

1. As for preventing war, there have been nearly 300 wars since 1945 and over 22 million deaths resulting from these wars. The U.N. has authorized military action to counter aggression just twice: North Korea's invasion of South Korea and Iraq's invasion of Kuwait.
2. The most urgent threat to international peace and security today is terrorism. Yet the U.N. cannot even agree upon a definition for terrorism—in large part because it counts terror-sponsoring states among its membership.
3. The U.N. counts the world's leading human rights violators and repressive governments among its membership. Worse, those members are disproportionately represented among the 53 countries elected to the U.N. Commission on Human Rights (UNCHR)—with Libya serving as chairman last year. I doubt the billions suffering from human rights abuses are comforted by U.N. efforts in this regard.
4. Equal rights for men and women are not observed among many U.N. members, particularly among Muslim nations.
5. As for advancing social progress, individual freedom, and the rule of law and improving living standards, Freedom House reports that a majority of U.N. members are not politically free and The Heritage Foundation and The Wall Street Journal revealed similar results among the U.N. members in terms of economic freedom.

Some have called on the U.S. to withdraw from the U.N. because of these flaws. I do not agree with this. Like it or not, other nations hold the U.N. in high esteem and it has become a central pillar of international relations and law. However, the U.S. would be better served by a U.N. that more closely adheres to its founding principles.

The most direct method for addressing the failures of the U.N. is to amend the Charter. I do not recommend this. Why? Consider the process set forth in Chapter 18 of the U.N. Charter, which states:

> Amendments to the present Charter shall come into force for all Members of the United Nations when they have been adopted by a vote of two thirds of the members of the General Assembly and ratified in accordance with their respective constitutional processes by two thirds of the Members of the United Nations, including all the permanent members of the Security Council.

Quite simply, opening up the Charter to amendment would be an invitation for log-rolling that would make Congress blush. Getting at least 128 U.N. members to agree to amendments and then getting their governments to ratify those amendments would require decades of work and would inevitably involve gross expansion of the U.N.'s authority, mandates, and power. This would not be in the interests of the United States and would inevitably aggravate the current problems of overreach, inefficiency, and duplication.

Worse, the quid pro quo for Charter reform would be likely to weaken America's power in the Security Council—a situation that would undermine the ability of the U.S. to protect its interests.

The remaining option is to work within the existing framework. Yet what to do and where to start? Past experience gives some clues.

One of the success stories of U.N. reform is the rejuvenation of the United Nations Educational, Scientific and Cultural Organization (UNESCO). This organization was deemed so irretrievably antithetical to U.S. interests that the Reagan Administration withdrew from it. This step, derided in many circles—especially within the U.N.—was critical to turning the organization around. It was so successful that President George W. Bush led the U.S. to rejoin UNESCO nearly 20 years after the U.S. first left. Whether this reform is lasting remains to be seen, but it is one of the few successful examples of reform in the U.N. system.

The U.S. should use this lesson and consider other candidates for withdrawal. The egregious behavior of the Human Rights Commission begs attention. Commission membership by Sudan, Cuba, China, and numerous other human rights violators tragically undermines the efforts of the organization and illustrates that U.N. member states do not take this issue as seriously as they should. The United States is faced with the sad situation of questioning if the cause of human rights is better served by participating in the Commission in order to champion the cause or by highlighting the complicity of the Commission in obscuring human rights abuses by publicly chastising the organization and refusing to lend it the credibility of U.S. membership.

Another step that the U.S. should take is to establish a Democracy Caucus and an Economic Freedom Caucus within the U.N. These groups would bring together countries that share common values on human rights, freedom of religion, equal rights, representative government, free trade, and economic freedom. As suggested by Kim Holmes, Assistant Secretary of State for International Organization Affairs, there are nations that agree with the U.S. on economic and political freedom, but who do not vote with the U.S. on these issues due to regional loyalties and other pressures.

However, members of the Caucus would be seen as supporting agreed principles rather than as supporting the U.S. Creating alternative groupings and voting blocs could serve U.S. interests by, hopefully, countering the efforts of a few key nations and establishing reliable allies to support efforts to expand freedom, basic rights, and the rule of law.

Another necessity is to reform the U.N. budget process. While the U.S. may have a technical veto due to the consensus requirement for budgets, it frequently fails to exert its veto due to concerns about the impact this could have on ongoing issues in the Security Council or the General Assembly. It makes no sense that Tuvalu—with its miniscule financial contribution—carries the same weight in budget decisions as does the U.S., Japan, or other large donors. The U.S. should lead an effort to get large contributors greater influence over budget decisions, though not necessarily by amending Article 18. Cooperation among large donors should be sufficient to enact change: After all, a handful of countries fund over 50 percent of the U.N. budget.

To Reform the U.N., Reform the Membership

In many ways the U.N. has fallen short of the hopes of its founders, not because of its staff, but because of its members. As discussed above, many U.N. members do not live up to the Charter's ideals. Unfortunately, over the years, the U.N. has regarded self-rule to be the main prerequisite for membership—rather than whether the proposed new member is a "peace-loving state [that is willing to] accept the obligations contained in the present Charter and, in the judgment of the Organization, are able and willing to carry out these obligations."

In reality, some U.N. members honor the Charter principles not at all. Yet they enjoy the privileges of U.N. membership and take that privilege for granted. For example, under what justification does North Korea merit U.N. membership? It is aggressive; a threat to international peace and security; a proliferator of weapons of mass destruction; and a repressive, undemocratic regime that brutalizes its own citizens. North Korea does not deserve membership alongside democratic, free countries in the U.N. that observe the founding principles of the organization.

Similarly, why should a country that continuously violates U.N. Security Council resolutions—such as Iraq in the 1990s—enjoy the privileges of U.N. membership? For that matter, why should a failed state like Somalia, which has no effective government, retain status as a U.N. member?

The U.N. needs to clean house by reprimanding those countries that habitually violate U.N. principles. The U.S. should raise the issue of ejecting from the organization the worst violators of U.N. principles. Some may suggest that this goes against the spirit of the U.N., but the procedures for revoking U.N. membership are set forth in Chapter 2 of the U.N. Charter, which states:

> A Member of the United Nations which has persistently violated the Principles contained in the present Charter may be expelled from the Organization by the General Assembly upon the recommendation of the Security Council.

Obviously, the drafters of the Charter envisioned the possibility of ejecting nations from the organization.

A two-thirds vote in the General Assembly may be difficult to achieve—as would a Security Council recommendation for the ejection of a member country—but the threat alone may encourage better behavior and may shame U.N. member nations into being more vocal and rigorous in support of freedom and human rights.

Conclusion

All nations use the U.N. to advance their national interests. The difference between the U.S. and other nations is that America has a vested interest in making the U.N. work. Otherwise, those problems normally assigned to the U.N. wind up on America's doorstep—mainly because no other nation has the capacity to do anything about them. America is better off with the U.N. heading up election monitoring campaigns, monitoring ceasefires, and rebuilding wrecked nations. Frankly, America is not very good at those tasks. That is no fault: America has rightly focused its efforts on larger security issues, warfare (when necessary), and preserving global security.

The United States should not hesitate to advance its interests by unapologetically pushing for fundamental change—even if that course is controversial. In the end, the efforts for reform I have outlined here are little more than insisting that the U.N. fulfill its mission. Central to this effort are getting rid of the rotten apples and allowing the U.N. to do its work as envisioned. I believe that these issues must be considered and an overarching vision set forth if reform efforts are to be consistent and effective.

Mary Robinson

NO

Relevance of the United Nations

In preparing my remarks for today, I have been reflecting on my last speech as U.N. High Commissioner for Human Rights in New York on 9th September last, when I addressed the NGO Annual Forum at the General Assembly on "Rebuilding Societies Emerging from Conflict: A Shared Responsibility." The United Nations has suffered quite a battering since then. Today, we are asked to reflect on its relevance in a changed world.

There is no need to be too defensive. For all its shortcomings, the U.N. is as relevant now as it was when created. A recent telling example can be found in the SARS [severe acute respiratory syndrome] epidemic. It has proved once again—if proof were necessary—that strong and effective global institutions such as the United Nations World Health Organization are essential to act in ways which complement and protect national governments. Led by Gro Brundtland, WHO coordinated global responses to the outbreak of the epidemic in terms of quarantine requirements and diagnosis. Laboratories were mobilized in 13 countries to identify the new virus and devise a diagnostic test. This broke the normal pattern in which laboratories compete against each other for commercial advantage—as has been the case with HIVAIDS. That SARS is now off the front pages shows that the international response has been quite effective.

Yet as we all know, some influential voices here in the United States argue that while the U.N. may still be relevant in addressing humanitarian crises or assisting developing countries, it is not up to the challenge of confronting today's threats to peace and security. President Bush made this point of view unmistakably clear when he said to the U.N. General Assembly in September of last year that the U.N.'s failure to confront Iraq would cause the world body to "fade into history as an ineffective, irrelevant debating society."

The scene was so different in the General Assembly just two years earlier. The adoption of the U.N. Millennium Declaration provided a sense of promise and shared commitment to international law and institutions at the start of the 21st century. But just one year and three days after this historic Declaration was adopted, the terrible events of September 11, 2001 shook the United States and the world. Since that day, the renewed commitments

Plenary Session on the Relevance of the United Nations, June 26–28, 2003.

which ushered in the new century have been overshadowed by the threats of terrorism, by fears and uncertainties about the future and by questions about the viability of open societies joined by international norms and values.

Some believe that multilateralism and international law are no longer relevant in a post 9/11 world. They contend that everything that has happened since that day of horrific violence is unprecedented—that new threats to domestic and international security require new strategies and new coalitions.

With the war in Iraq led by a coalition of the willing, and the rebuilding of that country under the control of the occupying powers, the international system's legitimacy and relevance have been put to yet another test.

In these worrying times, let us draw strength from the spirit that was present at the creation of the U.N.—a time which was no less fraught with uncertainty about the future than our own. Yet the conviction then was so different. Consider the words of U.S. President Harry Truman in his address in San Francisco to the closing session of the U.N. Conference, exactly 58 years ago this week.

Speaking about the U.N. Charter, he said:

> You have created a great instrument for peace and security and human progress in the world. The world must now use it! If we fail to use it, we shall betray all those who have died in order that we might meet here in freedom and safety to create it. If we seek to use it selfishly—for the advantage of any one nation or any small group of nations—we shall be equally guilty of that betrayal. The successful use of this instrument will require the united will and firm determination of the free peoples who have created it. The job will tax the moral strength and fibre of us all.
>
> We all have to recognize-no matter how great our strength—that we must deny ourselves the license to do always as we please. No one nation, no regional group, can or should expect, any special privilege which harms any other nation. If any nation would keep security for itself, it must be ready and willing to share security with all. That is the price which each nation will have to pay for world peace. Unless we are all willing to pay that price, no organization for world peace can accomplish its purpose.
>
> And what a reasonable price that is!"

What should the U.N. Charter's opening words—"We the peoples of the United Nations" mean today? Who speaks for whom in an increasingly interconnected yet divided world? How can people play a meaningful role, both through their local communities and through global networks, in taking forward the commitment to international institutions and the international legal framework which was affirmed by government leaders in the United Nations Millennium Declaration? If we want globalization to work for all the world's people—the overall priority identified in that Declaration—then multilateralism and respect for international law, in particular, international human rights law, must work as well.

In my view, it wasn't the inability of the Security Council to reach consensus on how to deal with the situation in Iraq, damaging as it was, that has posed the most serious threat to the future of the United Nations. There is something more fundamental that needs to be addressed. Namely, how do we ensure that the institutions of international governance, established more than 50 years ago, are seen today by the people of the world to be legitimate, accountable, transparent—democratic?

The question of international legitimacy has, I believe, taken on a new relevance in the context of the lack of security and continuing political instability in Iraq. There are more voices calling for a stronger United Nations involvement there. President Truman's words should guide us—we must use to the full the "great instrument for peace and security and human progress in the world."

These challenges to the legitimacy of the U.N. and other global governance institutions can be seen at three levels:

First, at the macro level of governance: globalization has shifted power from national to global levels with global institutions ill equipped to deal with current realities. Governance of the U.N. is still stuck in the geopolitics of 1945 and for it to be effective in a new world, it must be addressed.

Second, at the meso level of policy making processes: while there have been positive attempts to engage civil society, and thus enhance transparency, legitimacy and participation, there still remain several defects both with the content and context of global policy making processes.

Finally, at the micro level of operations: there is still much to do by all sides to operate more effectively and to meet goals even within the limitations of flawed governance and flawed policy making parameters.

Will the normative global system that restored peace and security after the Second World War be seen by future generations as an idealistic dream that was unable to respond to the realities of a changing international landscape? Or will it instead be reformed and adapted to remain the essential foundation of a more just and secure world based on respect for the international rule of law? The answer, of course, will depend on the choices we make, the priorities we set and the values we seek to uphold.

During my travels to over 80 countries as High Commissioner, I found that people believed overwhelmingly in the importance of the U.N. They recognized the vital work that its agencies and programs carried out in protecting refugees and improving public health, in supporting democratic development and seeking peaceful solutions to conflicts.

But I also found great skepticism. People viewed U.N. resolutions, declarations and treaties as commitments that are routinely ignored by governments or worse still, selectively implemented to benefit the strong at the exclusion of the weak. They felt that all the fine words about protecting human rights, eliminating poverty and ensuring sustainable development were only paper promises and that the U.N. was unable to hold its member states accountable for the commitments they had made.

Over recent months I have had opportunities to speak to a wide range of audiences here in the United States, including influential legal audiences such as the American Society of International Law and the American Law Institute. I have urged the importance of engagement by the United States in the international rule of law through support for the ICC, for the broader framework of international human rights law including economic, social and cultural rights, and the value of a broader approach to human security as advocated by the Commission on Human Security in its recent report "Human Security Now." The views I have heard are not, of course, a scientific survey, but I have been genuinely encouraged by the interest and responsiveness of those audiences to these issues.

We shouldn't forget that in some aspects of international law, real progress has been made. For example, governments take seriously the rules of the international trading system, despite growing concerns about lack of global fairness, and in the resolution of international trade disputes. This is in no small part because the U.S. has been fully engaged—both as an architect of the system, and as a party to a number of the disputes. Since 1995, 65 cases have been brought against the U.S., with findings in 22 cases that there had not been full compliance with WTO rules. In most of these cases, the U.S. has already acted effectively to change its practices—setting an important example to other countries.

There has also been major progress in developing international criminal law, and the administration of criminal justice, first through the International Criminal Tribunal for the former Yugoslavia, currently trying former President [Slobodan] Milosovic, and then through a second international tribunal to prosecute acts of genocide committed in Rwanda in 1994. The Convention against Torture makes the investigation and prosecution of torture a treaty obligation for states parties, who now include the U.S. It requires states to exercise universal jurisdiction over acts of torture—a provision most spectacularly applied by the House of Lords in the United Kingdom, when it ruled against Augusto Pinochet's claim of immunity from criminal prosecution in relation to the torture of political opponents while he was President of Chile.

Today, the new International Criminal Court builds on these achievements. The Court is an institutional recognition that certain crimes—because of their nature—affect the entire international community, and where they cannot be prosecuted by domestic courts, an international court must have jurisdiction. 139 states have signed, and so far, 89 have ratified the treaty and the Court is now meeting at The Hague.

It is a matter of great regret to supporters of the Court that the U.S. is not engaged in this great legal enterprise, that there is no U.S. judge, and that the decisions which will shape the Court's legal and procedural character for the duration of this new century are being taken without America. More seriously, the renewal earlier this month by the Security Council, at the insistence of the Bush administration, of a one-year exemption from ICC jurisdiction for American troops involved in U.N.-authorized military operations, and the pursuit of bilateral agreements seeking to immunize U.S. citizens from ICC scrutiny, risk undermining the legitimacy of the Court by bringing about a two-tiered system of justice. It is ironic that the country leading a war on terror is weakening the very institution which a recent New York Times editorial described as "a

U.S. 'ally' in its efforts to prevent the globe's most serious crimes and bring to trial those who commit them."

A key challenge for the U.N. is to find new and innovative ways of developing greater accountability for the decisions its member states make. This challenge cannot be met by the Secretary-General or the secretariat alone as important as they are in pushing for results. It cannot be imposed on any one state by another. Some say change is only possible if the structures of the U.N. itself, such as the Security Council, are reformed. These are important issues that should and are being discussed.

But I believe more can be done in the short term by drawing on the combined voice and influence of civil society at every level–local, national and international—in holding governments accountable for the commitments they have made under the U.N.

The need for greater involvement of civil society in the work of the organization has been a priority for some time now. In his initial U.N. reform proposals in 1997, Secretary General Kofi Annan noted that "civil society constitutes a major and increasingly important force in international life," but, he continued, ". . . Yet despite these growing manifestations of an ever more robust global civil society, the United Nations is at present inadequately equipped to engage civil society and make it a true partner in its work."

I saw for myself the growing importance of civil society during my time as High Commissioner. When I began my term, there was already a sophisticated network of groups around the world with a long track record in promoting human rights and speaking out on behalf of victims. I cannot count the number of times that these human rights activists provided me with valuable insights and information.

But I also witnessed the emergence of a powerful movement for change through which civil society groups in every region were using the tools of the legal commitments governments have made, under the six core international human rights instruments, to foster a deeper democratic discourse. For example, a sophisticated, literate women's movement world-wide is increasingly using the Convention on the Elimination of Discrimination against Women and its optional protocol to pin governments to their legal commitments under this treaty.

Another example can be seen in how the International Covenant on Economic, Social and Cultural Rights is being used by civil society in a growing number of countries to provide analysis of government spending on health care, education and access to clean water among other issues. Earlier this month I traveled to Thailand to participate in the launch of ESCR-Net, an international network on Economic, Social and Cultural Rights. This new network has brought together social movements and non governmental organizations working in human rights, development and the environment world wide all committed to connecting concrete local struggles for social justice with international standards and mechanisms for advancing human rights.

An international study just released by SustainAbility [an international group focusing on business strategies to support sustainable development] in partnership with the U.N. Global Compact and the U.N. Environment Program titled, "The 21st Century NGO: In the Market for Change" notes that increasing numbers of NGOs [nongovernment organizations, private policy-related groups] are making strategic decisions to engage with business and governments in efforts to achieve shared objectives. The report identifies some concern that this could compromise the independence of NGOs and leave them open to criticism that they are "selling out." I believe, however, that some of the larger international human rights NGOs, such as Amnesty and Human Rights Watch, who have recognized the importance of developing strategies to implement economic, social and cultural rights, are well positioned to illustrate that involvement of NGOs in wider partnerships can be independent, rigorous and principled.

The new project I have been developing since leaving the U.N.—the Ethical Globalization Initiative—seeks to build on this approach to civil society involvement in policy making at the national and international level. Our aim is to seek in a low-key, targeted way, to be a promoter of good practices or model projects of how human rights approaches can help produce ethical policies and processes at the national and global level. We also plan to be a "chorus leader," linking local activists and networks with academics and policy development, which together can produce the analysis and recommendations needed to influence decision makers at different levels in government, international organizations, the business sector and civil society.

To give an example, an issue we have identified to be tackled from a human rights perspective is health, access to life saving treatments and HIV/AIDS. Before going any further, allow me to say that I welcome President Bush's decision to give U.S. leadership in the fight against AIDS. The commitment of $15 billion by the U.S. government over the next five years to fight AIDS abroad through support to the Global AIDS Fund and other projects is vitally important and will set an example for other wealthy nations to follow. The President was right when he said in his State of the Union address that in an age of miraculous medicines, no person should have to hear the words— 'You've got AIDS. We can't help you. Go home and die.'

One of the first projects on this issue that we are developing in the Ethical Globalization Initiative, in cooperation with the Center for Research on Women, the International AIDS Trust and the Center for the Study of AIDS at the University of Pretoria in South Africa, is to engage with African parliamentarians, beginning with a meeting in Botswana this September, to reduce women's vulnerability and to combat stigma in the HIV/AIDS pandemic in Africa.

We hope to build greater understanding among African government leaders and AIDS experts that the disease could be more effectively addressed by emphasizing the extent to which it is also a women's rights issue both from the perspective of women as victims of the disease as well as primary caregivers for the sick and orphaned. As one slogan has put it—"The best investment in an AIDS vaccine is an investment in protecting women's rights."

We are also developing, in cooperation with the Respect Group in Europe, a new Business Leaders Initiative on Human Rights. It aims to involve senior business leaders from multi-national corporations in a consultative process with different stake holder groups to better define the extent of business responsibilities for human rights, particularly in countries facing problems of extreme poverty and deficient governance. Our aim is to support businesses committed to promoting human rights and avoiding practices which may lead to rights violations, while recognizing—indeed emphasizing—that the primary responsibility for human rights protection remains with governments.

But as we speak about the many worthy activities that diverse groups are carrying out in every part of the world, we also shouldn't forget that the global civil society movement today faces strong criticisms.

The new study on 21st Century NGOs emphasizes the importance of ensuring higher levels of transparency and disclosure around funding and effectiveness. It predicts that additional transparency and accountability will become prerequisites for NGO success in entering the mainstream through partnerships and will be crucial for retaining their position of trust. Others resent that trust and seek to undermine the influence of NGOs.

The backlash, particularly within this country, against the multilateralism of the U.N. and international mechanisms and standards is also being reflected in an assault on NGOs. Many of you will have read of a recent conference here in Washington entitled "Nongovernmental Organizations: The Growing Power of an Unelected Few" organized, as it happens, by two powerful NGOs! As a keynote speaker put it "The world is no longer divided between realists and idealists, but those who favor liberal internationalism and others who favor democratic sovereignty." NGOs, perceived to be liberal internationalists, came under sweeping attack—and do so on a new website called "NGO Watch" which seeks to monitor their activities.

A more productive approach to these questions is being pursued by the Geneva based International Council on Human Rights Policy—which has produced a draft report titled "Deserving Trust: Issues of Accountability for Human Rights NGOs." The report notes that:

> As governments are showing less willingness to be accountable to international institutions, and when transparency, consistency, and political accountability at international level appear to be weakening . . . NGOs have a choice in the way they respond: they can follow the example given by their governments, ignoring calls to become more accountable; or they can acknowledge that it is precisely in difficult times like this that it is important for NGOs to reset standards, and force states to again be accountable—not least by demonstrating accountability themselves.

The Council welcomes comments on the draft which is available at www.ichrp.org.

Another serious concern I feel it necessary to address is that attacks on the impartiality and neutrality of humanitarian NGOs threaten the future of multilateral, neutral and impartial humanitarian action. Firstly, we saw, in Afghanistan, U.S. military carrying out "humanitarian programmes"—often in civilian clothing yet carrying weapons—as part of a hearts and minds campaign, and thereby threatening the safety of neutral humanitarian action.

And just recently, the head of USAID, Andrew Natsios, has reportedly told U.S. NGOs which have received USAID money to promote humanitarian and rehabilitation projects in Iraq that they are, in essence, an arm of the U.S. Government.

Such actions dilute and distort a commitment to the fundamental humanitarian principles of humanity, neutrality and impartiality. They transform humanitarian programmes into extensions of military campaigns. This endangers the lives of humanitarian workers—be they U.N. or NGO—and compromises the work of NGOs and the U.N. agencies with whom they work.

It is vital therefore that the U.N. take a more active role in defending the integrity and independence of NGOs which are working in support of the goals of the U.N.: peace, development, human rights, environmental protection, women's rights and so on.

In the particular case of Iraq, the U.N. needs to recognize just how much its interests and those of humanitarian NGOs are inextricably intertwined. For both, there is an absolute imperative to ensure that neutrality and impartiality continue to be central values ensuring universal application of humanitarian principles. Humanitarian action cannot and must not be an extension of a political or military struggle, however well intentioned. We must ensure that we retain the right and the ability to reach all those in need regardless of race, religion, or political affiliation. I urge the U.N. system to recognize that defending space for NGOs is vital for defending its own future.

As many of you will know, as part of the next phase of his reform program, the Secretary-General recently established a panel of eminent persons, chaired by former President of Brazil Fernando Henrique Cardoso, to take stock of the Organization's experience in interacting with civil society and to recommend improvements for the future in order to make the interaction between civil society and the United Nations more meaningful.

As President Cardoso stressed in his initial paper for consideration by the panel, "A vibrant and forceful national civil society, working together with government, far from weakening democracy and good governance, increases the national resources invested in social development and strengthens the country's voice in global issues."

He also pointed out the growing recognition "that collaboration and partnership involving multiple actors increases the available stock of ideas, capacities and resources to deal with a given problem." I would encourage the U.N. Association of the U.S. to provide the Cardoso panel with your support and recommendations—setting out your vision of how the U.N. should work with citizens and organizations as the panel carries out its work over the remainder of the year.

The United Nations needs the voices and resources of civil society now more than ever before. That begins with each one of you and your roles as ambassadors for the U.N. in your own communities and through your own professional and personal networks. In taking on these duties, we should remember Eleanor Roosevelt's words about human rights: "Without concerned citizen action to uphold them close to home, we shall look in vain for progress in the larger world."

POSTSCRIPT

Is the United Nations Fundamentally Flawed?

Deciding about the worth of the UN has a great deal to do with what standard of evaluation you adopt. A recent and generally positive review of UN operations can be found in Thomas G. Weiss, David P. Forsythe, and Roger A. Coate, *The United Nations and Changing World Politics* (Westview Press, 2004). A more negative perspective is presented in Joshua Muravchik, "The Case Against the UN," *Commentary* (November 2004).

One evaluative standard is akin to asking whether a glass is half full or half empty. Undoubtedly, the UN has not come anywhere near achieving the lofty goals set out in its Charter. By this standard, the evaluative glass is at least half empty. However, it can also be rightly said that the UN has accomplished a great deal. Dozens of UN peacekeeping operations have been fielded, and some of them have made an important contribution. More on UN peacekeeping is available in Robert Lane Greene, "Blue Man Group," *New Republic* (August 27, 2003). For all the deprivations that remain, there is greater justice and better living conditions in the world now than existed than a few decades ago. Again, the UN can legitimately claim some of the credit. So from this perspective, the glass is at least fuller than it once was. A study that concentrates on the UN's efforts beyond security is Jean E. Krasno, ed., *The United Nations: Confronting the Challenges of a Global Society* (Lynne Rienner, 2004).

It is also important to decide whether to evaluate the UN from a global or a national perspective. President Bush argued that the UN was veering toward irrelevance because it did not agree with his desire to have it condemn Iraq and authorize the United States and other countries to wage war. From the perspective of U.S. national interests or at least Washington's preferences then (see Issue 10), the UN failed. But most of the other countries of the world opposed taking action at that point. And if their collective view constitutes the world's evaluation of its interest, then arguably the UN succeeded by resisting the superpower's pressure. Indeed, one could contend that if the UN failed, it was in that it could not prevent the United States and the so-called coalition of the willing from taking unilateral action.

Yet another standard of measurement has to do with the old adage about being careful when throwing stones in glass houses. The UN has sometimes wasted money, its workers have not always performed admirably, and in 2004 it received a black eye when reports of incompetence and even corruption in the oil-for-food program it had managed in connection with Iraq prior to the war. Yet it must be remembered that the annual UN regular and peacekeeping budget combined ($4.2 billion in 2003) were are equal to only 18 hours worth of spending by the U.S. government that year. So whatever waste may occur, the amount in dollars is relatively minor compared to ill-spent funds in wealthy countries. Additionally, all governments, including the U.S. government, suffer scandals. Yet there are few calls to fundamentally reorganize, abolish, or replace them. Certainly money and organizations should be well managed, but it is important when doing an evaluation to distinguish between fundamental structural flaws that promote problems and the occasional human failings of people who work in a structure.

It is also important to ask why the United Nations has not been able to accomplish more. Is the weakness in the organization, or is in a function of the UN being hamstrung by counties that refuse to support it when they disagree with its decisions. By the self-interest standard, the only good world body would be one that always did what the evaluator wanted. Of course, another standard would be to judge countries by how fully they cooperate with the UN, rather than evaluating the UN by how frequently it follow a country's wishes.

ISSUE 18

Should the United States Ratify the International Criminal Court Treaty?

YES: Lawyers Committee for Human Rights, from Statement Before the Committee on International Relations, U.S. House of Representatives (July 25, 2000)

NO: John R. Bolton, from Statement Before the Committee on International Relations, U.S. House of Representatives (July 25, 2000)

ISSUE SUMMARY

YES: The Lawyers Committee for Human Rights, in a statement submitted to the U.S. Congress, contends that the International Criminal Court (ICC) is an expression, in institutional form, of a global aspiration for justice.

NO: John R. Bolton, senior vice president of the American Enterprise Institute in Washington, D.C., contends that support for an international criminal court is based largely on naive emotion and that adhering to its provisions is not wise.

Historically, international law has focused primarily on the actions of and relations between states. More recently, the status and actions of individuals have become increasingly subject to international law.

The first significant step in this direction was evident in the Nuremberg and Tokyo war crimes trials after World War II. In these panels, prosecutors and judges from the victorious powers prosecuted and tried German and Japanese military and civilian leaders for waging aggressive war, for war crimes, and for crimes against humanity. Most of the accused were convicted; some were executed. There were no subsequent war crimes tribunals through the 1980s and into the mid-1990s. Then, however, separate international judicial tribunals' processes were established to deal with the Holocaust-like events in Bosnia and the genocidal massacres in Rwanda.

The 11-judge tribunal for the Balkans sits in The Hague, the Netherlands. The 6-judge Rwanda tribunal is located in Arusha, Tanzania. These tribunals have indicted numerous people for war crimes and have convicted and imprisoned a few of them. These actions have been applauded by those who believe that individuals should not escape punishment for crimes against humanity that they commit or order. But advocates of increased and forceful application of international law also feel that ad hoc tribunals are not enough.

Such advocates are convinced that the next step is the establishment of a permanent International Criminal Court (ICC) to prosecute and try individuals for war crimes and other crimes against humanity. The move for an ICC was given particular impetus when President Bill Clinton proposed just such a court in 1995. Just a year later, the United Nations convened a conference to lay out a blueprint for the ICC. Preliminary work led to the convening of a final conference in June 1998 to settle the details of the ICC. Delegates from most of the world's countries met in Rome, where their deliberations were watched and commented on by representatives of 236 nongovernmental organizations (NGOs). The negotiations were far from smooth. A block of about 50 countries informally led by Canada, which came to be known as the "like-minded group," favored establishing a court with broad and independent jurisdiction.

Other countries wanted to narrowly define the court's jurisdiction and to allow it to conduct only prosecutions that were referred to it by the UN Security Council (UNSC). The hesitant countries also wanted the court to be able to prosecute individuals only with the permission of the accused's home government, and they wanted the right to file treaty reservations exempting their citizens from prosecution in some circumstances. Somewhat ironically, given that President Clinton had been the impetus behind the launching of a conference to create the ICC, the United States was one of the principal countries favoring a highly restricted court. U.S. reluctance to support an expansive definition of the ICC's jurisdiction and independence rested on two concerns. One was the fear that U.S. personnel would be especially likely targets of politically motivated prosecutions. The second factor that gave the Clinton administration pause was the requirement that the Senate ratify the treaty. Senate Foreign Relations Committee chairman Jesse Helms has proclaimed that any treaty that gave the UN "a trapping of sovereignty" would be "dead on arrival" in the Senate.

In the following selections, the Lawyers Committee for Human Rights and John R. Bolton present their markedly differing views of the wisdom of founding an ICC. Both analyses agree that, if it works the way that it is intended to, the International Criminal Court will have a profound impact. Where they differ is on whether that impact will be positive or negative.

Statement of the Lawyers Committee for Human Rights

T he United States has compelling reasons to remain open to eventual cooperation with the International Criminal Court (ICC). United States interests may dictate such cooperation, even while the U.S. remains a non-party to the Rome Statute. . . . The following paper describes the U.S. interest in supporting the ICC.

> *I. A strong and independent International Criminal Court serves important national interests of the United States.*

At the end of World War Two, with much of Europe in ashes, some allied leaders urged that the leaders of the defeated Third Reich be summarily executed. The United States disagreed. U.S. leaders insisted that a larger and more valuable contribution to the peace could be made if the Nazis were individually charged and tried for violations of international law. The International Criminal Court is an expression, in institutional form, of an aspiration for justice with which the United States had been deeply identified ever since World War Two. It was created to advance objectives that are totally consistent with the long-term U.S. national interest in a peaceful, stable, democratic and integrated global system. And the Rome Treaty, in its final form, promised to advance that interest in the following ways:

- *First,* the treaty embodies deeply held American values. The establishment of the Court responds to the moral imperative of halting crimes that are an offense to our common humanity. The ICC promises to promote respect for human rights; advance the rule of law around the world, both domestically and internationally; reinforce the independence and effectiveness of national courts; and uphold the principle of equal accountability to international norms.
- *Second,* the ICC will help to deter future gross violations. It will not halt them completely, of course. But over time, its proceedings will cause prospective violators to think twice about the likelihood that they will face prosecution. This deterrent effect is already apparent in the former Yugoslavia. Even though leading architects of ethnic cleansing, such as

From U.S. House of Representatives. Committee on International Relations. *The International Criminal Court: A Threat to American Military Personnel?* Hearing, July 25, 2000. Washington, D.C.: U.S. Government Printing Office, 2000.

Radovan Karadzic and Ratko Mladic, have not been brought to trial, their indictment has limited their ability to act and has allowed more moderate political forces to emerge, reducing the risk to U.S. and other international peacekeepers still in Bosnia.

- *Third,* through this deterrent effect the ICC will contribute to a more stable and peaceful international order, and thus directly advance U.S. security interests. This is already true of the Yugoslav Tribunal, but it will be much more true of the ICC, because of its broader jurisdiction, its ability to respond to Security Council referrals, and the perception of its impartiality. The court will promote the U.S. interest in the preventing [of] regional conflicts that sap diplomatic energies and drain resources in the form of humanitarian relief and peacekeeping operations. Massive human rights violations almost always have larger ramifications in terms of international security and stability. These include widening armed conflict, refugee flows, international arms and drug trafficking, and other forms of organized crime, all of which involve both direct and indirect costs for the United States.

- *Fourth,* the ICC will reaffirm the importance of international law, including those laws that protect Americans overseas. For many people in the United States, "international law" is seen either as a utopian abstraction, or an unwelcome intrusion into our sovereign affairs. But as Abram Chayes, former Department of State Legal Adviser, remarked shortly before his death [in 2000], there is nothing utopian about international law in today's world. On the contrary, it is a matter of "hardheaded realism." Many nations who voted for the Rome Treaty had similar misgivings about its potential impact on their sovereignty. But they recognized that this kind of trade-off is the necessary price of securing a rule-based international order in the 21st century. France, for example, which participates extensively in international peacekeeping operations, made this calculation, joined the consensus in Rome and . . . ratified the treaty. The United States, likewise, should see the ICC as an integral part of an expanding international legal framework that also includes rules to stimulate and regulate the global economy, protect the environment, control the proliferation of weapons of mass destruction, and curb international criminal activity. The United States has long been a leading exponent, and will be a prime beneficiary, of this growing international system of cooperation.

II. The risks posed by the ICC to U.S. servicemen and officials are negligible in comparison to the benefits of the Court to United States' interests.

In assessing the U.S. government's concerns, it is important to bear in mind some basic threshold considerations about the ICC. Most fundamentally, it will be a court of last resort. It will have a narrow jurisdiction, and is intended to deal with only the most heinous crimes. The ICC will step in only where states are unwilling or unable to dispense justice. Indeed, that is its entire purpose: to ensure that the worst criminals do not go free to create further havoc just because their country of origin does not have a functioning legal system. The Court was designed with situations like Rwanda and Cambodia and Sierra Leone in mind, not to supplant sophisticated legal systems like those of the

United States. Furthermore, there are strict guidelines for the selection of ICC judges and prosecutors, as well as a set of internal checks and balances, that meet or exceed the highest existing international standards. The legal professionals who staff the Court will not waste their time in the pursuit of frivolous cases.

Second, the Court will only deal with genocide, war crimes and crimes against humanity, all of which are subject to a jurisdiction narrower than that available to domestic courts under international law. It will not be concerned with allegations of isolated atrocities, but only with the most egregious, planned and large-scale crimes.

Could a member of the U.S. armed forces face credible allegations of crimes of this magnitude? Genocide would seem to be out of the question. War crimes and crimes against humanity are more conceivable. The My Lai massacre in Vietnam revealed the bitter truth that evil knows no nationality: American soldiers can sometimes be capable of serious crimes. If such a crime were committed today, it would appear self-evident that the U.S. military justice system would investigate and prosecute the perpetrators, as it did at My Lai, whether or not an ICC existed. And if it were an isolated act, not committed in pursuit of a systematic plan or policy, it would not meet the threshold for ICC concern in any case.

Benign support by the United States for the ICC as a non-party to the Treaty would reaffirm the standing U.S. commitment to uphold the laws of war and could be offered in the knowledge that the Court would defer to the U.S. military justice system to carry out a good faith investigation in the unlikely event that an alleged crime by an American was brought to its attention. The marginal risk that is involved could then simply be treated as part of the ordinary calculus of conducting military operations, on a par with the risk of incurring casualties or the restraints imposed by the laws of war. The preparation and conduct of military action is all about risk assessment, and the marginal risk of exposure to ICC jurisdiction is far outweighed by the benefits of the Court for U.S. foreign policy.

> *III. The ICC provides an opportunity for the United States to reaffirm its leadership on the issue of international justice, which for so long has been a central goal of U.S. policy.*

We urge the United States to develop a long-term view of the benefits of the ICC. Such an approach would open the door to cooperation with the Court as a non-state party, and eventually to full U.S. participation. This policy shift should be based on the following five premises:

- **The creation of new international institutions requires concessions from all the participants.** As an international agreement, the Rome Statute bears the marks of many concessions to sovereign states—not least the United States. As such, the ICC will have a twofold virtue: it will be imbued with the flexibility of an international institution as well as with the rigor of a domestic criminal court. The risks involved in supporting the present ICC Treaty are more than outweighed by the expansion of an international legal framework that is congenial to U.S. interests and values.

- **The risks of U.S. exposure to ICC jurisdiction are in fact extremely limited, as a result of the extensive safeguards that are built into the Rome Treaty.** Those safeguards are there in large part because the United States insisted on their inclusion. The modest risks that remain can never be fully eliminated without compromising the core principles established at Nuremberg and undermining the basic effectiveness of an institution that can do much to advance U.S. interests. The best way to minimize any residual risk is to remain engaged with others in helping to shape the Court. The risks, in fact, will only be aggravated if the United States decides to withdraw from the ICC process. Joining the ICC, on the other hand, would allow the United States to help nominate, select and dismiss its judges and prosecutors, and so ensure that it operates to the highest standards of professional integrity. More broadly, the ICC's Assembly of States Parties would provide an ideal setting for the United States to demonstrate its leadership in the fight against impunity for the worst criminals.

- **The Pentagon's views, while important, should be balanced among other U.S. policy interests in reference to the ICC.** The U.S. military has an institutional interest in retaining the maximum degree of flexibility in its operational decisions. But this must be put in proper perspective by civilian authorities as they weigh the pros and cons of the ICC. Legislators and others who have so far remained on the sidelines of the ICC debate will have an important part to play in helping the Administration develop a broader approach to the ICC, one that puts long-term stewardship of the national interest into its proper perspective.

- **U.S. leadership requires working in close cooperation with our allies around the world.** It is tempting to believe that U.S. economic and military supremacy is now so absolute that the United States can go it alone and impose its will on the rest of the world. But the evolution of the ICC is a reminder that this kind of unilaterialism is not possible in today's more complex world. The United States has tried to impose its will on the ICC negotiations, and it has failed. In its repeated efforts to find a "fix," the United States has succeeded only in painting itself into a corner. Worse, it has disregarded one of the cardinal rules of diplomacy, which is never to commit all your resources to an outcome that is unattainable. Unable to offer credible carrots, decisive sticks, or viable legal arguments, the United States finds itself on what one scholar has called a "lonely legal ledge," able neither to advance nor to retreat. Asking for concessions it cannot win, in a process it can neither leave nor realistically oppose, the United States has so far resisted coming to terms with the limits of its ability to control the ICC process.

- **The costs of opposition to the Court are too high and would significantly damage the U.S. national interest.** Once the ICC is up and running, it seems highly unlikely that the United States would refuse to support the principle of accountability for the worst international crimes simply because the Court was the only viable means of upholding that principle. It is far more likely that a future U.S. administration will see the advantage in supporting the Court, if only as a matter of raw political calculus. Opposition to a functioning Court would undermine faith in a world based on justice and the rule of law and would shake one of the foundation stones on which the legitimacy of U.S. global leadership has rested since World War Two.

For the last half century, U.S. foreign policy has sought to balance military strength with the nurturing of an international system of cooperation based on democracy and the rule of law. It would be a serious mistake to imagine that victory in the Cold War means that the institutional part of this equation can now be abandoned, and that ad hoc applications of force should prevail over the consistent application of law.

Statement of John R. Bolton

Unfortunately, support for the ICC [International Criminal Court] concept is based largely on emotional appeals to an abstract ideal of an international judicial system, unsupported by any meaningful evidence, and running contrary to sound principles of international crisis resolution. Moreover, for some, faith in the ICC rests largely on an unstated agenda of creating ever-more-comprehensive international structures to bind nation states in general, and one nation state in particular. Regrettably, the Clinton Administration's naïve support for the concept of an ICC . . . left the U.S. in a worse position internationally than if we had simply declared our principled opposition in the first place.

Many people have been led astray by analogizing the ICC to the Nuremberg trials, and the mistaken notion that the ICC traces its intellectual lineage from those efforts. However, examining what actually happened at Nuremberg easily disproves this analysis, and demonstrates why the ICC as presently conceived can never perform effectively in the real world. Nuremberg occurred after complete and unambiguous military victories by allies who shared juridical and political norms, and a common vision for reconstructing the defeated Axis powers as democracies. The trials were intended as part of an overall process, at the conclusion of which the defeated states would acknowledge that the trials were prerequisites for their readmission to civilized circles. They were not just political "score settling," or continuing the war by other means. Moreover, the Nuremberg trials were effectively and honorably conducted. Just stating these circumstances shows how different was Nuremberg from so many contemporary circumstances, where not only is the military result ambiguous, but so is the political and where war crimes trials are seen simply as extensions of the military and political struggles under judicial cover.

Many ICC supporters believe simply that if you abhor genocide, war crimes and crimes against humanity, you should support the ICC. This logic is flatly wrong for three compelling reasons.

First, all available historical evidence demonstrates that the Court and the Prosecutor will not achieve their central goal—the deterrence of heinous crimes —because they do not (and should not) have sufficient authority in the real world. Beneath the optimistic rhetoric of the ICC's proponents, there is not a shred of evidence to support their deterrence theories. Instead, it is simply a near-religious article of faith. Rarely, if ever, has so sweeping a proposal for

From U.S. House of Representatives. Committee on International Relations. *The American Servicemembers' Protection Act of 2000.* Hearing, July 25, 2000. Washington, D.C.: U.S. Government Printing Office, 2000. Notes omitted.

restructuring international life had so little empirical evidence to support it. Once ICC advocate said in Rome that: "the certainty of punishment can be a powerful deterrent." I think that statement is correct, but, unfortunately, it has little or nothing to do with the ICC.

In many respects, the ICC's advocates fundamentally confuse the appropriate role of political and economic power; diplomatic efforts; military force and legal procedures. No one disputes that the barbarous actions under discussion are unacceptable to civilized peoples. The real issue is how and when to deal with these acts, and this is not simply, or even primarily, a legal exercise. The ICC's advocates make a fundamental error by trying to transform matters of international power and force into matters of law. Misunderstanding the appropriate roles of force, diplomacy and power in the world is not just bad analysis, but bad and potentially dangerous policy for the United States.

Recent history is unfortunately rife with cases where strong military force or the threat of force failed to deter aggression or gross abuses of human rights. Why we should believe that bewigged judges in The Hague will prevent what cold steel has failed to prevent remains entirely unexplained. Deterrence ultimately depends on perceived effectiveness, and the ICC is most unlikely to be that. In cases like Rwanda, where the West declined to intervene as crimes against humanity were occurring, why would the mere possibility of distant legal action deter a potential perpetrator? . . .

Moreover, the actual operations of the existing Yugoslav and Rwanda ("ICTR") tribunals have not been free from criticism, criticism that foretells in significant ways how an ICC might actually operate. A UN experts' study (known as the "Ackerman Report," after its chairman) noted considerable room for improvement in the work of the tribunals. . . .

[For example], ICC opponents have warned that it will be subjected to intense political pressures by parties to disputes seeking to use the tribunal to achieve their own non-judicial objectives, such as score-settling and gaining advantage in subsequent phases of the conflict. The Ackerman Report, in discussing the ICTY's quandary about whether to pursue "leadership" cases or low-level suspects, points out precisely how such political pressures work, and their consequences: "[u]navoidable early political pressures on the Office of the Prosecutor to act against perpetrators of war crimes . . . led to the first trials beginning in 1995 against relatively minor figures. And while important developments . . . have resulted from these cases, the cost has been high. Years have elapsed and not all of the cases have been completed." In short, political pressures on the Tribunals, to which they respond, are not phantom threats, but real. . . .

Second, the ICC's advocates mistakenly believe that the international search for "justice" is everywhere and always consistent with the attainable political resolution of serious political and military disputes, whether between or within states, and the reconciliation of hostile neighbors. In the real world, as opposed to theory, justice and reconciliation may be consistent—or they may not be. Our recent experiences in situations as diverse as Bosnia, Rwanda, South Africa, Cambodia and Iraq argue in favor of a case-by-case approach rather than the artificially imposed uniformity of the ICC.

For example, an important alternative is South Africa's Truth and Reconciliation Commission. After apartheid, the new government faced the difficulty of establishing truly democratic institutions, and dealing with earlier crimes. One option was certainly widespread prosecutions against those who committed human rights abuses. Instead, the new government decided to establish the Commission to deal with prior unlawful acts. Those who had committed human rights abuses may come before the Commission and confess their past misdeeds, and if fully truthful, can, in effect, receive pardons from prosecution.

I do not argue that the South African approach should be followed everywhere, or even necessarily that it is the correct solution for South Africa. But it is certainly a radically different approach from the regime envisioned by the ICC. . . .

Efforts to minimize or override the nation state through "international law" have found further expression in the expansive elaboration of the doctrine of "universal jurisdiction." Until recently an obscure, theoretical creature in the academic domain, the doctrine gained enormous public exposure (although very little scrutiny) during the efforts [of a Spanish judge] to extradite General Augusto Pinochet of Chile from the United Kingdom [to Spain].

Even defining "universal jurisdiction" is not easy because the idea is evolving so rapidly. . . . The idea was first associated with pirates. . . . Because pirates were beyond the control of any state and thus not subject to any existing criminal justice system, the idea developed that it was legitimate for any aggrieved party to deal with them. Such "jurisdiction" could be said to be "universal" because the crime of piracy was of concern to everyone, and because such jurisdiction did not comport with more traditional jurisdictional bases, such as territoriality or nationality. In a sense, the state that prosecuted pirates could be seen as vindicating the common interest of all states. (Slave trading is also frequently considered to be the subject of universal jurisdiction, following similar reasoning.)

. . . [This sense of universal jurisdiction] is a far cry from what "human rights" activists, NGOs [nongovernmental organizations], and academics . . . have in mind today. From a very narrow foundation, theorists have enlarged the concept of universal jurisdiction to cover far more activities, with far less historical or legal support, than arose earlier in the context of piracy. At the same time, they have omitted reference to the use of force, and substituted their preferred criminal prosecution. The proscribed roster of offenses now typically includes genocide, torture, war crimes and crimes against humanity, which are said to vest prosecutorial jurisdiction in all states.

Announcing his decision in the November 25, 1998 decision of *Ex parte Pinochet,* Lord Nicholls described the crimes of which the General stood accused by saying, "International law has made it plain that certain types of conduct . . . are not acceptable conduct on the part of anyone." Although that decision . . . did not actually rest on universal jurisdiction, Lord Nicholls in fact stated the doctrine's essential foundation.

The worst problem with universal jurisdiction is not its diaphanous legal footings but its fundamental inappropriateness in the realm of foreign policy. In effect (and in intention), the NGOs and theoreticians advocating the concept are misapplying legal forms in political or military contexts. What constitutes "crimes against humanity" and whether they should be prosecuted or otherwise handled—and by whom—are not questions to be left to lawyers and judges. To deal with them as such is, ironically, so bloodless as to divorce these crimes from reality. It is not merely naïve, but potentially dangerous, as Pinochet's case demonstrates.

Morally and politically, what Pinochet's regime did or did not do is primarily a question for Chile to resolve. Most assuredly, Pinochet is not, unlike a pirate or a slave trader, beyond the control of any state. Although many people around the world intensely dislike the solution that Chile adopted in order to restore constitutional and democratic rule in 1990, especially the various provisions for amnesty, the terms and implementation of that deal should be left to the Chileans themselves. They (and their democratically elected government) may continue to honor the deal, or they may choose to bring their own judicial proceedings against Pinochet. One may accept or reject the wisdom or morality of either course (and I would argue that they should uphold the deal), but it should be indisputable that the decision is principally theirs to make. The idea that Spain or any other country that subsequently filed extradition requests in the United Kingdom has an interest superior to that of Chile—and can thus effectively overturn the Chilean deal—is untenable. And yet, if the British had ultimately extradited Pinochet to Spain, that is exactly what would have happened. A Spanish magistrate operating completely outside the Chilean system will effectively have imposed his will on the Chilean people. One is sorely tempted to ask: Who elected him? If that is what "universal jurisdiction" means in practice (as opposed to the theoretical world of law reviews), it is hopelessly flawed.*

Spain *does* have a legitimate interest in justice on behalf of Spanish citizens who may have been held hostage, tortured, or murdered by the Pinochet regime. And the Spanish government may take whatever steps it ultimately considers to be in the best interest of Spanish citizens, but its recourse lies with the government of Chile, and certainly not with that of the United Kingdom. . . .

Because of the substantial publicity surrounding the Pinochet matter, we can expect copycat efforts covering a range of other "crimes against humanity" in the near future. But adding purported crimes (shocking though they may be) to the list of what triggers universal jurisdiction does not make the concept any more real. Nor does a flurry of law review articles (and there has been far more than a flurry) make concrete an abstract speculation. In fact, "universal jurisdiction" is conceptually circular: universal jurisdiction covers the most dastardly offenses; accordingly, if the offense is dastardly, there must be universal jurisdiction to prosecute it. Precisely because of this circularity, there is absolutely no limit to what creative imaginations can enlarge it to cover, and we can be sure that they are already hard at work. . . .

Third, tangible American interests are at risk. I believe that the ICC's most likely future is that it will be weak and ineffective, and eventually ignored, because [it was] naïvely conceived and executed. There is, of course, another

---

*Pinochet was ultimately returned to Chile, where he was charged with crimes and awaits trial. —Ed.

possibility: that the Court and the Prosecutor (either as established now, or as potentially enhanced) will be strong and effective. In that case, the U.S. may face a much more serious danger to our interests, if not immediately, then in the long run.

Although everyone commonly refers to the "Court" created at the 1998 Rome Conference, what the Conference actually did was to create not just a Court, but also a powerful and unaccountable piece of an "executive" branch: the Prosecutor. Let there be no mistake: our main concern from the U.S. perspective is not that the Prosecutor will indict the occasional U.S. soldier who violates our own laws and values, and his or her military training and doctrine, by allegedly committing a war crime. Our main concern should be for the President, the Cabinet officers on the National Security Council, and other senior leaders responsible for our defense and foreign policy. They are the real potential targets of the ICC's politically unaccountable Prosecutor.

One problem is the crisis of legitimacy we face now in international organizations dealing with human rights and legal norms. Their record is, to say the least, not encouraging. The International Court of Justice and the UN Human Rights Commission are held in *very* low esteem, and not just in the U.S. ICC supporters deliberately chose to establish it independently of the ICJ to avoid its baggage.

Next is the overwhelming repudiation by the Rome Conference of the American position supporting even a minimal role for the Security Council. Alone among UN governing bodies, the Security Council does enjoy a significant level of legitimacy in America. And yet it was precisely the Council where the U.S. found the greatest resistance to its position. The Council has primacy in the UN for "international peace and security," in all their manifestations, and it is now passing strange that the Council and the ICC are to operate virtually independently of one another. The implicit weakening of the Security Council is a fundamental *new* problem created by the ICC, and an important reason why the ICC should be rejected. The Council now risks both having the ICC interfering in its ongoing work, and even more confusion among the appropriate roles of law, politics and power in settling international disputes.

The ICC has its own problems of legitimacy. Its components do not fit into a coherent international structure that clearly delineates how laws are made, adjudicated and enforced, subject to popular accountability, and structured to protect liberty. Just being "out there" in the international system is unacceptable, and, indeed, almost irrational unless one understands the hidden agenda of many NGOs supporting the ICC. There is real vagueness over the ICC's substantive jurisdiction, although one thing is emphatically clear: this is *not* a court of limited jurisdiction. . . .

Examples of vagueness in key elements of the Statute's text include:

- "Genocide," as defined by the Rome Conference is inconsistent with the Senate reservations attached to the underlying Genocide Convention, and the Rome Statute is not subject to reservations.
- "War crimes" have enormous definitional problems concerning civilian targets. Would the United States, for example, have been guilty of "war

crimes" for its WWII bombing campaigns, and use of atomic weapons, under the Rome Statute?

- What does the Statute mean by phrases like "knowledge" of "incidental loss of life or injury to civilians"? "long-term and severe damage to the natural environment"? "clearly excessive" damage?

Apart from problems with existing provisions, and the uncertain development of customary international law, there are many other "crimes" on the waiting list: aggression, terrorism, embargoes (courtesy of Cuba), drug trafficking, etc. The Court's potential jurisdiction is enormous. Article 119 provides: "any dispute concerning the judicial functions of the Court shall be settled by the decision of the Court."

Consider one recent example of the use of force, the NATO air campaign over former Yugoslavia. Although most Americans did not question the international "legality" of NATO's actions, that view was not uniformly held elsewhere. During the NATO air war, Secretary General Kofi Annan expressed the predominant view that "unless the Security Council is restored to its preeminent position as the sole source of legitimacy on the use of force, we are on a dangerous path to anarchy." . . .

Implicitly, therefore, in Annan's view, NATO's failure to obtain Council authorization made its actions illegitimate, which is what those pursuing the hidden agenda want to hear: while one cannot stop the United States from using force because it is so big and powerful, one can ensure that it is illegitimate absent Security Council authorization, *and thus a possible target of action by the ICC Prosecutor.* . . .

Many hope to change [U.S. military] behavior as much as the international "rules" themselves, through the threat of prosecution. They seek to constrain military options, and thus lower the potential effectiveness of such actions, or raise the costs to successively more unacceptable levels by increasing the legal risks and liabilities perceived by top American and allied civilian and military planners undertaking military action.

. . . Amnesty [International] asserted . . . that "NATO forces violated the laws of war leading to cases of unlawful killing of civilians." The NGO complained loudly about NATO attacks on a "civilian" television transmitter in Belgrade, even though it served the Milosevic regime's propaganda purposes. Similarly, Human Rights Watch . . . concluded that NATO violated international law, but stopped short of labeling its actions as "war crimes." In another recent report, this NGO announced its opposition to the sale of American air-to-ground missiles to Israel because of the Israeli "war crime" of attacking Lebanese electrical power stations. Of course, much the same could also be said about American air attacks during the Persian Gulf War, aimed at destroying critical communications and transportation infrastructure inside Iraq, in order to deny it to Saddam's military. If these targets are now "off limits," the American military will be far weaker than it would otherwise be. . . .

What to do next is obviously the critical question. Whether the ICC survives and flourishes depends in large measure on the United States. We should not allow this act of sentimentality masquerading as policy to achieve indirectly what was rejected in Rome. We should oppose any suggestion that we cooperate, help fund, and generally support the work of the Court and Prosecutor. We should isolate and ignore the ICC.

Specifically, I have long proposed for the United States a policy of "Three Noes" toward the ICC: (1) no financial support, directly or indirectly; (2) no collaboration; and (3) no further negotiations with other governments to "improve" the Statute. . . . This approach is likely to maximize the chances that the ICC will wither and collapse, which should be our objective. The ICC is a fundamentally bad idea. It cannot be improved by technical fixes as the years pass. . . . We have alternative approaches and methods consistent with American national interests, as I have previously outlined, and we should follow them.

POSTSCRIPT

Should the United States Ratify the International Criminal Court Treaty?

After the statements by the Lawyers Committee for Human Rights and Bolton were given, the prospects for U.S. adherence to the ICC treaty dimmed even further. In the waning days of his administration, President Clinton directed a State Department representative to sign the ICC treaty on behalf of the United States. That act was mostly symbolic, however, because at the same time Clinton warned that the "United States should have the chance to observe and assess the functioning of the Court, over time, before choosing to become subject to its jurisdiction. Given these concerns, I will not, and do not recommend that my successor submit the Treaty to the Senate for advice and consent until our fundamental concerns are satisfied." In fact, Clinton's cautious advice to the next president was superfluous because the election of George W. Bush in 2000 brought into the Oval Office a president who agreed fully with the opponents of the ICC. In May 2002 the administration announced that it would not even continue to work toward a revision of the treaty that would satisfy Washington. "It's over. We're washing our hands of it," Pierre-Richard Prosper, the State Department's ambassador-at-large for war-crimes issues, said of the ICC.

Despite the U.S. stance, the ICC became a reality on April 12, 2002, when the number of countries ratifying the treaty passed 60, the minimum required for the ICC treaty to take effect. "The long-held dream of a permanent international criminal court will now be realized," Secretary-General Kofi Annan of the United Nations proclaimed. By March 2003 the number of accessions to the ICC treaty stood at 89, with ratification pending in another 50 signatory countries. More on the ICC can be found in William A. Schabas, *An Introduction to the International Criminal Court* (Cambridge University Press, 2001) and Bruce Broomhall, *International Criminal Justice and the International Criminal Court: Between State Consent and the Rule of Law* (Oxford University Press, 2003).

Some of the basic provisions of the ICC are the following:

1. The court's jurisdiction includes genocide and a range of other crimes committed during international and internal wars. Such crimes must be "widespread and systematic" and committed as part of "state, organization, or group policy," not just as individual acts.
2. Except for genocide and complaints brought by the United Nations Security Council (UNSC), the ICC will not be able to prosecute alleged crimes unless either the state of nationality of the accused or the state where the crimes took place has ratified the treaty.

3. Original signatories will have a one-time ability to "opt out" of the court's jurisdiction for war crimes, but not genocide, for a period of seven years.
4. The UNSC can delay one prosecution for one year. The vote to delay will not be subject to veto.
5. The ICC will only be able to try cases when national courts have failed to work.

The irony of the U.S. opposition to the ICC is that the ICC treaty is, in part, an American product. The United States was a major force behind the convening of a conference to draft a treaty, with President Clinton issuing a clarion call during one commencement address for a permanent court that could try and punish those who committed war crimes and other abominations. Once at the conference in Rome, however, the United States retreated from Clinton's rhetorical position and sought to eliminate virtually any possibility that an American civilian or military leader would ever stand before the ICC's bar of justice. For more on the U.S. attitude toward the ICC, see Sarah Sewall and Carl Kaysen, eds., *The United States and the International Criminal Court* (Rowman & Littlefield, 2000).

Technically, U.S. ratification of the ICC treaty and support of the ICC once it begins is not necessary for the court to function. But, in reality, the United States is the world's hegemonic power, and U.S. opposition to the court will almost certainly hinder its operations and could prevent the full establishment of the ICC.

Two Web sites that provide more information on the ICC are the UN site http://www.un.org/law/icc/index.html and the site of the Coalition for the International Criminal Court at http://www.iccnow.org. Finally, Henry Kissinger expresses wariness of the ICC in "The Pitfalls of Universal Jurisdiction," *Foreign Affairs* (July/August 2001).

ISSUE 19

Is the Convention on the Elimination of All Forms of Discrimination Against Women Worthy of Support?

YES: Harold Hongju Koh, from Statement Before the Committee on Foreign Relations, U.S. Senate (June 13, 2002)

NO: Christina P. Hoff-Sommers, from Statement Before the Committee on Foreign Relations, U.S. Senate (June 13, 2002)

ISSUE SUMMARY

YES: Harold Hongju Koh, the Gerard C. and Bernice Latrobe Smith Professor of International Law at Yale University and former assistant secretary of state for human rights and democracy, contends that the United States cannot be a global leader championing progress for women's human rights around the world unless it is also a party to the global women's treaty.

NO: Christina P. Hoff-Sommers resident scholar at the American Enterprise Institute, Washington, D.C., tells Congress that the United States can and should help women everywhere to achieve the kind of equity American women have. She maintains, however, that ratifying the CEDAW is the wrong way to pursue that goal.

Females constitute about half the world population, but they are a distinct economic-political-social minority because of the wide gap in societal power and resources between women and men. Women constitute 70 percent of the world's poor and two-thirds of the world's literates. They occupy only 14 percent of the managerial jobs constitute less than 40 percent of the world's professional and technical workers, and garner a mere 35 percent of the earned income in the world.

Women are also disadvantaged politically in late 2003, only 10 women were serving as presidents or prime ministers of their countries; women make up just 8 percent of all national cabinet officers; and only one of every seven national legislators is a woman.

On average, life for women is not only harder and more poorly compensated than it is for men, it is also more dangerous. "The most painful devaluation of women," the UN reports, "is the physical and psychological violence that stalks them from cradle to grave." Signs of violence against women include the fact that about 80 percent of the world's refugees are women and their children. Other assaults on women arguably constitute a form of genocide. According the United Nations Children's Fund, "In many countries, boys get better care and better food than girls. As a result, an estimated one million girls die each year because they were born female."

None of these economic, social, and political inequities is new. Indeed, the global pattern of discrimination against women is an ancient story. What is new is the global effort to recognize the abuses that occur and to ameliorate and someday end them.

To help accomplish that goal, the UN General Assembly in 1979 voted by 130 to 0 to put the Convention on the Elimination of All Forms of Discrimination Against Women (CEDAW) before the world's countries for adoption. Supporters hailed the treaty as a path breaking step on behalf of advancing the status of women by defining their rights on the international level. Many countries agreed, and by September 1981 enough countries had signed and ratified CEDAW to put it into effect. Women's rights had not been specifically and fully addressed in any other treaty before CEDAW.

Countries that legally adhere to the convention agree to undertake a series of measure to end all the various forms of discrimination against women. Doing so entails accepting the principle of legal gender equality and ensuring the practice of gender equality by abolishing all discriminatory laws, enacting laws that prohibit discrimination against women, and establishing agencies to promote and protect women's rights.

President Jimmy Carter signed CEDAW on behalf of the United States in 1980 and submitted it to the Senate for ratification. However, he was soon thereafter defeated for reelection, and the treaty languished in legislative limbo through the presidencies of Ronald Reagan and George H. W. Bush. By contrast, President Bill Clinton made an effort to move CEDAW forward. The Senate Committee on Foreign Relations held hearings on the pact in 1994 and issued a report recommending ratification. There the effort on behalf of CEDAW stalled. The measure did not come up for debate or a vote in the Senate. There was little chance that the measure would garner the required votes. Therefore, rather than risk the defeat of CEDAW, its advocates did not press the issue to a vote.

The testimony that forms the two readings here came during new hearings in 2002 that were part of yet another effort to gain Senate approval of CEDAW. In the first reading, Harold Hongju Koh maintains that failure to ratify the treaty will hamper and undermine U.S. efforts to fight for democracy and human rights around the world. Christine P. Hoff-Sommers disagrees. She tells senators that the UN has often used human rights doctrines to score points against Western democracies while carefully refraining from censuring countries that notoriously abuse the rights of their citizens and that CEDAW would probably be similarly used to criticize the United States if it were to ratify the treaty.

Harold Hongju Koh  **YES**

Statement Before the Committee on Foreign Relations, U.S. Senate

In his [2002] State of the Union address, President George W. Bush . . . announced that "America will always stand for the non-negotiable demands of human dignity: the rule of law; limits on the power of the state; respect for women; private property; free speech; equal justice; and religious tolerance" (emphasis added). I can imagine no more fitting way for this Administration and this Senate to answer that demand than by moving quickly to ratify this treaty for the rights of women. . . .

My main message today is that this commitment should not stop at the water's edge. Particularly after September 11, America cannot be a world leader in guaranteeing progress for women's human rights, whether in Afghanistan, here in the United States, or around the world, unless it is also a party to the global women's treaty.

Let me first review the background and history of CEDAW [Convention to Eliminate All Forms of Discrimination Against Women]; second, explain why ratifying that treaty would further our national commitments to eliminating gender discrimination, without jeopardizing our national interests; and third, explain why some concerns occasionally voiced about our ratification of this treaty are, upon examination, completely unfounded.

First, some history. The United Nations Charter reaffirms both the faith of the peoples of the United Nations "in the equal rights of men and women," Preamble, and their determination to promote respect for human rights "for all without distinction as to race, sex, language, or religion." In 1948, the Universal Declaration of Human Rights similarly declared that "everyone" is entitled to the rights declared there "without distinction of any kind, such as race, colour, (or) sex . . ." In 1975, a global call for an international convention specifically to implement those commitments emerged from the First World Conference on Women in Mexico City. But until 1979, when the General Assembly adopted the CEDAW, there was no convention that addressed comprehensively women's rights within political, social, economic, cultural, and family life. After years of drafting, the United Nations adopted the Convention on the Elimination of All Forms of Discrimination Against Women in December 18, 1979, and the Convention entered into force in September 1981.

Committee on Foreign Relations, U.S. Senate, June 13, 2002.

In the more than two decades since, 169 nations other than our own have become parties to the Convention. Only nineteen United Nations member states have not. That list includes such countries as Afghanistan, Bahrain, Iran, Somalia, Sudan, Syria, Qatar, and the United Arab Emirates. To put it another way, the United States is now the only established industrialized democracy in the world that has not yet ratified the CEDAW treaty. Frankly, Senators, this is a national disgrace for a country that views itself as a world leader on human rights.

Why should the United States ratify this treaty? For two simple reasons. First, ratification would make an important global statement regarding the seriousness of our national commitment to these issues. Second, ratification would have a major impact in ensuring both the appearance and the reality that our national practices fully satisfy or exceed international standards.

The CEDAW treaty has been accurately described as an international bill of rights for women. The CEDAW simply affirms that women, like the rest of the human race, have an inalienable right to live and work free of discrimination. The Convention affirms the rights of all women to exercise on an equal basis their "human rights and fundamental freedoms in the political, economic, social, cultural, civil or any other field."

The treaty defines and condemns discrimination against women and announces an agenda for national action to end such discrimination. By ratifying the treaty, states do nothing more than commit themselves to undertaking "appropriate measures" toward ending discrimination against women, steps that our country has already begun in numerous walks of life. The CEDAW then lays a foundation for realizing equality between women and men in these countries by ensuring women's equal access to, and equal opportunities in, public and political life—including the right to vote, to stand for election, to represent their governments at an international level, and to enjoy equal rights "before the law" as well as equal rights in education, employment, health care, marriage and family relations, and other areas of economic and social life. The Convention directs States Parties to "take into account the particular problems faced by rural women," and permits parties to take "temporary special measures aimed at accelerating de facto equality" between men and women, a provision analogous to one also found in the Convention on the Elimination of All Forms of Racial Discrimination, which our country has already ratified.

Ratifying this treaty would send the world the message that we consider eradication of these various forms of discrimination to be solemn, universal obligations. The violent human rights abuses we recently witnessed against women in Afghanistan, Bosnia, Haiti, Kosovo, and Rwanda painfully remind us of the need for all nations to join together to intensify efforts to protect women's rights as human rights. At the State Department, where I supervised the production of the annual country reports on human rights conditions worldwide, I found that a country's ratification of the CEDAW is one of the surest indicators of the strength of its commitment to internalize the universal norm of gender equality into its domestic laws.

Let me emphasize that in light of our ongoing national efforts to address gender equality through state and national legislation, executive action, and judicial decisions, the legal requirements imposed by ratifying this treaty would not be burdensome. Numerous countries with far less impressive practices regarding gender equality than the United States have ratified the treaty, including countries whom we would never consider our equals on such matters, including Iraq, Kuwait, North Korea, and Saudi Arabia.

At the same time, from my direct experience as America's chief human rights official, I can testify that our continuing failure to ratify CEDAW has reduced our global standing, damaged our diplomatic relations, and hindered our ability to lead in the international human rights community. Nations that are otherwise our allies, with strong rule-of-law traditions, histories, and political cultures, simply cannot understand why we have failed to take the obvious step of ratifying this convention. In particular, our European and Latin American allies regularly question and criticize our isolation from this treaty framework both in public diplomatic settings and private diplomatic meetings.

Our nonratification has led our allies and adversaries alike to challenge our claim of moral leadership in international human rights, a devastating challenge in this post-September 11 environment. Even more troubling, I have found, our exclusion from this treaty has provided anti-American diplomatic ammunition to countries who have exhibited far worse record on human rights generally, and women's rights in particular. Persisting in the aberrant practice of nonratification will only further our diplomatic isolation and inevitably harm our other United States foreign policy interests.

Treaty ratification would be far more than just a paper act. The treaty has demonstrated its value as an important policy tool to promote equal rights in many of the foreign countries that have ratified the CEDAW. As a recent, comprehensive world survey issued by the United Nations Development Fund for Women chronicles, numerous countries around the world have experienced positive gains directly attributable to their ratification and implementation of the CEDAW. CEDAW has been empowering women around the globe to change constitutions, pass new legislation, and influence court decisions in their countries. Ratification of the CEDAW by the United States would similarly make clear our national commitment to ensure the equal and nondiscriminatory treatment of American women in such areas as civil and political rights, education, employment, and property rights.

Most fundamentally, ratification of CEDAW would further our national interests. Secretary of State Colin Powell put it well when he said earlier this year: "The worldwide advancement of women's issues is not only in keeping with the deeply held values of the American people; it is strongly in our national interest as well. . . . Women's issues affect not only women; they have profound implications for all humankind. Women's issues are human rights issues. . . . We, as a world community, cannot even begin to tackle the array of problems and challenges confronting us without the full and equal participation of women in all aspects of life."

After careful study, I have found nothing in the substantive provisions of this treaty that even arguably jeopardizes our national interests. Those treaty provisions are entirely consistent with the letter and spirit of the United States Constitution and laws, both state and federal. The United States can and should accept virtually all of CEDAW's obligations and undertakings without qualification. Regrettably, the Administration has not provided a witness here today to set forth its views on the ratification of this treaty. Although past Administrations have proposed that ratification be accompanied by certain reservations, declarations, and understandings, only one of those understandings, relating to limitations of free speech, expression and association, seems to me advisable to protect the integrity of our national law.

Finally, let me address some myths and fallacies that have been circulated about the likely impact of United States ratification of the CEDAW. The most common include the following:

First, that CEDAW supports abortion rights by promoting access to "family planning." This is flatly untrue. There is absolutely no provision in CEDAW that mandates abortion or contraceptives on demand, sex education without parental involvement, or other controversial reproductive rights issues. CEDAW does not create any international right to abortion. To the contrary, on its face, the CEDAW treaty itself is neutral on abortion, allowing policies in this area to be set by signatory states and seeking to ensure equal access for men and women to health care services and family planning information. In fact, several countries in which abortion is illegal—among them Ireland, Rwanda, and Burkina Faso—have ratified CEDAW.

A second fallacy is that CEDAW ratification would somehow undermine the American family by redefining traditional gender roles with regard to the upbringing of children. In fact, CEDAW does not contain any provisions seeking to regulate any constitutionally protected interests with respect to family life. The treaty only requires that parties undertake to adopt measures "prohibiting all discrimination against women" and to "embody the principle of the equality of men in women" in national laws "to ensure, through law and other appropriate means, the practical realization of this principle." How best to implement that obligation consistent with existing United States constitutional protections—which as you know, limit the government's power to interfere in family matters, including most parental decisions regarding childrearing—is left for each country to decide for itself.

Third, some have falsely suggested that ratification of CEDAW would require decriminalization of prostitution. Again, the text of the treaty is to the contrary. CEDAW's Article 6 specifically states that countries that have ratified CEDAW "shall take all appropriate measures, including legislation, to suppress all forms of traffic in women and exploitation of prostitution in women."

Fourth, some claim that if CEDAW were U.S. law, it would outlaw single-sex education and require censorship of school textbooks. In fact, nothing in CEDAW mandates abolition of single-sex education. As one way to encourage equal access to quality education for all children, Article 10 requires parties to take all appropriate measures to eliminate "any stereotyped concept of the roles of men and women at all levels and in all forms of education by encouraging [not requiring] coeducation and other types of education which will help to achieve this aim . . . ," (emphasis added) including, presumably, single-sex education that teaches principles of gender equality. CEDAW also encourages the development of equal education material for students of both genders. This provision is plainly designed not to disrupt educational traditions in countries like ours, but rather, to address those many countries in the world (like Afghanistan during Taliban rule) in which educational facilities for girls are either nonexistent or remain separate and unequal.

Fifth, some have suggested that U.S. ratification of CEDAW would require the legalization of same-sex marriage. Whatever view one may hold regarding the desirability of same-sex marriage, this treaty plainly contains no such requirement. Article 10 of CEDAW requires only elimination of discrimination directed against women "in all matters related to marriage and family relations." Thus, for example, the practice of polygamy is inconsistent with the CEDAW because it undermines women's equality with men and potentially fosters severe financial inequities. Article 10 would neither require nor bar any national laws regarding same-sex marriage, which by their very nature, would apply equally to men and women.

Finally, and most pervasively, opponents of CEDAW have claimed that U.S. ratification would diminish our national sovereignty and states' rights by superseding or overriding our national, state or local laws. Given the broad compatibility between the treaty requirements and our existing national laws, however, very few occasions will arise in which this is even arguably an issue. Moreover, the treaty generally requires States to use "appropriate measures" to implement the nondiscrimination principle, which by its terms accords some discretion to member countries to determine what is "appropriate" under the national circumstances. Finally, the Senate is, of course, free to address any material discrepancies between national law and the treaty by placing understandings upon its advice and consent, along the lines of the "freedom of speech" understanding discussed above, or by the Congress passing implementing legislation—as it has done, for example, to effectuate the Genocide Convention—specifying the precise ways in which the Federal legislature will carry out our international obligations under this treaty.

Ironically, many of the unfounded claims about the likely effects of CEDAW ratification have been asserted by self-proclaimed advocates of states' rights. In fact, within our own country, the emerging trend has been the opposite. Broad sentiment has been emerging at both the state and local level to incorporate the CEDAW requirements into local law. As I speak, governmental bodies in some fifteen states and Guam, sixteen counties and forty-two cities have adopted resolutions or instruments endorsing CEDAW or adopting it on behalf of their jurisdictions. Far from CEDAW imposing unwanted obligations on local governments, local governments are in fact responding to the demands of their citizens, who have become impatient at the lack of federal action to implement these universal norms into American law.

Particularly in a time of terror, promoting human rights and eradicating discrimination should not be partisan issues. As President Bush recently reminded us, the United States cannot fight a war on terrorism alone; it needs cooperation not only from its current allies, but also from the rest of the world. "We have a great opportunity during this time of war," he said, "to lead the world toward the values that will bring lasting peace . . . [such as] the non-negotiable demands of human dignity [that include] respect for women. . . ." First Lady Laura Bush echoed that sentiment on International Women's Day 2002, when she said, "People around the world are looking closely at the roles that women play in society. And Afghanistan under the Taliban gave the world a sobering example of a country where women were denied their rights and their place in society. . . . Today, the world is helping Afghan women return to the lives that they once knew. . . . Our dedication to respect and protect women's rights in all countries must continue if we are to achieve a peaceful, prosperous world. . . . Together, the United States, the United Nations and all of our allies will prove that the forces of terror can't stop the momentum of freedom."

The world looks to America for leadership on human rights, both in our domestic practices and in our international commitments. Ours is a nation conceived in liberty and dedicated to the proposition that all human beings— not just men—are created equal. Our country has fought a civil war and a centuries-long social struggle to eliminate racial discrimination. It is critically important that we seize this opportunity to announce unequivocally to the world that we, of all nations, insist on the equality of all human beings, regardless of gender.

Senators, in closing let me say how much United States ratification of this important treaty means to every American. My mother, Hesung Chun Koh, came to this country more than fifty years ago from the Republic of Korea and found equal opportunity here as a naturalized American citizen. My wife, Mary-Christy Fisher, is a natural-born American citizen and lawyer of Irish and British heritage. I am the father of a young American, Emily Koh, who will turn sixteen years old in ten days' time.

Although I have tried, I simply cannot give my daughter any good reason why her grandmother and mother would have been protected by CEDAW in their ancestral countries, but that she is not protected by it in the United States, which professes to be a world leader in the promotion of women's rights and gender equality. I cannot explain to her why this country we love, and which I have served as Assistant Secretary of State for Human Rights, has for so long failed to ratify the authoritative human rights treaty that sets the universal standard on women's equality. Finally, I cannot explain why, by not ratifying, the United States chooses to keep company with such countries as Afghanistan, Iran, Sudan, and Syria, in which human rights and women's rights have been brutally repressed.

The choice is simple. Our continuing failure to ratify this treaty will hamper and undermine our efforts to fight for democracy and human rights around the world. Ratification now of the CEDAW treaty would be both prudent foreign policy and simple justice.

Christina P. Hoff-Sommers  **NO**

Statement Before the Committee on Foreign Relations, U.S. Senate

The Case Against Ratifying the United Nations Convention on the Elimination of All Forms of Discrimination Against Women (CEDAW)

Although I shall be arguing that we should not ratify the CEDAW convention, I want first to speak as a feminist who would very much like to see a realistic international effort for securing women's rights.

American women have been the beneficiaries of two major waves of feminism. In the First Wave, led by the great foremothers, Elizabeth Cady Stanton and Susan B. Anthony, women won basic political and legal rights, including the right to vote. The Second Wave, which came in the sixties and early seventies, advanced women economically and socially. Employers could no longer legally restrict a job to one sex. A company could no longer refuse to hire a woman because she had children. Such laws have been critical to the well-being and success of American women and most of the reforms of the First and Second Waves are appropriate and necessary for women everywhere.

With this historical progress, American women have achieved virtual equality with men. There are still some unresolved equity issues, but overall, we are now among the freest and most liberated women in the world. In some ways, we are not merely doing as well as men—we are doing better. We live longer, we are better educated, we have more choices on how to lead our lives. By any reasonable measure, equity feminism is the great American success story.

When I lecture about the history of the women's movement on college campuses, students often ask what's next for the Third Wave. My answer is always the same; we have to help women in other parts of the world secure the freedoms we now take for granted. There are countries, especially in Africa and Asia, where women have not yet had their Elizabeth Cady Stanton and Susan B. Anthony; as for second wave reforms, they are light-years away from them.

American women have much to tell the women of the world. We can and should help women everywhere to achieve the kind of equity we have

Committee on Foreign Relations, U.S. Senate, June 13, 2002.

here. But joining the CEDAW convention is the wrong way to do that. I have several reasons for opposing ratification of this treaty. I will focus here on two or three that I regard as decisive.

The CEDAW convention has many admirable and sound goals that any person of conscience must support. But it was formulated in the 1970s and it promotes several reforms that we now know to be harmful. These programs looked promising, exciting and progressive in 1975, but since then we have come to realize that they undermine economic prosperity. Article 11, for example, calls for governments to set wages. It demands "The right to equal remuneration . . . in respect of work of equal value." This is the policy we call "comparable worth." Americans have rightly rejected comparable worth as unjust and unworkable at home. So, why should we advocate it for women anywhere?

Article 11 also demands that governments provide paid maternity leave, and provide the "necessary supporting social services to enable parents to combine family obligation with work responsibility and participation in public life . . . through the establishment and development of a network of childcare facilities." All very salutary, except that experience shows that such programs tend to burden a country's economy to everyone's detriment. American women have benefited from a free, open and economically dynamic society: shouldn't we be promoting policies that bring these advantages to needy women everywhere?

The treaty includes several sweeping demands that are socially divisive and likely to create unnecessary misery. Article 5, for example, calls for all governments to "modify the social and cultural patterns of conduct of men and women with a view to achieving the elimination of prejudices and all other practices which are based on . . . Stereotyped roles for men and women." What exactly does this provision entail? Of course, some gender stereotypes are destructive and prejudicial and we must call disparaging attention to them. (Typical examples include generalizations that women are irrational, that they are less intelligent than men, that they are politically immature, etc.) But, other male/female stereotypes are descriptively true. In the 1970s, many feminists believed that truly liberated men and women would become more and more alike—that a gender-just society would eventually become androgynous. Gender was supposedly an artificial social construction that gave men the advantage. Well, today, only a handful of scholars in Women's Studies programs still believe that.

A growing body of research in neuroscience, endocrinology, and psychology over the past 40 years provides evidence that there is a biological basis for many sex differences in aptitudes and preferences. Males have better spatial reasoning skills, females better verbal skills. Males are greater risk takers, females are more nurturing. (There are exceptions, but these are the rules.) As the Rutgers University anthropologist Lionel Tiger has said, "Biology is not destiny, but it is good statistical probability." Unfortunately, much in CEDAW is premised on the false idea that all gender preferences are socially constructed.

Of course, in recognizing the obvious differences between men and women, I am not for one moment suggesting that women should be prevented from pursuing their goals in any field they choose; but I am suggesting we should not expect or aim at parity in all fields. More women than men will continue to want to stay at home with small children and pursue careers in fields like early childhood education or psychology; men will continue to be heavily represented in fields like helicopter mechanics and hydraulic engineering.

A few years ago I took part in a television debate with celebrity lawyer, Gloria Allred. Ms. Allred was representing a 14-year-old girl who was suing the Boy Scouts of America for excluding girls. Allred characterized same-sex scout troops as a form of "gender apartheid." She spoke of the need to "socialize" boys to play with dolls so they could be more nurturing and less fractious. CEDAW will give all the Ms. Allred's in this country a treaty of their own to create mischief.

Consider, for example, how hard-liners could deploy Article 10 of the treaty: It calls for the "elimination of stereotyped concepts of the roles of men and woman at all levels in all forms of education . . . in particular, by the revision of textbooks and school programs." Our textbooks and school materials cannot endure any more political corrections. *The New York Times* recently ran a story about how politics of textbook revisions is now out of control: great works of literature were recently scanned for insensitivity and altered by censors before intense lobbying eliminated the practice. The CEDAW Treaty demands this kind of textual revision—which amount to censorship inconsistent with American civil liberties.

Can there be anyone in the United States, apart from a small coterie of feminists activists and academics, who would favor empowering a committee of foreign bureaucrats to oversee American social mores—or intrude into public education by distorting the textbooks our children read?

The treaty could do us harm by promoting male/female resentments and divisions at a time when the country badly needs social unity. Most American women feel blessed to live in a country where, for the most part, the men are fair-minded, decent and supportive of women in their quest for equality. We are proud and grateful to be part of a society that has afforded us unprecedented freedoms and opportunities. But this very favorable view of American men and of American society is not shared by the hard-line feminists in our universities. These activists/scholars tend to take a dim view of American society, routinely referring to it as a "patriarchy," a "male hegemony," a culture that keeps women socially subordinate. One leading textbook in women's studies talks of an epidemic of gender "terrorism" plaguing the average American women. Another calls the United States a "Rape Culture." Now, Bosnia, for a time, was truly a rape culture. Afghanistan, under the Taliban, routinely practiced gender terrorism. To apply such terms to the United States is ludicrous.

The activists and scholars who characterize America as a sexist society sincerely believe we are in a gender war. In all wars, the first casualty is truth. Too much of what we hear from contemporary women's organizations is outrageously false. Too much of what passes as gender scholarship is ideological and factually wrong: American men are depicted as violent predators and American women their hapless victims. If you ask me to reduce the philosophy of academic feminism to a single phrase it be this one: Women are from Venus, Men are from Hell.

For the past decade, moderate feminist academics like myself, and a growing number of dissidents scholars such as Camille Paglia (University of the Arts), Daphne Patai (University Of Massachusetts), Betsy Fox-Genovese (Emory), Noretta Koertge (University of Indiana), Judith Kleinfeld (University of Alaska), Jennifer Braceras (Harvard Law)—to name only a few—have been hard at work correcting the misinformation, challenging the naive hostility to the free market system, and calling for an end to the male bashing-rhetoric that is standard fare at most of our colleges and universities. We have made slow but steady progress in opening up the national discussion on gender to diverse perspectives, but thinking on these matters on campus and in the major feminist organizations remains dismayingly rigid and intolerant. For the time being, the organized women's movement in this country is dominated by ideological gender theorists and by well-intentioned, but misinformed, women's groups that take what these theorists say seriously.

Now what does this have to do with CEDAW? If the United States signs the treaty, it would dramatically increase the power of the misguided gender scholars. The treaty calls for the elimination of sexism. Reasonable people believe that our American society has already achieved this goal in most of the ways that count. If you compare us with the rest of the world, we are a shining example of gender equity. Unfortunately, most campus theorists do not agree with that. They believe that American women live in a male supremacist society; and they can cite twenty years of feminist "scholarship" to persuade themselves and us that they are right. What they actually cite is a body of statistically challenged gender ideology.

This treaty in conjunction with the counterfeit feminist research could be a most toxic combination. If CEDAW is ratified, expect more rancor, more lawsuits, and more divisiveness. Gender bureaucrats from the United Nations will join the feminist ideologues and the United States will be subject to relentless legal assaults for alleged violations of the treaty.

The United Nations has a history of using its human rights doctrines and commissions for scoring points against Western democracies—all the while carefully refraining from censuring countries that notoriously abuse the rights of their citizens. The United States was banished from the Commission on Human Rights for a year. The UN's 2001 Conference against Racism in Durban, South Africa turned into a shameful anti-Semitic condemnation of Israel. There is no reason to believe that the CEDAW would not be used in a highly political way as well.

Women in the developing countries need help. We are morally bound to assist them in ways that are constructive and that reflect ideals of fairness and common sense that have lifted American women to a level of freedom and unprecedented in human history. CEDAW is not the way.

POSTSCRIPT

Is the Convention on the Elimination of All Forms of Discrimination Against Women Worthy of Support?

The concentrated effort to promote women's rights internationally within the context of advancing globalism dates back only to 1975, which the United Nations declared the International Women's Year. There have been many changes that benefit women since that time, but those changes have only begun to ease the problems have advocates of women's rights argue need to be addressed.

CEDAW has been a keystone of the international effort to promote women's rights. By late 2004, CEDAW had been formally adhered to by 179 countries, 94 percent of all those in the world. The United States remains one of the few of countries that have not accepted the treaty. Following the hearings in 2002, the Senate's Committee on Foreign Relations Committee voted twelve to seven to recommend that the Senate ratify CEDAW. Two Republicans and all nine of the committee's Democrats favored ratification. The balance of the committee's Republican members cast all the seven votes against. One again, however, the measure stalled. Senator Sam Brownback (R-KS), one of the seven committee members who opposed CEDAW, explained, that "many undecided senators heard from many constituents who were opposed to [CEDAW], so at this point it appears we have stopped [it] for this year. Hopefully the election will place us in a stronger position when the new Congress convenes in January." As it turned out, Brownback's hopes were realized. The Republicans became the majority party in the Senate after the 2002 elections. This change of political power dimmed the conventions prospects in the Senate, and the reelection of President Bush in 2004 along with an even stronger majority in the Senate for the Republicans makes it unlikely the Senate will soon reconsider, much less ratify, CEDAW.

In the longer term, however, the political wheel will certainly turn yet again, and CEDAW's chances will improve either there are more Democrats in the Senate to join with those Republicans who favor the measure or because more Republicans will come to support CEDAW. So the issue is far from dead.

The heart of the opposition to CEDAW is entwined with the status of U.S. sovereignty and abortion rights policy. Opponents believe that U.S. adherence to CEDAW would lessen U.S. independence and might also be used to supplement national laws and court decisions that favor the pro-abortion, or pro-choice, position. Expressing this view is John Fonte, "Democracy's Trojan Horse," *The National Interest* (Summer 2004). Also on this topic, read Sandra F. VanBurkleo,

Belonging to the World: Women's Rights and American Constitutional Culture (Oxford University Press, 2000).

To further inform your views on this debate, read the text of CEDAW, which can be found on the Web site of the UN's Division for the Advancement of Women (DAW) at `http://www.un.org/womenwatch/daw/cedaw/cedaw.htm`. DAW favors CEDAW, as does the group Working Group on Ratification of the U.N. Convention on the Elimination of All Forms of Discrimination Against Women, whose Web site is at `http://www.womenstreaty.org/`. A group opposed to CEDAW is Women for Faith and Family at `http://www.wff.org/CEDAW.html`.

EarthTrends

An environmental information portal of the World Resources Institute sponsored by such diverse organizations as the Dutch government, the UN Development Programme, and the World Bank

        http://earthtrends.wri.org/

Worldwatch Institute

The environmental-activist Worldwatch Institute offers a range of information and commentary on environmental, social, and economic trends. The institute's work revolves around the transition to an environmentally sustainable and socially just society—and how to achieve it.

        http://www.worldwatch.org

The Common-Sense Environmentalist's Suite

The organizations listed on this links page offer research and commentary on environmental topics that take a skeptical view of many projectons on looming environmental disaster.

        http://www.heartland.org/archives/
            suites/environment/links.htm

PART 6

People and the Environment

*W*hen *all is said and done, policy is, or at least ought to be, about values. That is, how do we want our world to be? There are choices to make about what to do (and what not to do). It would be easy if these choices were clearly good versus evil. But things are not usually that simple, and the issue in this part shows the disparity of opinions regarding the current state of the environment.*

- Do Environmentalists Overstate Their Case?

ISSUE 20

Do Environmentalists Overstate Their Case?

YES: Bjørn Lomborg, from "Debating 'The Skeptical Environmentalist,'" A Debate Held at the Graduate Center of the City University of New York (April 9, 2002)

NO: Fred Krupp, from "Debating 'The Skeptical Environmentalist,'" A Debate Held at the Graduate Center of the City University of New York (April 9, 2002)

ISSUE SUMMARY

YES: Professor of statistics Bjørn Lomborg argues that it is a myth that the world is in deep trouble on a range of environmental issues and that drastic action must be taken immediately to avoid an ecological catastrophe.

NO: Fred Krupp, president of Environmental Defense, asserts that although Lomborg's message is alluring because it says we can relax, the reality is that there are serious problems that, if not addressed, will have a deleterious effect on the global environment.

We live in an era of almost incomprehensible technological boom. In a very short time—less than a long lifetime in many cases—technology has brought some amazing things. If you talked to a 100-year-old person, and there are many, he or she would remember a time before airplanes, before automobiles were common, before air conditioning, before electric refrigerators, and before medicines that could control polio and a host of other deadly diseases were available. A centenarian would also remember when the world's population was 25 percent of what it is today, when uranium was considered to be useless, and when mentioning ozone depletion, acid rain, or global warming would have engendered uncomprehending stares.

There are three points to bear in mind here. One is that technology and economic development are a proverbial two-edged sword. Most people in the economically developed countries (EDCs) and even many people in the less developed countries (LDCs) have benefited mightily from modern technology. For these people, life is longer, easier, and filled with material riches that were the stuff of science fiction not long ago. Yet we are also endangered by the byproducts of progress. There is a burgeoning world population that is now over 6 billion people. Resources are being consumed at an exponential rate. There are many who fear global warming. Acid rain is damaging forests. And extinction claims an alarming array of species of flora and fauna yearly, perhaps daily.

The second notable point is that most of this has occurred very rapidly. Most environmentalists date the escalation of the rate of environmental degradation to the start of the Industrial Revolution in the mid-eighteenth century. Since then, they say, the speed of change has increased steadily, and it is probable that between 80 and 90 percent of all technological advancement has occurred within the last 100 years—that is, the last 2.9 percent of humankind's 3,500 years of recorded history.

The speed of change is important because if there are pressing problems, then they must be addressed quickly. From many people's point of view, the globe cannot stand 100 more years of progress like that of the last century.

In many ways, the issues revolve around whether or not environmental safety requires us to drastically alter some of our consumption patterns; to pay more in taxes and higher prices for technologies that clean the environment; to use more expensive or less satisfactory substitutes for products that threaten the environment or scarce resources; and to alter (some might say lessen) our lifestyles by conserving energy.

Sustainability is one term that is important to this debate. Sustainable development means progress that occurs without further damaging the ecosystem. *Carrying capacity* is another key term. The question is whether or not there is some finite limit to the number of people that the Earth can accommodate. Carrying capacity is about more than just numbers. It also involves how carefully people manage the planet's resources—their lifestyles. If you live to be 100, you may well share the Earth with a world population of 10 billion. Can the world carry 10 billion people if we continue to use resources as rapaciously as we do today? Probably not. Can 10 billion environmentally careful people survive? Maybe.

The third notable point is that individual countries and the global community have collectively begun to try to figure out how to protect the environment while maintaining—indeed increasing and spreading—economic prosperity as well. A significant number of global conferences have been held during the past decade on one or another environmental issue. Data show that there has been progress in some areas but little or no progress in others.

In the following selection, Bjørn Lomborg argues that good progress is being made and will continue to be made and that we should not overreact to predictions of an environmental doomsday. His central assertion is that almost every economic and social change or trend has been positive, as long as the matter is viewed over a reasonably long period of time. In the second selection, Fred Krupp maintains that what is currently being done on many fronts is too little, too late, and he warns of the costs of inaction.

Bjørn Lomborg  **YES**

Debating "The Skeptical Environmentalist"

Let me just say there are 2 things that I try to say, and these are like the only two take-home points. One is, we need to remove our myths. Nobody can disagree with that, right. But we need to understand doomsday is actually not now. It's not like we have to act in desperation. This is important, because it means that we can start focusing on making the best possible decisions, or as politicians love to point out, there's only one bag of money but there's lots and lots of good purposes and the basic idea being here if we feel like we've been painted into a corner, we're desperate and we're willing to do pretty much anything, also making really bad decisions. If people come up to you with a gun to your head and say, "Give me your money," you don't stand around and say, "Oh, or would I like to buy a toaster?" You do what he says. That's the basic point. So, what I want to point out is that things are actually getting better—it doesn't mean that there are no problems—and that means that we can also start prioritizing. And that's what I'll get back to at the very end.

Basically, have things been getting better? Yes, on most of the important points we have actually got more leisure time, greater security, fewer accidents, more education, more amenities, higher income, fewer starving, more food, and a healthier and longer life. And this is not only true for the industrialized world, but perhaps more importantly also for the developing world.

The point here is to say, what I'm basically trying to do is take the best and very uncontroversial data we have from for instance the UN organizations—this is not the only thing there is, but at least it should make you somewhat less skeptical of the data that I have produced. What I've tried to show you here is the calories per capita per day in the world, for both the developing and developed world. What you can't see up there in the developed world is that we have enough calories, right? Because our problem is possibly getting too much. But what is really important is to look at the developing world. . . . If you look at the developing world we've gone from in 1961 which is the first data that the UN makes, 1932 calories per person per day in the developing world, that's on average just a little above what it takes to sustain your life. To in 1998, which is the last data that we have, to about 2650 calories, that's an increase of almost 40%—that's a dramatic increase.

From Bjørn Lomborg, "Debating 'The Skeptical Environmentalist,'" A Debate Held at the Graduate Center of the City University of New York (April 9, 2002). Copyright © 2002 by Bjørn Lomborg. Reprinted by permission of Environmental Defense. http://www.environmentaldefense.org.

And what I'm saying here, and I'm going to say this again and again, this means things have been getting better. I'm not saying that that means things are fine. There's a very, very big difference. Getting better is a scientific discussion that's basically a question of, is this line going up or down?" Whereas, "It's good enough, they don't need any more"? I'm not saying that. That's a political judgement. I would definitely say they need more, so they can decide themselves whether they want to be fat or not. So the idea here is to say, things have actually been getting better. And also note that the UN actually predicts that despite the fact that there are going to be even more people here on Earth, they will continue to get even more food. So we actually end up in 2030 with a situation where they'll have about 3000 calories per person per day or be at the same level as the developed world in 1960. This means it's better and it's getting even better, but it doesn't mean it's fine. I'm just making the scientific point of saying the data actually moves in the right direction.

Now of course you've also heard and I hear this all the time, "There are lies, damn lies, and statistics." But basically you also need to understand it's the only way we can understand our world, is to look the statistics. But of course we shouldn't just take a look at averages, because this could actually hide some very important differences. It's very unlikely it's one person eating all the calories, but it could be the middle class that's eating up this stuff, so on average they have much more food but it still could mean that there's a lot of poor people who don't get fed and possibly even more. But that's not the case. The UN made its first estimate of how many people are starving in the third world in 1970; the answer was back then about 1.2 billion people were starving, or the equivalent of 35 percent of all people in the developing world—35%, that's more than one out of three—today it's down to 18% in 2030; the UN expects it to be down to 6%. This points out both the fact that things are getting better— it's a lot better when it's only 6% starving than 35% starving—but it still means that in 2030, there'll be 400 million people starving needlessly, because we could easily feed them. It's only a question of allowing them the possibility to make enough money so that they could actually feed themselves. So the idea here is to say I'm trying to make the complex—or at least for the press the complex— point of saying, things are getting better but we can still do more.

Let me go through some of the other issue areas. . . .

Basically one of the arguments that I very often hear is people say, "Sure, Bjørn, you may be right when you're talking about money, but that's because you're *only* talking about money. But where does that help us if we're really undermining our future and our kids' future with pollution? Basically, oh sure we'll get more and more money but we're going to cough all the way to the bank." That's an important point. But I try and say, let's look at the most important pollutant, namely air pollution. The US EPA [Environmental Protection Agency] estimates that anywhere from 86 to 96 percent of all social benefits that stem from regulating pollution, any kind of pollution, comes from regulating one single pollutant—particulate pollution. That means that's the most important thing of all to look at when we're talking about pollution. [For the United Kingdom] we have the data back from 1585 until today. Of course, we love the fact that there've actually been people out measuring it in 1585. We don't.

This is based on models, based on very meticulous descriptions of what was imported into London. But basically this is the best data that we have and most of the EPAs and pretty much all over the world accept this data, and then it's correlated with the data we actually have measured from the early '20s on. If you look at smoke, which is by far most important, particulate pollution, you have an increase since 1585 up until about 1890 and from then on a decline down to today; air pollution by the most important factor is down below what it was in 1585. Americans actually believe air pollution is getting worse in their country; it's not true. This is true for all developed countries; things have been getting better at least for the last 30 or 40 years. When we look at the UK data where we actually have data back from 1585, we can actually say, "No, air pollution is an old phenomenon, and it's been getting better for the last 110 years. The air has never been cleaner in London since 1585, in medieval times."

That's an important point because it means we're not painted into a corner, it is not such that air pollution is taking over and breaking down the world. However, that does not mean that we shouldn't do something about it. Actually, it turns out that because particulate pollution is such an important problem, cutting this even further is a very good idea. So not only can we say yes, air pollution has been dropping dramatically, but we should do even more about it. One of the very obvious ways especially in Europe where we have lots of diesel cars, and diesel cars contribute by far the most particulate pollution in the UK, they constitute about 6% of the car pool, but make up 92% of all small particles so yes, we should fit them with filters. It's a very cost-efficient way, and it's probably also one of the best investments that we can make at all, not only in the environmental area but in any area whatsoever. So the idea here is again to say yes, things have been declining, we have not painted ourselves into a corner but we can still do even more. But it's important we do it because it's a good investment, not because we fear the world is coming to an end.

Let me point out one more thing—this is true for the developed world; it's is not true for the developing world. If you live in Beijing or Mexico City things are getting worse, but that's really because they do exactly the same thing as we did. This is one of the World Bank's analyses that—they pretty much show up any way you make it—if you put income out this way and then you put problems with particulate pollution out this way, you basically see first it gets worse and then it gets better. Really, it's no big surprise, it's exactly what we saw with London and it also makes conceptual sense; first you don't have industry, you don't have pollution but you don't have any money either, right? And then you get industrialization and you say cool, I can buy fruit for my kid, give him an education, buy stuff for myself and then never mind our cough. That's what we did. That's what London did. And it's only when you get sufficiently rich that you start saying, "Hmm, it'd be nice to cough a little less," and then we buy some environment. That's what we've got. And it makes good sense. Environment, in this sense, is a luxury good. When you don't know where your next meal comes from you don't care about the environment 10 or 100 years down the line. However if you actually say, "Now I'm sufficiently rich," if we make the developing countries sufficiently rich, they will also start to worry much more about the environment.

So, let me just make two more points here. One is global warming, obviously because it's one of the most important areas of environmental discussion, and then I'm going to finally talk about what are the consequences of us not prioritizing correctly.

First of all, about global warming, what should we do. There's a lot of discussion about global warming—is it happening, all that stuff. Let's just point out global warming is important. I certainly think carbon dioxide does increase warming and I think we need to take our departure as the best possible scientific data we have from the IPCC [Intergovernmental Panel on Climate Change], the UN climate panel, and it doesn't mean it's infallible but it's the best we have. Now, the point here is to say global warming, the total cost will probably be somewhere around 5 to 8 trillion dollars. This is not a trivial amount of money. It means global warming will make great damage to the world. You also need to put it in perspective, that the total worth of the 21st century is about 900 trillion dollars, so we're talking about a 0.5 percent problem. There are not many other problems that reach that scale, but it is not going to drive us to the poor house no matter what we do.

We need to say it's an important problem, but it is not a problem that in any way will damage our future dramatically. It doesn't mean that we shouldn't handle it carefully, because it's an important problem. And so the idea is to say how important, what about the future of carbon dioxide emissions and what should we do. Let me just say very quickly that everyone is worried about should we run out of oil, should we run out of gas or coal or all these things. Sheik Yamani, the guy who founded OPEC [Organization of Petroleum Exporting Countries], loves to say, "The Oil Age isn't going to come to an end because of lack of oil, just as the Stone Age didn't come to the end because of lack of stone." It wasn't like people said, "Oh God, we're out of flint!" Right? They did it because it made good sense, and we're going to do exactly the same with our energy supply; eventually we'll move to renewables [nonfossil energy sources, such as wind and solar power].

Renewables have been dropping in price about 50% per decade over the last 30 years. Even if they continue at a much lower rate, to about 30% per decade, they'll become competitive before mid-century, and that means certainly we will not be using massive amounts of fossil fuel by the end of the century. It means global warming will be a limited problem, that's not the same as saying it won't be a *big* problem, but it will be a limited problem, probably 2 to 3 degrees centigrade temperature increase. Now at that rate, the main problems will occur in the developing countries. Actually the UN IPCC second summary document said, in what was later mangled by a politician, that it is not going to harm the developed world, it *will* harm the developing world, with a median temperature increase of 2 to 3 degrees. And that's an important part. . . . [B]asically, let's just point out it's going to harm by far the most the developing world.

And then we have to ask ourselves: Is it really a good idea what we're talking about doing right now, namely Kyoto [a 1997 treaty aimed at reducing emissions of greenhouse gases]? . . . [F]rom 1990 up through 2100 what's going to happen if we don't do something . . . we're going to have a temperature increase of 2.1 degrees centigrade. If we do Kyoto, it's not like it's going to stop global warming, it's going to simply slow it down slightly . . . —this is totally uncontroversial. All models show this. . . . it will go down to 1.9 degrees, or to put it more clearly, the temperature that we would have had in 2094, we have now postponed until 2100. In other words, we've bought the world six years.

Of course if Kyoto was cheap or something, maybe that would be a good idea, it's something good. But not very good. Basically what we're saying is the fact the guy in Bangladesh, who has to move because his house got flooded in 2100, he only has to move now in 2106. It's a little good but not very much good, right? On the other hand, the cost of Kyoto is going to be anywhere from 150 to 350 billion dollars a year. That's three to seven times the global development aid to the third world. Is that a good investment?

Well actually, all cost/benefit analyses show that it's a very, very bad investment. Just to give you a sense of the cost of this, the cost of Kyoto for one year in 2010, for just that one year cost, we could solve the single biggest problem in the world, once and for all. We could give clean drinking water and sanitation to every single human being on earth once and for all. It would save 2 million lives each year, in fact, half a billion lives each year. We have to ask ourselves, wouldn't that be a better way of helping the developing world? Actually the UN estimates that for 70–80 billion dollars—much less than the cost of Kyoto—we could permanently solve all the basic problems of the developing world: it could include clean drinking water, sanitation, basic health care and education to every single human being on earth. Wouldn't we do better by doing that? And again the idea here is to say, we should not allow ourselves to be painted into a corner to believe we have to do something. If it actually makes sense, we should do not only something that sounds good, but also actually *does* good.

. . . [T]he Harvard Center for Risk Analysis . . . showed all the data publicly available from the US on all legislation which had as its primary purpose to save human life. Notice a lot of environmental legislation does not have as its primary purpose to save human life—if you talk about saving the Bengal tiger for instance, it probably has the opposite effect! The idea here is to say we're talking about all the costs of the legislation that tries to save human life and then they compare what is the median or the typical cost of saving one human life one year. . . . [T]he biggest study we have in the world basically [says] that in the health area it costs 19 thousand dollars to save one human life one year. You can see the other areas—and what we basically have out here is the environment area costing $4.2 million to save one human life a year. And that's when the purpose was to save human life. . . . It's not the same thing as saying there are no good investments to be made in the environmental area. It's simply to say that on average we over-worry about the environmental area—and it does have consequences because if we over-worry about some areas we end up under-worrying about other areas, and that's my last point, that's the reason why we need to focus on what is the real state of the world. Things are actually getting better and better and they're likely to do so in the future. This does not mean that there are no problems and that we don't need to worry, but it means that what we need to understand, the problems are getting smaller, and that means that we have to start focusing and prioritizing correctly, so that we not only make sure that we make a better world—we probably will no matter how stupidly we act—but that we make an even better world for our kids and grand-kids and that involves both knowing how the world looks, and how we should prioritize correctly.

Fred Krupp

 **NO**

Debating "The Skeptical Environmentalist"

I think this debate is timely because now at the beginning of the 21st century scientists have told us that there are some big risks that are posed by climate change, the loss of species and the sorry state of our oceans. It's timely as well because Bjørn [Lomborg] has been on tour saying there really is not much to worry about, a message that's been trumpeted in the media throughout the world. This is important because if Bjørn's right, we can all relax, take a deep breath and reallocate our time and energies to other problems, and the news media or corporate and citizen leaders can also reallocate their time and resources as well.

And it's important though, because if he's wrong, such complacency can have a big cost. . . . I want to stick to the topic of "Is our environmental future secure," and not get into all the graphs and charts that Bjørn has put up. But I do want to point out that others have written articles about those and I commend them to you: *Science, Scientific American,* Union of Concerned Scientists. . . . But basically instead of debating the charts and the graphs, I intend to challenge Bjørn's conclusions. My own answer to this important question, "Is our environmental future secure" is no. *Not yet.* It depends on us, on what we do, on the actions we take. If we make the right choices things can turn out OK, but it could also turn out the other way.

I'll support this answer with three points. First, my involvement with environmental policy over the last 30 years, which teaches me that the progress we have made comes not just from the accumulation of wealth and reaching a certain GDP [gross domestic product] level, but from citizen and government action. Bjørn tells us that progress in the past has come from wealth, and asserts that more wealth will solve any problems in the future. Second, I look to the best scientific assessment and see that yes, important progress has been made. But many important challenges remain that demand urgent attention and good policy choices. Bjørn seems to argue that these problems are largely being solved on their own. Since world scientific bodies and our own National Academy certainly see these problems, I think it's fair, even important to ask why we should rely on Bjørn's views instead. And finally, my third point is that real world experience shows that we can make progress keeping an eye toward keeping cost down, carefully choosing strong scientific standards and incentive-based policies. Given the stakes involved, not to continue to forge ahead on big problems like climate change would be like encouraging my teenage son to go around driving his car without insurance or seat belts. The risks are just too great.

OK. First, my point that environmental progress is not automatic. Others have argued that environmental progress is the inevitable outcome of wealth. Their argument is first we grow, then we clean up. Or as [Bjørn believes], the environment is a luxury good. But this argument neglects to point out the role of citizen and government action in achieving that progress. Take the case . . . of air pollution in London. It didn't just get better; London and the UK passed a series of laws *requiring* that smoke and pollution levels come down. Some of these laws were passed after an infamous inversion that killed many, many folks. Look at the history of air pollution in our own country, or cleaner, high mileage cars. Time and again the prevailing sentiment of industry was against passing a tough Clean Air Act, fighting acid rain controls, and campaigning furiously against cleaner exhaust pipes and higher mileage standards. Opponents claimed each new regulatory proposal would bankrupt them. They haven't.

Regulations to reduce sulfur dioxide pollution largely generated by the burning of coal in power plants were projected by industry to cost as much as $2000 for each ton of sulfur removed from their smokestacks. The nation chose instead a new performance-based approach to cleaning up, giving industry flexibility, but requiring results. And today removing a ton of sulfur costs less than $200, a tenth of the doomsayers' predictions. Reductions of 50% in the annual emissions were required, and so far we've reduced pollution by even more. This happy story was not an automatic happening that was triggered by some level of GDP. It was the result of a campaign waged by many environmental groups (including Environmental Defense) and then of the choices made by our legislators. The victory on sulfur and car exhaust and the Clean Air Act produced much of the air pollution improvements Bjørn points to.

The same citizen and government action was required to take lead out of gasoline, to ban DDT and other pesticides to protect our wetlands, to restore the ozone layer, and to clean up water pollution. Now in addition, other countries really don't need to spend their money cleaning up after the fact when they have the option of doing things right beforehand. China, for example: they don't accept the idea that they need to be consigned to some environmental hell. Their citizens are way down on the ladder in terms of per capita income but they're already demanding a cleanup. And their government is beginning an acid rain control program modeled after our own. Why should they wait?

Second, while I agree that progress has been made on many issues, many important problems remain. It's not true that we're making progress on most of the important problems. Sure, we have made progress on emissions from power plants. But acid rain continues to strip our soils of essential elements and kill our lakes. We need to further cut sulfur emissions. Yes we have cleaned up car exhaust, but with more cars and trucks driving more miles, smog and particulate pollution remain a huge issue both in our country and abroad. Here in this country, 15,000 premature deaths are ascribed by the scientists to our air pollution. And if you look south in Central and South America, 200,000 deaths a year are attributed by the scientists to particulate pollution.

On biodiversity, I think it's fair to say the very web of life is unraveling. For example, let's talk about what happens right before extinction: endangerment. Based on very solid data the World Conservation Union has discovered that of the world's birds, 1 out of every 8 is endangered. The same ratio applies to the world's plant life: 1 out of 8 endangered. These are troubling numbers and they are not the result of natural die-offs. These species are endangered because of what people are doing to their ecosystems here and now. Fortunately, we're talking about endangered species, not extinct species, so there is still time to save some of them. But to deny there's a problem is just plain unscientific.

On the problem of global warming, this needs to be a huge priority because the stakes are so high. It threatens our ecosystem. The parks and reserves that we set aside are at risk, our oceans, our coral reefs. What we know for certain is that the earth is warming, that sea level has been rising partly as a result of this warming, and that human activity has been contributing significantly. Glaciers are melting, the permafrost in Alaska is melting, causing power lines and phone lines to begin to topple. And . . . in New York City we've already seen samplings of the kinds of intense storms and sea level increase that may well inundate some of our own major airports, subways and highways. Our own National Academy has confirmed that this is a very real problem.

Finally, let's get to that question of costs and benefits and buying insurance, with special attention to climate change. Bjørn has [said] that it's not worth the cost to aggressively tackle climate change, but the problem will self-correct as renewables magically become competitive. I disagree. Just looking here at the US, as long as we continue to subsidize the burning of fossil fuels by a 3 to 1 ratio compared to renewables, and as long as we allow the true cost of burning fossil fuels, the true cost to our health, and our environment, to basically frankly be borne by all of us, it's not likely that renewables will be able to play the role that he predicts. Without changes in policy, it just won't happen.

Rather than citing theories about what the cost of reducing emissions will be, the truth is we can rarely know with precision what the costs or benefits of the decision will be, but time and time again we find that the projections of costs have been systematically overestimated. In particular the economic models just cited about climate tell Bjørn that we should only be prepared to spend a small amount on this critical problem. But those models don't include the potentially catastrophic but unknown cost of the collapse of the world's oceans' heat circulation patterns, or the episodic events like the flooding of the Mississippi in 1995, or the intense nor'easter that hit New York in 1992. The models don't account for heat waves that have killed thousands of people. And most importantly, the models assign no monetary value to the loss of natural systems.

As for our experience so far in climate change let's look at the facts. Environmental Defense has been working over the last few years with a number of multinational corporations to reduce emissions. The results of those efforts are now beginning to come in. For example, four years ago, BP [British Petroleum] committed to reduce their worldwide emissions by 10% below 1990 levels by the year 2010. . . . [E]ight years ahead of schedule, they announced that they had already achieved that level of reduction at no net costs whatsoever. Not every experience will turn out this way. But this is a real world example, not some theoretical projection.

[Bjørn] says Kyoto [1997 treaty aimed at reducing emissions of greenhouse gases] will cost too much for too little benefit, but the Kyoto agreement actually is fundamentally structured to minimize cost with flexible mechanisms, incentives for action, and opportunities to invest in reduction strategies like carbon sequestration. And as to the benefits of Kyoto, [Bjørn's evidence] went out to 2100 showing very little benefit. Kyoto only covers reductions mandated up to the year 2012, and everyone in the process anticipates that then subsequent reductions will be required. As for the benefits again, he hasn't taken into account, the models don't take into account the catastrophic storms that climate change can cause, and moreover, the limited economic models don't calculate in what are called co-benefits, the value of protecting forest ecosystems, coral reefs, and these can make a world of difference in the calculations. Simply put, the idea that we should rely on renewables and put the whole future of our world on the cost of renewables coming down stakes a pretty big bet on faith that that will happen. Doesn't it make more sense that we purchase insurance and make sure we put into place policies that will make it happen?

In closing, let me just say my first point, progress has been the result of deliberate decisions often by government spurred on by nongovernmental groups campaigning for action; second, sure, some important problems are getting better, but several important ones are getting worse; and third, while costs are important and they need to be considered, we need to be very wary of the way cost benefit analysis as it's historically been practiced, systematically exaggerating costs and not reliably totaling up the benefits. Time and time again, these projections have been proven wrong. Will our environment be secure? As I indicated I believe the answer is no. Not yet. We can't be sure our environmental future is secure but we can act to secure it. I'm actually quite optimistic about the outcome. But I try to anchor my optimism in realism; that is, in order to solve problems we have to first recognize them, we have to admit they exist, and then we have to understand what has worked before and aggressively apply it to those lessons. Using this approach gives me cause for both hope and concern, and I think that's the combination that can yield the most progress.

POSTSCRIPT

Do Environmentalists Overstate Their Case?

Y ou can't have your cake and eat it too" is a trite phrase. Such bits of folk wisdom, though, often get to be trite because there is a kernel of truth to them that is worth repeating. The environment is akin to our common cake. People have been consuming it gluttonously during the past century, and most experts agree that this cannot go on any longer. The question is whether or not we have to go on a bread-and-water diet. Also, can the world's less developed countries be asked to forgo cake when the developed countries have already consumed so much?

Some, such as Krupp, say we have exceeded the boundaries of responsibility. Lomborg and others are more optimistic. They contend that we have or can develop the technology and the environmental-use policies to continue to develop and to enhance the existence of less developed countries while protecting—even improving—the environment. They may be correct, and it certainly is more comforting to believe Lomborg's optimistic view than to accept Krupp's dire outlook. More on Krupp and his organization, Environmental Defense, is available at the organization's home page at http://www.environmentaldefense.org/home.cfm. Also see Lomborg's personal Web page at http://www.lomborg.com and his book *The Skeptical Environmentalist: Measuring the Real State of the World* (Cambridge University Press, 2001).

Even if Lomborg is correct, it is important not to ignore the costs of sustainable development. They can be substantial. Because of population and economic development patterns, the less developed countries require particular care and assistance. It is easy to preach about not cutting down Brazilian rain forests or not poaching cheetah skins in Kenya. But what do you say to the poor Brazilian who is trying to scratch out a living by clearing cropland or grazing land? What do you tell the equally poor Kenyan who is trying to earn a few dollars in order to supply food for his family? Questions such as these have brought environmental issues much closer to the forefront of world political concerns.

The point is that environmental protection is not cost free. This is because environmentally safe production, consumption, and waste disposal techniques are frequently much more expensive than current processes. It is also because poorer people will generally do what they must to survive, whether it is environmentally safe or not. Moreover, the less developed countries have precious few financial resources to devote to developing, constructing, and implementing environmentally safe processes. Therefore, if the changes that need to occur are going to be put in place before further massive environmental degradation occurs, there will have to be a massive flow of expensive technology and financial assistance from the developed to the less developed countries.

The debate over the state of the environment is complex. One good place to begin researching it is with the United Nations Environment Programme's report *Global Environment Outlook 3* (Earthscan, 2002). To keep abreast of the current thinking on the environment, read John L. Allen, ed., *Annual Editions: Environment 03/04* (McGraw-Hill/Dushkin, 2003). Finally, to learn more about global environmental politics, consult Lorraine Elliott, *The Global Politics of the Environment* (New York University Press, 2003).

Contributors to This Volume

EDITORS

JOHN T. ROURKE, Ph.D., is a professor of political science at the University of Connecticut. He has written numerous articles and papers, and he is the author of *Congress and the Presidency in U.S. Foreign Policymaking* (Westview Press, 1985); *The United States, the Soviet Union, and China: Comparative Foreign Policymaking and Implementation* (Brooks/Cole, 1989); and *International Politics on the World Stage*, 10th ed. (McGraw-Hill, 2004). He is also coauthor, with Ralph G. Carter and Mark A. Boyer, of *Making American Foreign Policy*, 2d ed. (Brown & Benchmark, 1996).

STAFF

Larry Loeppke	Managing Editor
Jill Peter	Senior Developmental Editor
Nichole Altman	Developmental Editor
Beth Kundert	Production Manager
Jane Mohr	Project Manager
Tara McDermott	Design Coordinator
Bonnie Coakley	Editorial Assistant
Lori Church	Permissions

AUTHORS

LEON ARON is director of Russian studies at the American Enterprise Institute. He has a Ph.D. in sociology from Columbia University.

WALDEN BELLO is executive director of Focus on the Global South, based in Thailand, and a professor of sociology and public administration at the University of the Philippines. He is also former chair of the board of Greenpeace South East Asia, visiting professor in Southeast Asian Studies at the University of California at Los Angeles, and board member of Food First, the International Forum on Globalisation, and the Transnational Institute. He became a political activist in the 1970s opposing the regime of Ferdinand Marcos. He has a Ph.D. in sociology from Princeton University. The Belgian newspaper *Le Soir* recently called Bello "the most respected anti-globalization thinker in Asia."

BRUCE BERKOWITZ is a research fellow at the Hoover Institution at Stanford University. He has also served as a consultant to the Department of Defense and the Intelligence Community, and lectures on national security issues at the National Defense University. He has also been an international affairs fellow at the Council on Foreign Relations (1985–86), a visiting fellow at the Brookings Institution, on the staff of the U.S. Senate Select Committee on Intelligence, and with the Central Intelligence Agency. He has a Ph.D. in political science from the University of Rochester.

P. J. BERLYN is an author of Israelite history and culture and a former associate editor for the *Jewish Bible Quarterly* in Jerusalem. She has also worked for the Council on Foreign Relations in New York, as well as its journal *Foreign Affairs*. She is coauthor, with Shimon Bakon, of *Shani—Her Adventures Beyond the Sambatyon* (En-Gedi Books, 2000).

JOHN R. BOLTON is senior vice president of the American Enterprise Institute. During the presidential administration of George Bush, he served as the assistant secretary of state for international organization affairs. A Yale University–educated lawyer, Bolton has held a variety of posts in both the Reagan and Bush administrations.

JOSÉ BOVÉ, a French farmer and unionist, is an anti-globalization activist. He is best known for leading a group of activists that ransacked a McDonald's outlet in France in 1999 to protest industrialized food.

ROBERT BYRD is a democratic U.S. senator from West Virginia. He was first elected to the Senate in 1958, is the longest-serving member of the Senate, and is the ranking Democrat on the Senate Committee on Appropriations. He has a J.D. from American University.

RICHARD CHENEY is vice president of the United States. He has also served as White House chief of staff for President Gerald Ford, as Secretary of Defense to President George H. W. Bush, and as a member of the U.S. House of Representatives from Wyoming. He is the recipient of the Presidential Medal of Freedom.

ARIEL COHEN is a research fellow in Russian and Eurasian studies in the Kathryn and Shelby Cullom Davis Institute for International Studies at The Heritage Foundation. He has a law degree from Bar Ilan University Law School in Tel Aviv and a Ph.D. from the Fletcher School of Law and Diplomacy at Tufts University

COMMISSION FOR ASSISTANCE TO A FREE CUBA was established in October 2003 by President George W. Bush to develop a comprehensive U.S. policy toward Cuba. Its chair was Secretary of State Colin Powell, and its membership included representatives of most Cabinet departments and several leading independent agencies. Assistant Secretary of State for Western Hemisphere Affairs Roger Noriega oversaw the commission's day-to-day operations.

WILLIAM EASTERLY is senior adviser in the Development Research Group at the World Bank and the author of *The Elusive Quest for Growth: Economists' Adventures and Misadventures in the Tropics* (MIT Press, 2001).

ROBERT J. EINHORN is a senior adviser in the CSIS International Security Program of the Center for Strategic and International Studies. Prior to that, he served in numerous government positions, including assistant secretary for nonproliferation at the U.S. Department of State. He has an M.A. from the Woodrow Wilson School of Public and International Affairs, Princeton University.

DOUGLAS J. FEITH is the U.S. undersecretary of defense for policy. Prior to that he was the managing attorney of the Washington, D.C., law firm of Feith & Zell, deputy assistant secretary of defense for negotiations policy, and a staff member of the National Security Council specializing in Middle East affairs. He has a J.D. from the Georgetown University.

NIALL FERGUSON is the Herzog Professor of History at New York University's Stern School of Business and senior fellow at the Hoover Institution at Stanford University. He has published works on the nineteenth- and twentieth-century European political and financial history.

JULIA GALEOTA is a high school senior at Holton Arms School in McLean, Virginia. Her essay won first place in the thirteen-to-seventeen-year-old age category of the 2004 *Humanist* Essay Contest for Young Women and Men of North America and was published in the *Humanist*.

JOHN GERSHMAN is a senior analyst at the Interhemispheric Resource Center and the co-director of Foreign Policy in Focus. He has also been a research associate at the Institute for Health and Social Justice Research Associate and a visiting doctoral candidate at Princeton University's Woodrow Wilson School.

ROMILLY GREENHILL is the Jubilee Research economist at the New Economics Foundation, a radical think tank that creates practical and enterprising solutions to the social, environmental, and economic challenges facing local, regional, national, and global economies. She spent 2 years as an Overseas Development Institute fellow in Uganda.

HIGH-LEVEL PANEL ON THREATS, CHALLENGES, AND CHANGE was established in November 2003 by Secretary-General Kofi Annan to examine the major threats and challenges the world faces in the broad field of peace and security. The panel was chaired by Anand Panyarachun, former prime minister of Thailand. Other panel members included Robert Badinter (France, member of the French Senate and former Minister of Justice), João Clemente Baena Soares (Brazil, former secretary-general of the Organization of American States), Gro Harlem Brundtland (Norway, former prime minister of Norway and former director-general of the World Health Organization), Mary Chinery-Hesse (Ghana, vice-chairman, National Development Planning Commission of Ghana and former deputy director-general, International Labour Organization), Gareth Evans (Australia, president of the International Crisis Group and former minister for foreign affairs of Australia), David Hannay (United Kingdom, former ambassador of the United Kingdom to the United Nations), Enrique Iglesias (Uruguay, president of the Inter-American Development Bank), Amre Moussa (Egypt, secretary-general of the League of Arab States), Satish Nambiar (India, former lt. general in the Indian Army and force commander of UN peacekeeping in Croatia), Sadako Ogata (Japan, former United Nations high commissioner for refugees), Yevgeny Primakov (Russia, former prime minister of Russia), Qian Qichen (China, former vice prime minister and minister for foreign affairs of China), Nafis Sadik (Pakistan, former executive director of the United Nations Population Fund), Salim Ahmed Salim (Tanzania, former secretary-general of the Organization of African Unity), and Brent Scowcroft (United States, former lt. general in the U.S. Air Force and national security adviser to President George H. W. Bush).

CHRISTINA P. HOFF-SOMMERS is a resident scholar at the American Enterprise Institute. She is also chairman, Board of Academic Advisors, Independent Women's Forum and has served on the faculty of Clark University and the University of Massachusetts at Boston. She has a Ph.D. in philosophy from Brandeis University.

HAROLD HONGJU KOH is the dean and Gerard C. and Bernice Latrobe Smith Professor of International Law, Yale University. He is also a former U.S. assistant secretary of state for democracy, human rights and labor; attorney-advisor, Office of Legal Counsel, U.S. Department of Justice; and law clerk for U.S. Supreme Court Justice Harry A. Blackmun. He hasd a J.D. from Harvard University.

JOHN C. HULSMAN is Research Fellow, Kathryn and Shelby Cullom Davis Institute for International Studies. He has a Ph.D. in modern history and international relations from the University of St. Andrews, Scotland.

LOUIS JANOWSKI is a former U.S. diplomat with service in Vietnam, France, Ethiopia, Saudi Arabia, and Kenya.

A. ELIZABETH JONES is a career foreign service office currently serving as the U.S. assistant secretary of state for European and Eurasian affairs. Prior to that, she served as senior advisor for Caspian Basin energy diplomacy principal deputy assistant secretary of state in the Bureau of Near Eastern Affairs, and as ambassador to Kazakhstan.

STEVEN L. KENNY is a colonel in the U.S. Army.

STEPHEN D. KRASNER is the Graham H. Stuart Professor of Political Science at Stanford University. Krasner earned his Ph.D. from Harvard University. He served as editor for *International Organization* from 1987 to 1992, and he is the author of *Sovereignty: Organized Hypocrisy* (Princeton University Press, 1999).

WILLIAM KRISTOL is the editor of the Washington-based political magazine, *The Weekly Standard*. Prior to that he headed the Project for the Republican Future and was chief of staff to Vice President Dan Quayle. He has taught politics at the University of Pennsylvania and Harvard's Kennedy School of Government.

ANNE O. KRUEGER is the first deputy managing director of the International Monetary Fund. Prior to taking this position, she was the Herald L. and Caroline L. Ritch Professor in Humanities and Sciences in the department of economics at Stanford University. She was also director of Stanford's Center for Research on Economic Development and Policy Reform and a senior fellow of the Hoover Institution. From 1982 to 1986, Krueger was the World Bank's vice president for economics and research. She earned her Ph.D. in economics from the University of Wisconsin.

FRED KRUPP is president of Environmental Defense. He has also been a member of President Bill Clinton's Advisory Committee for Trade Policy and Negotiations, a member of Clinton's Commission on Sustainable Development, and an advisory board member of the Environmental Media Association. Identified as a key figure behind the congressional passage of the 1990 Clean Air Act, Krupp also led the Environmental Defense Fund's delegation for the Buenos Aires climate change negotiations. He holds a J.D. from the University of Michigan.

LAWYERS COMMITTEE FOR HUMAN RIGHTS is a New York–based civil rights advocacy group. The committee seeks to influence the U.S. government to promote the rule of law in both its foreign and domestic policy and presses for greater integration of human rights into the work of the UN and the World Bank. The committee works to protect refugees through the representation of asylum seekers and by challenging legal restrictions on the rights of refugees in the United States and around the world.

PHILIPPE LEGRAIN is chief economist of Britain in Europe. He was previously special adviser to the director-general of the World Trade Organization and the trade and economics correspondent for *The Economist*. He has a master's degree in economics from the London School of Economics. He has also written for the *Financial Times, The Wall Street Journal Europe, The Times, The Guardian, The Independent, New Statesman, Prospect,* and *The Ecologist,* as well as *The New Republic, Foreign Policy,* and the *Chronicle Review.*

BJØRN LOMBORG is an associate professor of statistics in the department of political science at the University of Aarhus in Denmark and a frequent participant in topical coverage in the European media. His areas of professional interest include the simulation of strategies in collective action dilemmas, the use of surveys in public administration, and the use of statistics in the environmental arena. In February 2002, Lomborg was named director of Denmark's national Environmental Assessment Institute. He earned his Ph.D. from the University of Copenhagen in 1994.

JOHAN NORBERG is a fellow at Timbro, a policy institute in Stockholm, Sweden. He has an M.A. from Stockholm University, Sweden.

MARGOT PATTERSON is a senior writer for *National Catholic Reporter.*

WILLIAM NORMAN GRIGG is senior editor of *The New American* and host of the John Birch Society's biweekly audio commentary *Review of the News Online.* He is the author of a number of books, including *Global Gun Grab: The United Nations Campaign to Disarm Americans* (John Birch Society, 2001).

WILLIAM RATLIFF is a research fellow and curator of the Americas Collection at the Hoover Institution. Before joining the Hoover Institution, he was a columnist and chief editorial writer for *Peninsula Times Tribune.* He has a Ph.D. in Latin American and Chinese history from the University of Washington in Seattle and has taught at Stanford, San Francisco State University, the University of San Francisco, and Tunghai University in Taiwan.

MARY ROBINSON is United Nations high commissioner for human rights. Prior to that, she was president of Ireland, a member of Ireland's senate, and Reid Professor of Constitutional Law at Trinity College, Dublin. She has law degrees from the King's Inns in Dublin and from Harvard University.

BRETT D. SCHAEFER is Jay Kingham Fellow in International Regulatory Affairs at The Heritage Foundation. Prior to that, he was assistant for International Criminal Court Policy at the U.S. Department of Defense. He has an M.A. in international development economics from the School of International Service at The American University in Washington, D.C.

ROSEMARY E. SHINKO teaches in the department of political science at the University of Connecticut, where she is working toward her Ph.D. in international relations and political theory.

MICHAEL D. SWAINE came to the Carnegie Endowment. Earlier he was research director of the RAND Center for Asia-Pacific Policy. He has a Ph.D. from Harvard University and has held positions at the University of California, Berkeley, and Harvard University.

YEVGENY VOLK is director of the Moscow office of The Heritage Foundation and president of The Hayek Foundation in Moscow. He is a former deputy director for the Russian Institute for Strategic Studies, advisor to the Committee on Defense and Security Issues for the Supreme Soviet of the Russian Federation, and staff member of the Soviet Ministry of Foreign Affairs. He has a Ph.D. from the Moscow State Institute of International Relations.

KIMBERLY WEIR is an assistant professor of political science at Northern Kentucky University. She was director of the GlobalEd High School Simulation Project, which was created to systematically and scientifically evaluate perceived gender differences in leadership and decision-making styles and values and in approaches to technology. She holds a Ph.D. in political science from the University of Connecticut and an M.A. from Villanova University.

Index